Green Energy and Technology

Climate change, environmental impact and the limited natural resources urge scientific research and novel technical solutions. The monograph series Green Energy and Technology serves as a publishing platform for scientific and technological approaches to "green"—i.e. environmentally friendly and sustainable—technologies. While a focus lies on energy and power supply, it also covers "green" solutions in industrial engineering and engineering design. Green Energy and Technology addresses researchers, advanced students, technical consultants as well as decision makers in industries and politics. Hence, the level of presentation spans from instructional to highly technical.

Indexed in Scopus.

Indexed in Ei Compendex.

Hairus Abdullah
Editor

Photocatalytic Activities for Environmental Remediation and Energy Conversion

Springer

Editor
Hairus Abdullah
Department of Industrial Engineering
Universitas Prima Indonesia
Medan, Sumatera Utara, Indonesia

ISSN 1865-3529 ISSN 1865-3537 (electronic)
Green Energy and Technology
ISBN 978-981-19-6747-4 ISBN 978-981-19-6748-1 (eBook)
https://doi.org/10.1007/978-981-19-6748-1

This Springer imprint is published by the registered company Springer Nature Singapore Pte Ltd.
The registered company address is: 152 Beach Road, #21-01/04 Gateway East, Singapore 189721, Singapore

Preface

It is a great pleasure to introduce a special issue of Springer Book containing promising photocatalysis methods for environmental remediation and energy conversion. Authors with educational backgrounds in material science and chemical engineering have contributed to the chapters in this book. The chapters have reviewed some crucial applications in photocatalytic methods for organic pollutant degradation, air purifier, nitrogen fixation to generate ammonia, CO_2 reduction reactions, and water splitting. Chapter "Photocatalytic and Adsorptive Removal of Liquid Textile Industrial Waste with Carbon-Based Nanomaterials" discusses the promising carbonaceous materials with different semiconductor nanomaterials for degrading organic pollutants such as textile dyes. The chapter also indicates the benefits of the photocatalysis method and the drawbacks of traditional wastewater treatments in industries. Chapter "Syntheses and Applications of Nanomaterials-Based Photocatalysts for Air Purification" shows the prospective photocatalytic treatment of polluted air using filters with different semiconductors. Some critical works on famous TiO_2, C_3N_4,ZnO, ZnS, and WO_3 have been reviewed. The essential ideas of photocatalysis heterojunctions with nitrogen reduction reaction mechanism, ammonia production techniques, and the existing advantages are mainly discussed in Chapter "Photocatalytic Nitrogen Fixation on Semiconductor Materials: Fundamentals, Latest Advances, and Future Perspective". Some of the thermodynamic and reaction kinetics of photocatalytic CO_2 reduction using different types of photocatalysts for CO_2 reduction are discussed in depth by considering metallic, non-metallic, and composite systems in Chapter "A Recent Review on Photocatalytic CO_2 Reduction in Generating Sustainable Carbon-Based Fuels". Furthermore, Chapter "Semiconductor-Based Photocatalytic Oxygen Evolution Performance for Water Splitting: Light-Driven Energy Conversion and Storage" introduces the difficulties in oxygen evolution reaction (OER) in a water-splitting (WS) process, which are also thoroughly discussed since OER is a sluggish process that determines the whole WS reaction efficiency. Based on the discussion on this issue, it is expected to stimulate further discussion and research in nanomaterials to advance the photocatalysis field with more applications.

Medan-North Sumatera, Indonesia Dr. Hairus Abdullah

Contents

Contributors

Abdullah Hairus Department of Materials Science and Engineering, National Taiwan University of Science and Technology, Taipei, Taiwan, ROC;
Department of Industrial Engineering, Universitas Prima Indonesia, Medan, Indonesia

Ahmed Mohamed Tarek Department of Materials Science and Engineering, National Taiwan University of Science and Technology, Taipei, Taiwan

Amanah Nur Layli Department of Materials Science and Engineering, National Taiwan University of Science and Technology, Taipei, Taiwan, ROC

Ambar Prawesti Department of Materials Science and Engineering, National Taiwan University of Science and Technology, Taipei, Taiwan, ROC

Aprilio Muhammad Razaki Department of Materials Science and Engineering, National Taiwan University of Science and Technology, Taipei, Taiwan, ROC

Assa Muhammad Hawary Department of Materials Science and Engineering, National Taiwan University of Science and Technology, Taipei, Taiwan, ROC

Astriani Rapita Department of Materials Science and Engineering, National Taiwan University of Science and Technology, Taipei, Taiwan, ROC

Aziz Galang Dhaifullah Abdul Department of Materials Science and Engineering, National Taiwan University of Science and Technology, Taipei, Taiwan, ROC

Bafe Woldesenbet Department of Materials Science and Engineering, National Taiwan University of Science and Technology, Taipei, Taiwan, ROC

Bekele Kebede Department of Materials Science and Engineering, National Taiwan University of Science and Technology, Taipei, Taiwan, ROC

Catherine Stefani Department of Materials Science and Engineering, National Taiwan University of Science and Technology, Taipei, Taiwan, ROC

Chang Li-Hsiang Department of Materials Science and Engineering, National Taiwan University of Science and Technology, Taipei, Taiwan

Cheng Yichia Department of Materials Science and Engineering, National Taiwan University of Science and Technology, Taipei, Taiwan, ROC

Chiu Yun Xuan Department of Materials Science and Engineering, National Taiwan University of Science and Technology, Taipei, Taiwan, ROC

Damastuti Retno Department of Materials Science and Engineering, National Taiwan University of Science and Technology, Taipei, Taiwan, ROC

Ernawati Lusi Department of Chemical Engineering, Institut Teknologi Kalimantan, Balikpapan, East Kalimantan, Indonesia

Fadhilah Gianna Department of Chemical Engineering, National Taiwan University of Science and Technology, Taipei, Taiwan

Fariz Farghani Department of Materials Science and Engineering, National Taiwan University of Science and Technology, Taipei, Taiwan, ROC

Gemechu Habib Department of Materials Science and Engineering, National Taiwan University of Science and Technology, Taipei, Taiwan, ROC

Gemeda Tadele Negash Department of Materials Science and Engineering, National Taiwan University of Science and Technology, Taipei, Taiwan;
Department of Chemical Engineering, School of Mechanical, Chemical and Material Engineering, Adama Science and Technology University, Adama, Ethiopia

Ginting Riski Titian Department of Electrical Engineering, Universitas Prima Indonesia, Medan, Indonesia

Haliq Ridhwan Department of Materials Science and Engineering, National Taiwan University of Science and Technology, Taipei, Taiwan, ROC;
Department of Mechanical Engineering, Institut Teknologi Kalimantan, Balikpapan, Indonesia

Huang Ting-Chen Department of Materials Science and Engineering, National Taiwan University of Science and Technology, Taipei, Taiwan, ROC

Khan Uzair Ali Department of Materials Science and Engineering, National Taiwan University of Science and Technology, Taipei, Taiwan, ROC

Laksono Andromeda Dwi Department of Materials Science and Engineering, National Taiwan University of Science and Technology, Taipei, Taiwan, ROC;
Study Program of Materials and Metallurgical Engineering, Institut Teknologi Kalimantan, Balikpapan, East Kalimantan, Indonesia

Liang Yu Tse Department of Materials Science and Engineering, National Taiwan University of Science and Technology, Taipei, Taiwan

Lin Jia Hung Department of Materials Science and Engineering, National Taiwan University of Science and Technology, Taipei, Taiwan, ROC

Meutia Farah Department of Materials Science and Engineering, National Taiwan University of Science and Technology, Taipei, Taiwan, ROC

Nugroho Agung Department of Chemical Engineering, Universitas Pertamina, Simprug Kebayoran Lama, Jakarta, Indonesia

Phan Van Hoang Khang Department of Materials Science and Engineering, National Taiwan University of Science and Technology, Taipei, Taiwan

Prasetyo Fery Department of Chemical Engineering, National Taiwan University of Science and Technology, Taipei, Taiwan

Putra Dwi Fortuna Anjusa Department of Materials Science and Engineering, National Taiwan University of Science and Technology, Taipei, Taiwan, ROC

Qadir Abdul Department of Materials Science and Engineering, National Taiwan University of Science and Technology, Taipei, Taiwan, ROC

Tarigan Olivia Christy Department of Mechanical Engineering, National Taiwan University of Science and Technology, Taipei, Taiwan, ROC

Urgesa Merga Hailemariam Department of Materials Science and Engineering, National Taiwan University of Science and Technology, Taipei, Taiwan, ROC

Wijaya Francisca Tania Deviani Department of Materials Science and Engineering, National Taiwan University of Science and Technology, Taipei, Taiwan, ROC

Photocatalytic and Adsorptive Removal of Liquid Textile Industrial Waste with Carbon-Based Nanomaterials

Andromeda Dwi Laksono, Retno Damastuti, Nur Layli Amanah, Muhammad Hawary Assa, Yichia Cheng, Lusi Ernawati, Agung Nugroho, and Hairus Abdullah

Abstract Textile industry wastewater is considered one of the most terrible sources of pollution to our precious aquatic environment because it is mutagenic, carcinogenic, cytotoxic, and other genotoxic to natural compounds. This literature review collects and investigates the latest available research on photocatalytic wastewater or organic pollutant degradation using carbon-based materials in textile industrial wastewater treatment. Materials such as carbon have been used in several studies to enhance photocatalytic performance due to superior properties such as small particle size, high specific surface area-to-volume ratio, high reactivity, excellent chemistry, high thermal stability, wide availability, and catalytic potential on the nanoscale. Heterojunction formation is one of the most effective methods for increasing charge separation efficiency and reducing photogenerated electron–hole pair recombination. There are many ways to increase the photocatalytic performance for carbon-based material, such as heterojunction formation, using g-C_3N_4 or zinc oxide/carbon nanocomposite as a photocatalyst, surface modification to improve electron transport efficiency in semiconductor particles, using a green hydrothermal method to synthesize a hierarchy of graphene oxide/zinc nanocomposites, and more. Interestingly, the membrane separation technique explains the removal and recovery of synthetic

A. D. Laksono (✉) · R. Damastuti · N. L. Amanah · M. H. Assa · Y. Cheng · H. Abdullah
Department of Materials Science and Engineering, National Taiwan University of Science and Technology, Taipei 10672, Taiwan, ROC
e-mail: andromeda@lecturer.itk.ac.id

A. D. Laksono
Study Program of Materials and Metallurgical Engineering, Institut Teknologi Kalimantan, Soekarno-Hatta Street Km. 15, Karang Joang, Balikpapan, East Kalimantan 76127, Indonesia

L. Ernawati
Department of Chemical Engineering, Institut Teknologi Kalimantan, Soekarno-Hatta Street Km. 15, Karang Joang, Balikpapan, East Kalimantan 76127, Indonesia

A. Nugroho
Department of Chemical Engineering, Universitas Pertamina, Simprug Kebayoran Lama, Jalan Teuku Nyak Arief, Jakarta 12220, Indonesia

H. Abdullah
Department of Industrial Engineering, Universitas Prima Indonesia, Medan, Indonesia

H. Abdullah (ed.), *Photocatalytic Activities for Environmental Remediation and Energy Conversion*, Green Energy and Technology,
https://doi.org/10.1007/978-981-19-6748-1_1

colors from wastewater, which is a potential technical route. Functional composite membranes integrating photoelectrocatalysis and membrane filtration have been the subject of recent research. Finally, carbon-based materials are prospective to improve photocatalytic reaction due to their adsorption capability.

Keywords Wastewater · Photocatalysis · Adsorption · Industry waste · Textile

1 Introduction

1.1 Textile Industry

Contamination of water is presently one of the most severe problems of the world caused by textile producers' inability to dispose of their effluent correctly. The textile industry has a significant role in the world economy and pollution in several locations, including China and the South African estuaries [1]. Dye-containing sewage is a significant pollution problem of the environment that also influences human beings since textile firms create a substantial quantity of vivid wastewater that is tinted, carrying toxins that stay in the body for a long time [2, 3]. Every year, approximately around the world, 7×10^7 tons of synthetic dyes are produced, with the textile sector using over 10,000 tons of these hues [4]. According to their application, dyes, origin, and structure are frequently categorized into many groups [5]. The most often used synthetic dyes in textile manufacturing include reactive, mordant, azo, direct, dispersion, sulfide dyes, base, and acid (Fig. 1). Organic and inorganic fibers used in the textile manufacturing business include cotton, silk, wool, acrylic, polyamide, and polyester [6]. Furthermore, textile companies use a range of highly harmful intoxicants during the manufacturing procedure, including size, softening, desizing, whitening, and finishing agents [7]. On the other hand, a direct discharge of textile dye wastewater into aquatic environments such as ponds, lakes, streams, and rivers causes significant ecotoxicological problems and toxic effects on biological creatures [8, 9].

Textile industry wastewater is regarded as one of the most horrific sources of pollution to our valuable water and soil environment. It is mutagenic, carcinogenic, cytotoxic, and genotoxic causing other hopeless hazards to natural compounds. The textile industry wastewater contains various pollutants (colors), phenol, surfactants, formaldehyde, phthalates, pentachlorophenol, perfluorooctanoic acid destroyer (PFOA), and several heavy metals such as lead (Pb), arsenic (As), cadmium (Cd), nickel (Ni), zinc (Zn), and chromium (Cr). The wastewater is available in a high variety, high pH, biochemical oxygen demand (BOD), chemical oxygen demand (COD), total suspended solids (TSS), total dissolved solids (TDS), sulfuric acid salt, chlorides, and total organic carbon (TOC). Appropriate treatment is expected to safeguard population health and the environment before critical discharge into water bodies.

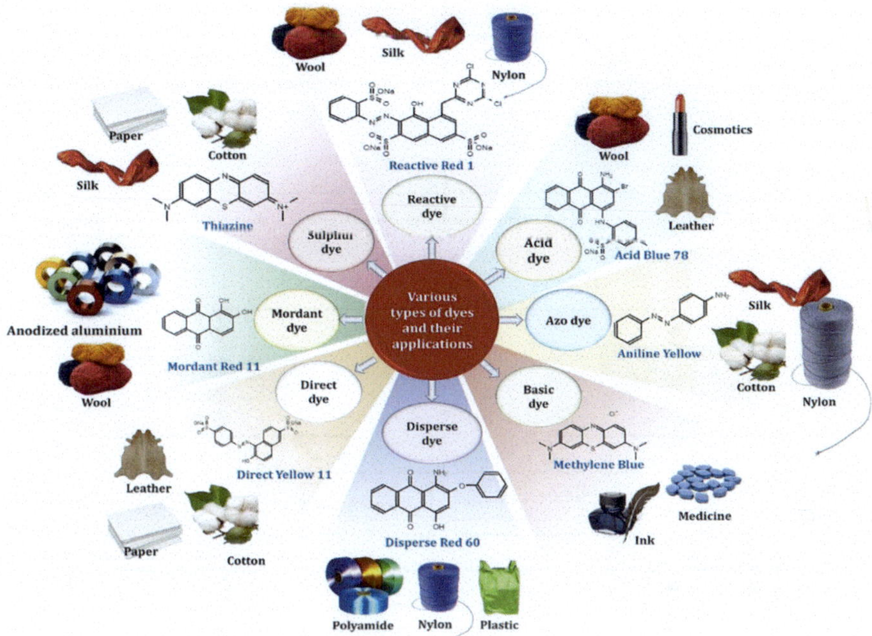

Fig. 1 The industrial potential of several dye groups. Reproduced with permission. [8] Copyright 2022, Elsevier (License Number: 5323471112278)

Treatment of textile industry wastewater is a crucial test since there is no particularly effective treatment for textile industry wastewater. It is critical to nurture innovative, modest, and biological system-friendly advancements to adapt to textile industry wastewater truly. This review article is undoubtedly wrapped in the altered material current cycles, the time of sewage, its propensity and designed course of action, regular consequences and wellness threats, and treatment options available for textile industry wastewater treatment. It also depicts a variety of top to bottom methodologies, snags, major focal points, and prospects for textile industry wastewater impurities and their metabolites identification and presentation [10].

The textile industry influences the environment in water pollution since it produces significant volumes of discharges mixed with diverse colors. Effective treatment processes employing specialized adsorbents or other advanced technologies are necessary to remove pollutants before discharging wastewater into waterways; they must be treated.

Standard and engineered colors are two sorts of colors utilized in material materials, as seen in Fig. 2. The most recent tendency is to discard traditional colors due to population expansion [11].

To minimize ecosystem degradation and promote sustainability, suitable and efficient administration solutions are essential because of the enormous quantity of

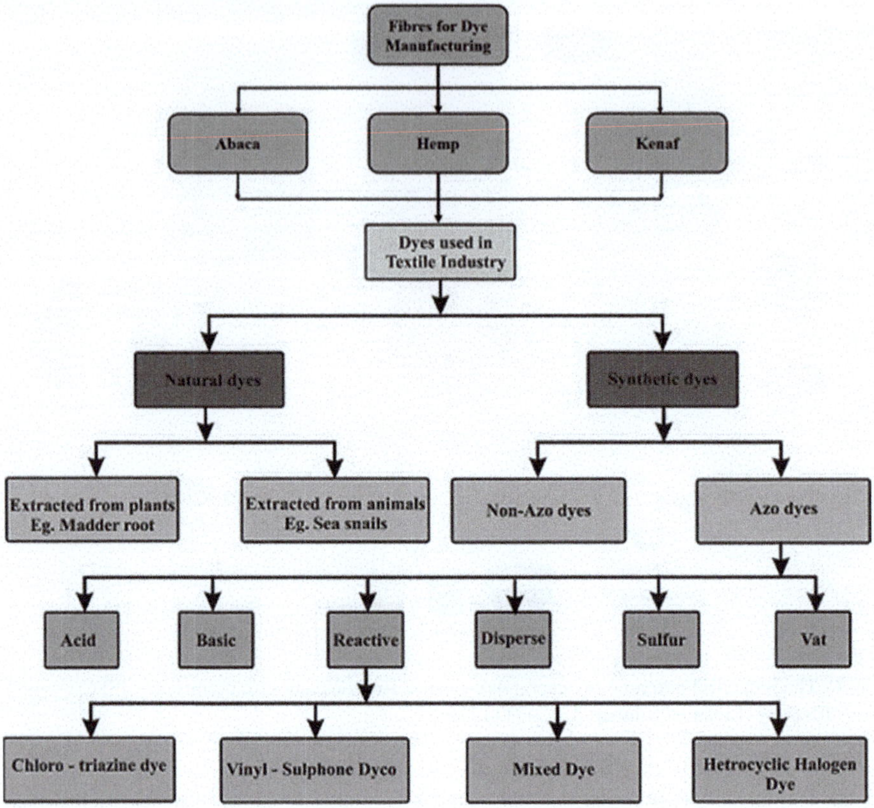

Fig. 2 Textile colors produced from distinct fibers are classified and utilized in the textile industry [11]

textile enterprises and dye wastewater. When choosing the optimum treatment technique, however, the occurrence of various organic and inorganic compounds, their toxicity, and the applicable discharge onto the environment requirements must all be considered [12, 13].

Azo hues create a wide range of different forces and types. Corrosive hues are ideal for silk, nylon, and fleece textures. Cowhide, cosmetic care goods, printing, and culinary items are all made using it. When metal-complex colors, commonly used in fleece textures, react with water, they cause cancer [14]. Exact colors that are solvents in water have been used on rayon, nylon, and cotton fabrics. Basic tones are frequently used in several types of nylon and polyester texture materials. They are also utilized in medications as a killing specialty. Receptive hues can benefit both silk and celluloid fiber materials [11].

Various wastewater treatment procedures, including natural, synthetic, and genuine medications, are used to remove material colors. Figure 3 depicts a few of the

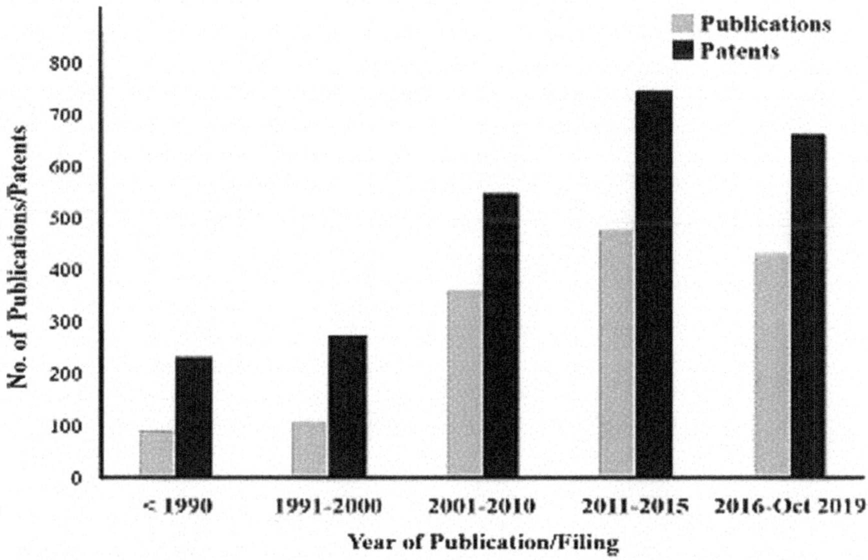

Fig. 3 The publication/patent filing is prevalent in the field of textile effluent treatment [11]

most well-known methods for removing colors from material effluents using various medications. The schemes still work out the arrangements and types of wastewater in the schemes [11].

Due to the massive water quantities needed in dyeing operations, the textile industry generates a huge amount of dye effluent [15]. A variety of treatment strategies for removing dye from wastewater samples have been studied. These are chemical oxidation [16, 17], adsorption [18, 19], chemical reduction [20], photodegradation [21], electrochemical oxidation [22], coagulation–flocculation [23], membrane separation [24, 25], Fenton oxidation [25, 26], and biological methods [27]. Each one of these strategies has advantages and disadvantages. The adsorption method has garnered special attention due to its simplicity, cheap cost, and excellent substance recovery [28].

Textile effluent is hazardous wastewater that includes toxic complex components that can cause cancer, among other problems. If not adequately handled, those toxic components will have a substantial environmental impact, harming human health and aquatic ecosystems. The current research examines the chemicals and dyes used in the textile sector. It emphasizes the traditional treatment methods for removing them from industrial wastewaters, including physical, chemical, hybrid, and biological treatments. There is also a look at remote management solutions for remote water quality monitoring and control based on the Internet of Things. The lack of a single platform for integrating old treatment methods with modern technologies has been identified as a significant stumbling block to successful sustainable production. Consequently, this study aims to increase the number of feasible solutions by presenting cost-effective, high-performance combination approaches that promise future industrial applications (Fig. 4) [11].

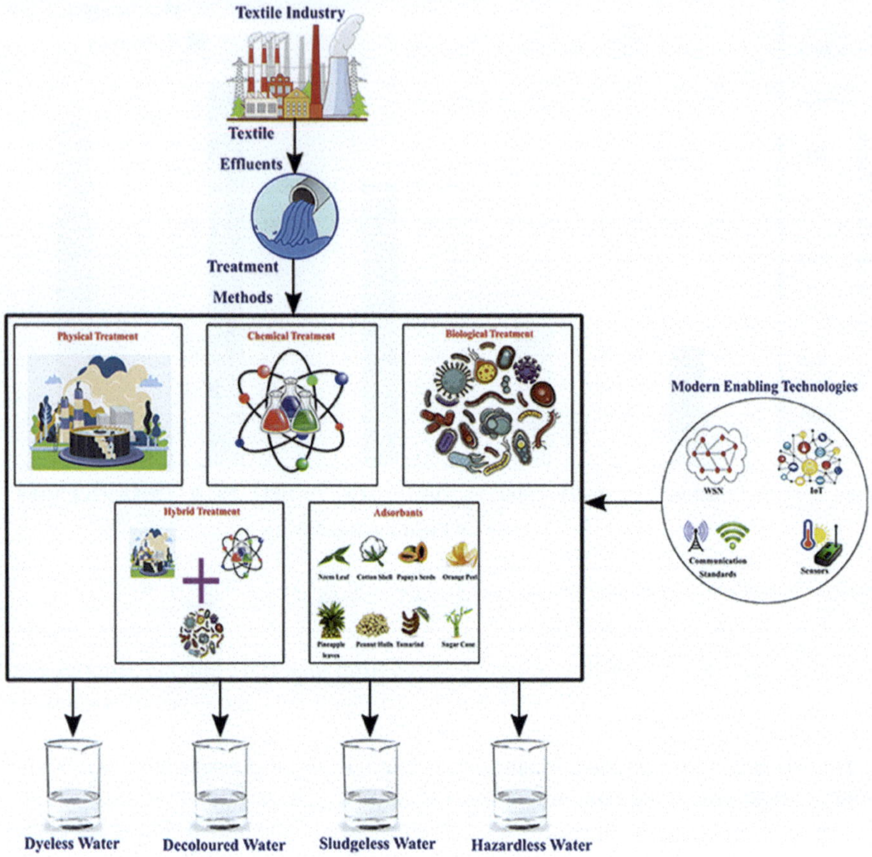

Fig. 4 Dye removal from textile effluents treatment techniques [11]

1.2 Chemical Treatment

Chemicals such as PAC, BacEnz, and poly are used in the chemical treatment of dye effluent. The chemical PAC has the virtue of settling sludge quickly in borosilicate beakers. Sludge and dye wastewater are separated after chemical treatment [29]. Photocatalytic reactions were also discovered to be responsible for dye removal. The photocatalytic process focuses on the formation of the bandgap, surface area, and electron–hole pairs for the breakdown of dyes. It has a larger surface area and a better adsorption capacity—experiments on flow rate, COD, pH, and other operational variables [30]. The jar test determines how well the coagulation process is working. In addition, the operating costs of the different advanced oxidation processes are calculated. Textile effluents treated using Poly Aluminum Chloride (PAC) are compact, easy to regulate, inexpensive, and compact.

Pollutants often identified in textile wastewater include the following:

1. Concentrations of non biodegradable organics or inorganics like metals, dyes, phenols, pesticides, phosphates, and surfactants.
2. Elevated concentrations of total suspended solids (TSS) [31].
3. High biological oxygen demand (BOD) [32].
4. High chemical oxygen demand (COD) [33].
5. High levels of total dissolved solids (TDS) [34].

As a result, appropriate treatment is required before ultimate discharge into water bodies to preserve human and environmental health.

1.3 Biological Treatment

Biological treatment is used to achieve depigmentation [35]. The treatment necessitates the right proportioning of materials such as cow dung, jaggery, and stirring of the material to develop bacteria. The method is low-cost, employs a fundamental technology, and improves the decolorization effect. Biological therapy is thought to be the most effective way of decolorization. Chemicals like biocides and stain repellents are utilized to soften and moisten materials in wastewater [11].

1.4 Physical Treatment

The reverse osmosis (RO) process involves moving water from a higher concentration to a lower concentration. Purified water with a pH of 7 was obtained due to this treatment, and it was safe to drink. The wastewater undergoes five phases, extracting pure water using a reverse osmosis technique [11].

Dye-containing wastewater, adsorption, ion exchange, and membrane filtration are among the physical processes used, with high clearance effectiveness ranging from 85 to 95% [36]. Physical methods have many benefits, incorporating a straightforward design, ease of use, fewer chemical requirements, cheap cost, and no obstruction from dangerous compounds [37, 38]. These technologies are typically not chosen due to several restrictions, such as hazardous by-products, sludge production, and limited application [5]. In addition, BOD, COD, high temperatures, pH, heavy metals, and color. Their use in textile wastewater treatment is usually hampered.

1.5 Adsorption

There are two forms of adsorption: physisorption and chemisorption. This categorization is determined by how the dye molecules are adsorbed onto the surface of the adsorbent [39]. Several forces may be present during dye adsorption, including van der Waals forces, hydrophobic and electrostatic interactions, and hydrogen bonding [40]. The adsorption process offers several benefits, including the flexibility to reuse the adsorbent, high efficiency, and the capacity to quickly remove the color from wastewater [5, 41]. Adsorbents, which are often porous to measure the overall surface area necessary for the rapid and effective adsorption of pigment molecules from wastewater, are at the center of the adsorption idea [36, 42]. Adsorbents such as zeolites, alumina, silica gel, and activated carbon have been used to remove colors from wastewater. On the other hand, activated carbon is an adsorbent regularly employed on a vast basis [43]. For Basic Fuchsin (15 mg/L; pH 9.0), Crystal Violet (15 mg/L; pH 7.5), and Methylene Blue (15 mg/L; pH 8.0), the Biopolymer/ZSM–5 zeolite adsorbent was used to treat dye-containing wastewater, with decolorization percentages of 881.2, 775.3, and 86.6, respectively [44]. Up to 90% Congo Red decolorization was obtained using ZnO@Ze composite particles as an adsorbent at a dye concentration of 25–500 mg/L and an adsorbent dosage range of 0.025–0.1 g/L [45].

1.6 The Need for Water Consumption in the Textile Industry

The textile industry is the world's third greatest consumer of freshwater resources, contributing significantly to the business industry. Water is required in the manufacturing business at many phases of the production process. The average textile industry requires between 230 and 270 t of water to make 1 t of finished textile fabric [46]. By 2050, global water consumption for manufacturing enterprises is predicted to increase by 400% [47]. Water is used only in textile processing, such as bleaching, desizing, finishing, scouring, and dyeing [48].

One of the most significant sectors contributing to the exploitation of water resources to the point of exhaustion worldwide is the textile industry [49]. Increased manufacturing and other operations connected to the process directly impact human health and environmental resources, necessitating a cost-effective and efficient solution using current and practical methods [50]. As the uncertainty and hazards associated with overstressed natural resources develop, an increasing emphasis has been placed on the notions of "cleaner production" and "greener economy" firms in the industrial sector, which have a direct detrimental influence on economic profitability and growth [51]. In addition, the over-reliance on groundwater, the scarcity of freshwater supply sources, and the textile industries becoming more numerous make the current state of water management critical in the textile business [47].

The textile sector is regarded as one of the most severe environmental risks. The many steps of the industrial textile production process include pre-treatment, dyeing, printing, and finishing. These industrial procedures consume a lot of energy and resources and produce much trash. To give cloth desired attributes, the textile industry uses a variety of chemicals, dyes, and other materials, and these activities generate much waste. Sewage is of such excellent quality that it cannot be utilized for different reasons and, if not adequately treated, it might cause environmental hazards [52].

1.7 What is the Danger?

Complex compounds, bases, acids, colors, inorganic salts, and heavy metals abound in textile industry wastewater [53]. Textile waste is characterized by the US Environmental Protection Agency (USEPA) as a large-volume, dispersible harmful pollutant that is difficult to address [54]. Pigments are the principal and stubborn toxins that are harmful and inherently carcinogenic among the different contaminants found in the effluent [55]. The release of raw sewage into freshwater bodies or sewer systems reduces dissolved oxygen, increases turbidity, hinders the biological treatment process, and unsightly hurts the overall water quality of river and the surrounding environment [56, 57]. It is hazardous to biotic and abiotic environmental components, including water, air, aquatic life, terrestrial, and soil [38, 58, 59].

Textile effluent contributes to severe environmental deterioration as well as human sickness. Across 40% of all colorants in worldwide use include organically bonded chlorine, a recognized carcinogen. Due to their toxic effect and sensitivity to various disinfectants, they are not recommended for usage; organic compounds found in textile effluent constitute a significant source of concern [60]. Even at low concentrations, the occurrence of dyes in drinking water is damaging to humans and aquatic life. Furthermore, the dyes are hazardous, carcinogenic, and challenging to remove efficiently. They are incredibly stable, nonbiodegradable, and common contaminants. Humans get allergic reactions, dermatitis, skin irritation, cancer, and mutations due to this substance [61].

1.8 Toxicity and the Impact of Textile Dyes

The untreated effluents of the textile sector contain a variety of organic pollutants, the most frequent of which are fabric dyes [62]. Azo dyes, which structurally incorporate one or more azo groups, are the most extensively used in the textile industry, accounting for more than 60% of all textile dyes [63]. Inefficient textile dyeing techniques result in 15 ~ 50% of azo dyes not bonded to fibers and textiles disposed in wastewater [64]. The untreated azo dyes are released to the environment directly into water bodies and pose severe ecotoxicity concerns, which harm the environment and living things. For farmers in developing nations, if the unclean industry effluent-containing wastewater is used to irrigate the land, it will cause soil quality and crop germination percentages to suffer [8, 65, 66].

Because of its water-intensive nature, the textile sector is among the most environmentally hazardous, resulting in the depletion of freshwater and groundwater. The researchers were forced to evaluate the suitability of various textile wastewater treatment techniques for effective effluent disposal and reuse due to the rising scarcity of water resources and deterioration of water quality. Textile wastewater can be treated in a variety of ways. The four types of processes are hybrid, biological, chemical, and physical [38]. Most of these processes differ in their core operating principles and are compared in performance, costs, and advantages and disadvantages.

Furthermore, the construction of textile wastewater treatment facilities requires extensive infrastructure, uses much energy during operation, degrades ecosystem hydrology, and leaves a carbon imprint [67]. The effluent treatment reduces pollutants and makes it suitable for reuse and disposal, producing hazardous sludge, which is expensive. As a result, selecting a long-term textile wastewater treatment technique is a complex and demanding task where decision-makers must resolve several conflicts. The appropriate textile wastewater treatment may be a multi-criteria decision-making difficulty [59].

1.9 Textile Dyes Have an Impact on Both Aquatic and Terrestrial Habitats

Even though liquid covers over 70% of the earth surface, a shortage of clean, high-quality drinking water is one of humanity's most serious challenges today. Industry, agriculture, and homes consume 19, 70, and 11% of freshwater resources, respectively [68, 69]. People have negatively interacted with natural settings. The textile sector undoubtedly contributes to the deterioration of water quality, which will continue to poison existence eternally by depositing toxins in terrestrial ecosystems. The textile industry contribution and marine environments are significant to the decline of the quality of water, which will pollute nature indefinitely [70].

Large volumes of treated effluent, including organic substances and other organic pollutants, are treated in the textile business. Because azo dyes break down pre- or post-disposal, treated wastewater may include amino acids that are more hazardous than the original chemicals. However, the health risks associated with untreated wastewater have numerous detrimental effects on aquatic creatures and the environment [40, 71].

1.9.1 Impact of Textile Dyes on Human Health

Textile dyes have been connected to various human and animal ailments because they are poisonous and potentially carcinogenic [72, 73]. Textile pigments can cause everything from dermatitis to central nervous system problems. The replacement of the enzyme co-factor, which leads the enzymes to become inactive, might be the source of these problems [74].

Several dyes, notably azo dyes, have the potential to be mutagenic. Sudan I, an azo-lipophilic dye used in the textile and food sectors, was enzymatically transformed by intestinal microbes into carcinogenic aromatic amines. Cancer can be caused by both the dye and its metabolites [75]. Azure B (a cationic dye that involves inserting into DNA and the double RNA helical structures) is a significant metabolite of methylene blue [76].

1.9.2 Solution to Use Carbon-Based Photocatalyst

Flocculation and coagulation, anaerobic biological treatment, membrane separation, sonolysis, UV/ozone treatment, photocatalysis, and biodegradation are only a few wastewater treatment procedures now accessible [61].

The only way to avoid a future energy crisis and environmental deterioration is to convert solar energy into sustainable, safe, and cost-effective. Water splitting, for example, is a clean technology. Graphitic carbon nitride has an appealing band structure, is chemically stable, is abundant on Earth, and is relatively inexpensive to manufacture, making it ideal for water splitting to make hydrogen. We sought to critically assess the current progress and future development of several water splitting systems that produce hydrogen using graphitic carbon nitride (g-C_3N_4).

Hydroelectricity, wind, solar light, biomass (solar drying, solar cooking), and geothermal energy are only some environmentally beneficial energy sources [77]. It is prudent to convert solar energy to chemical energy through photosynthesis to ease future energy crises and environmental deterioration. Sunlight harvesting may create ecologically benign hydrogen, which produces water or water vapor when burned with oxygen. As a result, making hydrogen as an energy source can assist in meeting future energy demands while also addressing environmental concerns [78].

Hydrogen with a heat value of 120–142 MJ/kg derived from solar energy is considered clean energy [79]. Hydrogen is being produced by over 44.5 million tons worldwide, and it will be the dominant energy source until 2080. Although there are other methods for photocatalytic water, producing hydrogen splitting is a promising method that has attracted much attention due to its various environmental and energy applications. Artificial photosynthesis is one of the few environmentally friendly options for dealing with future energy and environmental challenges [78].

One of the most critical areas is creating highly efficient semiconductor photocatalysts because of the increasing need for photocatalytic removal of harmful dyes. These colors are carcinogenic and poisonous, endangering both humans and aquatic life. Rhodamine B (RhB) and other cationic dyes have been recognized as a significant cause of discharge from various businesses [80]. Conventional approaches have been devised to partially degrade risky pigments in wastewater. In other situations, secondary pollutant production was also discovered [80, 81]. Fenton oxidation is one of many dye wastewater treatment processes [82, 83], physical adsorption [84, 85], biodegradation [86], and heterogeneous photocatalysis oxidation [87] have risen in popularity recently.

Physical adsorption only serves to increase the concentration of contaminants [88]. It is also challenging to find a good adsorbent. However, Fenton oxidation has indeed been demonstrated to treat colored wastewaters successfully. However, the expense is not monetary and the uses of iron sludge are limited [89]. Bio-degradation procedures are a less effective due to dye biological persistence [90]. Compared to other dye wastewater treatment processes, photocatalytic activity oxidation is now recognized as a viable choice due to its environmental contribution and maintenance ease [91]. However, the present efficiency of photocatalytic oxidation is insufficient for industrial use. For decades, scientists have been fascinated by the photocatalysts powered by visible light for environmental remediation [92, 93].

2 Carbon-Based Research for Wastewater Remediation

2.1 Carbon-Based Photocatalyst

This systematic review assembles and investigates available research on photocatalytic wastewater or organic pollutant degradation employing carbon-based materials in industrial textile wastewater treatment [94]. The textile sector is the most significant contributor to wastewater and pollution [95]. Many researchers used photodegradation of a specific pollutant to test the performance of their manufactured photocatalyst materials [96]. Material like carbon has been used in several studies to increase the photocatalytic performance of photocatalytic systems [97] because of eminent traits such as small particle size, high specific surface area-to-volume ratio, high reactivity, excellent chemical and high thermal stability, wide availability, and catalytic potential at the nanoscale [98] (Fig. 5).

As examples, Carbon Quantum Dots (CQDs) have eminent physicochemical properties, such as photoluminescence, light absorption, up-conversion luminescence properties, quantum effect, and non-toxicity [99]. On the other hand, Carbon Nitride

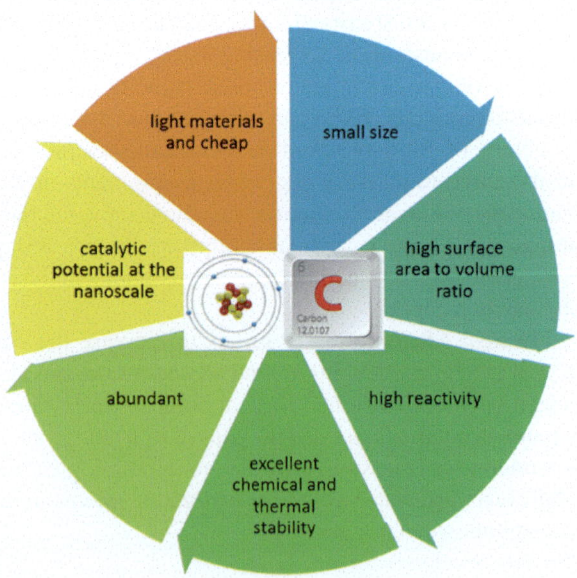

Fig. 5 Advantages of carbon materials as a supporting material of semiconductor photocatalysts

(CN) demonstrated vigorous photocatalytic activity, providing hydrogen generation and pollutant degrading activities while maintaining high cycle stability [100]. Carbon Fiber (CF) is employed because it is efficient, stable, and reusable with a simple reuse procedure [101]. Many more carbon-based materials can be used to develop the limitations of photocatalytic systems, including the restricted use of fast charge recombination, visible light, and poor movement ability of the photogenerated holes and electrons [102].

This section discusses the simple and standard synthetic methods recently used to synthesize carbon-based photocatalysts, including sol–gel, precipitation method, hydrothermal, solvothermal, thermal polymerization, polycondensation, spray pyrolysis, and electrospinning.

Chemical synthesis provides various benefits for photocatalyst materials synthesis, including the ability to size-controlled, distributed structure shape, and large-scale production potential, suitable particle size, unagglomerated nanoparticles, composition, and easy handling. As a result, the preparation procedure determines the morphology and cation order of photocatalytic materials [103]. Because it creates particles with appropriate shape and homogeneous size, the sol–gel approach is the most common and viable wet-chemical synthesis process for making homogenous nanocomposite. By modifying simple process factors like pH and annealing temperature and introducing reaction modifiers, it is easy to control the morphology and size [104]. To generate an inorganic bridge network, alkoxide-based metal precursors (MOR) are hydrolyzed and polycondensed in a sol–gel process. The term "sol" refers to solvated metal precursors that form in a colloidal suspension. The gel converts the produced into a homogenous liquid phase; sol is a network structure. Nanocomposite precursors are metal-based alkoxide components encased in various reactive ligands [105, 106].

In a typical synthesis, Fe_2TiO_5/C hollow spheres can be prepared via sol–gel method, as shown in Fig. 6. First, 0.24 mol C_2H_5ONa and 0.08 mol $FeCl_3$ were dissolved with 36 mL anhydrous ethanol for 2 h with vigorous continuous stirring. After forming $Fe(OC_2H_5)_3$, filtration was used to remove the by-product of NaCl. A precursor was prepared by reacting 0.08 mol of $Ti(OC_4H_9)_4$ and then adding it to the solution by mixing it using magnetic stirring for 1 h at 40 °C. The sol was then refluxed with continuous stirring in an oil bath at 80 °C for 24 h, resulting in the development of wet gel. A xerogel was obtained when the wet gel was dried at 110 °C for 6 h. Finally, the xerogel was calcined in air at 200–900 °C (heating rate 55 °C/min) [107].

The wet-chemical technique to produce mixed metal oxide nanocomposites is a precipitation technique. It is a bottom-up preparation method. The precipitation process forms nanoparticles from solution; the process is entirely dependent on the existence of specific nuclei. Various amounts of ionic metals present in dissolution, obtained precipitation, reaction conditions (pH of the solution), and the presence of additional dopants may influence the reaction [108]. Figure 7 shows a typical co-precipitation method for micro- and nanoparticle synthesis. Coprecipitation has three primary mechanisms: inclusion, occlusion, and adsorption. An inclusion (formation in the crystalline lattice) happens when a contaminant occupies a lattice site in the

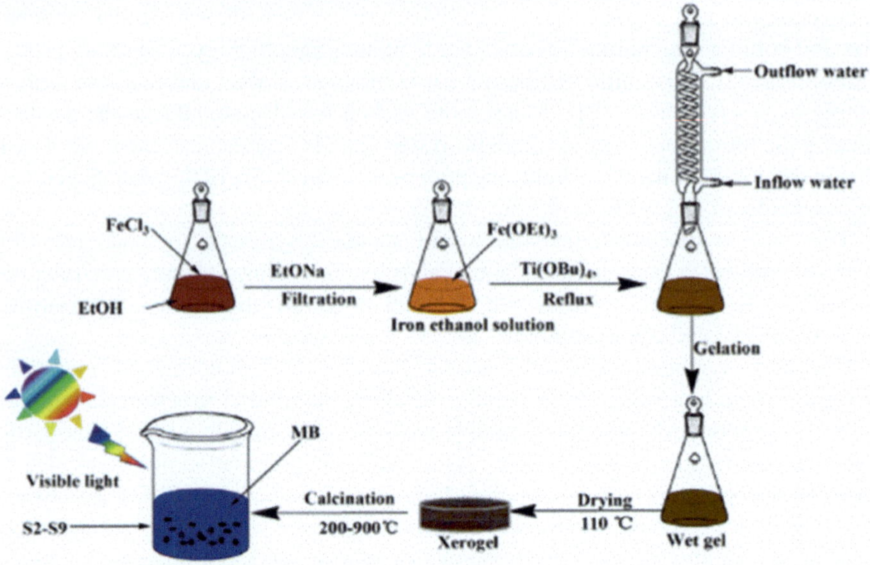

Fig. 6 Sample preparation process via sol–gel method [107]

matrix structure, producing a crystallographic defect. The formation occurs when the atomic radii and charge of dopants are present close to its matrix. An adsorbent surface is a dopant that is lightly or firmly attached (adsorbed) to the matrix surface. When the adsorbed species becomes physically stuck inside the crystal, this is referred to as an occlusion.

Hydrothermal refers to an environment operated at high temperatures and high pressures involving water in a chemical reaction for the synthesis of material. The equipment includes an autoclave made of high-quality stainless steel and a Teflon coating, which can protect the reactants from contaminants and environmental reactions. Water is used in the hydrothermal process to synthesize nanocomposites with unique morphologies; water can operate as a catalyst as well as a reducing agent at high temperatures (over 100 °C) and pressure (approximately a few atmospheric pressure) [109].

In particular research work, Perumal et al. used hydrothermal-assisted precipitation to create a g-C_3N_4/BiOBr/CdS photocatalyst (Fig. 8). The structure and morphology of the as-prepared photocatalyst were described. The photocatalytic performance of the visible-light-driven g-C_3N_4/BiOBr/CdS photocatalyst toward TC (Tetracycline) degradation was improved. The g-C_3N_4/BiOBr/CdS photocatalyst had the best photocatalytic degradation rate performance, outperforming 8.95, 3.11, and 3.18 times compared to g-C_3N_4, BiOBr, and BiOBr/CdS composite, respectively, with a high stability. The improved photocatalytic effectiveness was primarily due to a fast charge transfer at the g-C_3N_4/BiOBr/CdS interfaces, which boosts photogenerated charge-carrier separation efficiency and improves optical absorption [110].

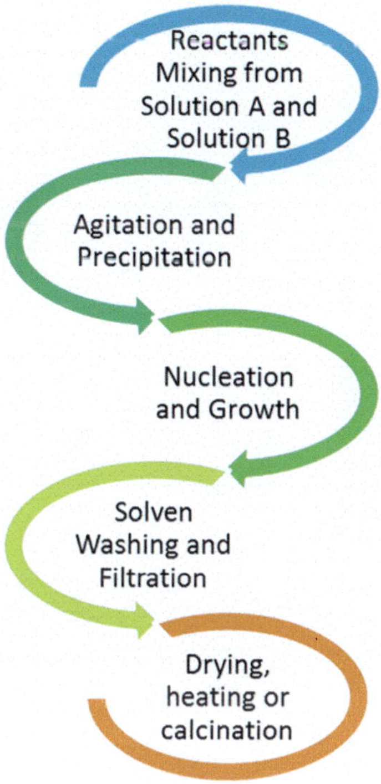

Fig. 7 Schematic diagram of co-precipitation

Chemical synthesis (solvothermal synthesis), on the other hand, is an effective method for obtaining nanostructural materials with shape control. The results are impacted by more grain, scattering limits, reducing thermal conductivity, and boosting quantum confinement. The solvothermal process in a stainless-steel autoclave with high pressure is required to induce a chemical reaction. The primary benefit of this method is the risen temperature and pressure at a critical level influence the material solubility [111, 112].

Yuan Guan et al. (Fig. 9) prepared carbon quantum dots (CQDs)/Bi/BiOBr with a facile one-pot solvothermal method. It has shown an efficient ternary heterojunction photocatalyst to degrade methyl orange (MO). For the first time, ethylene glycol (EG) is utilized as a solvent and carbon source. While BiOBr is synthesized at 190 °C for 3 h, EG is used to create CQDs using the bottom-up technique. CQDs and Bi are controllably deposited on top of BiOBr microspheres by varying the solvothermal temperature and duration. CQDs, BiOBr, and Bi interfaces are coupled through ohmic junctions with low contact impedance by varying Fermi levels and work functions [113].

The solvothermal method is very similar to the hydrothermal route and the difference is that the reaction medium is not an aqueous solution. Weili Yu et al. synthesized $Bi_2W_{20}Mo_{1.x}O_6$ by solvothermal assisted with calcination. The photocatalyst

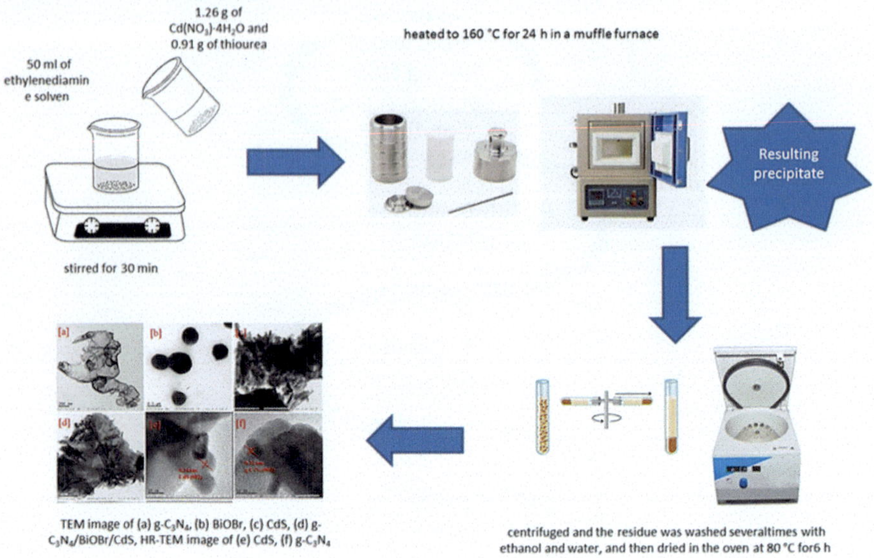

Fig. 8 Hydrothermal-assisted precipitation synthesis of highly stable g-C_3N_4/BiOBr/CdS photocatalyst. Reproduced with permission. [110] Copyright 2022, Elsevier (License Number: 5316450032079)

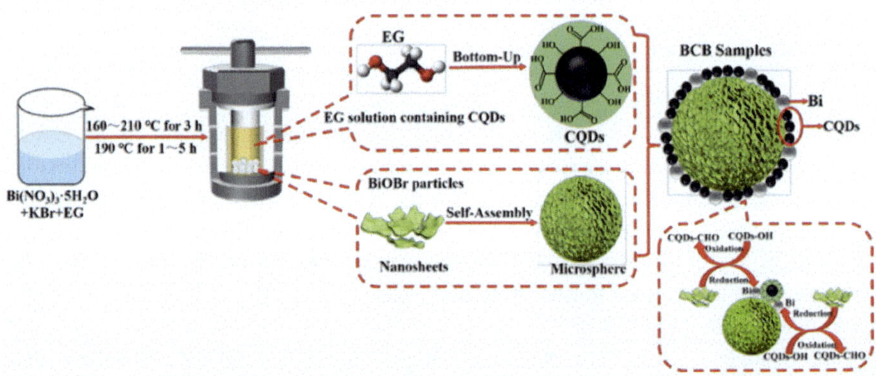

Fig. 9 Schematic illustration of the preparation process for BCB samples using solvothermal. Reproduced with permission. [113] Copyright 2022, Elsevier (License Number: 5316860555938)

was applied for bisphenol A (BPA) degradation under simulated sunlight irradiation [114]. On the other hand, calcination is the thermal treatment of a solid chemical compound (e.g., mixed carbonate ores) in which the mixture is heated to high temperatures without melting under a limited supply of ambient oxygen (i.e., the gaseous O_2 fraction of air), typically to remove impurities or volatile substances and to undergo thermal decomposition (Fig. 10).

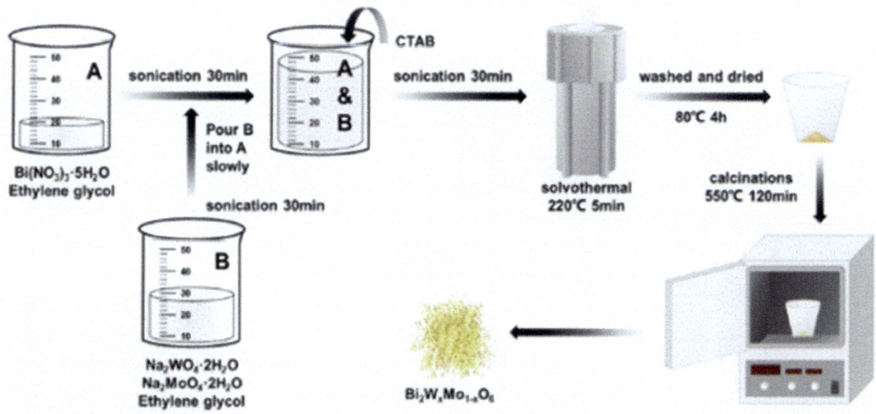

Fig. 10 Schematic drawing for the fabrication process of $Bi_2W_{20}Mo_{1.x}O_6$. Reproduced with permission. [114] Copyright 2022, Elsevier (License Number: 5316830459077)

Some reactions involve organic materials during the preparation. In the case of polymerization, polycondensation occurs when monomers with two or more reactive functional groups (e.g., hydroxyl, carboxyl, and amino) condense with one other. The following are the primary characteristics of polycondensation: (i) the monomer must have two reactive functional groups to continue polymerization; (ii) polymerization occurs through sequential interactions between reactive functional groups; (iii) in polymer formation, just one reaction (in this case, a condensation reaction) occurs between two functional groups; and (iv) if the polymer still has two reactive functional groups at the end of the chain, it is "active" rather than "dead," as in chain polymerization [115, 116]. Rui Zhang et al. used the thermal polycondensation method to synthesize graphite phase carbon nitride (g-C_3N_4) (Fig. 11). The g-C_3N_4 is coupled with Bi_2WO_6 and activated carbon fiber membrane (ACF) as a thin-film photocatalyst can be utilized as a reusable photocatalyst with a vast surface area to remove water antibiotics in the environment [117].

Furthermore, spray pyrolysis is another typical method of depositing a thin coating on a heated surface by spraying a precursor solution, where the chemical elements react to obtain a chemical product. The chemical reactants are designed with a particular deposition temperature to vaporize the volatile products except for the target molecules. In a droplet forming system, the precursor solution is atomized. Spray pyrolysis occurs at a somewhat high temperature; the vapor phase may be carried in the air to produce transparent thin films of oxides such as F-doped SnO_2, TiO_2, graphene oxide, ZnO, and other compact layers on the surface of a transparent conductor. Spray pyrolysis was selected as a processing method since it is easy and reasonably priced (especially concerning equipment cost) [118].

The spray pyrolysis technique is a thermal deposition process employing metal precursors. Several processes are involved in the reaction procedure, including.

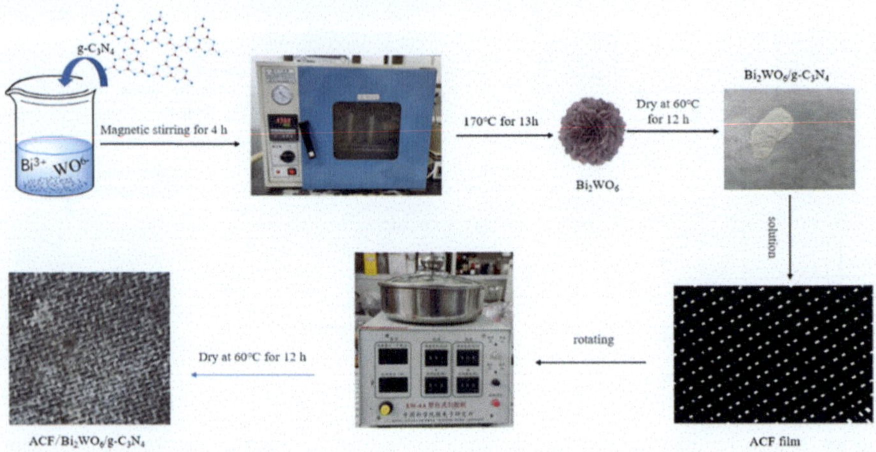

Fig. 11 Bi_2WO_6/g-C_3N_4/ACF preparation process. Reproduced with permission. [117] Copyright 2022, Elsevier (License Number: 5316980791297)

1. Precursor or starting materials solution loading in a tiny needle to generate minute droplets called as atomizer.
2. Droplet spraying on the top of the targeted thermal substrate.
3. Fabrication experiment and deposition materials of mixed metal oxide films onto thermal barrier substrates by thermal decomposition.

To produce CNTs/Cu_2O-CuO composite powder with uniform dispersed CNTs, the spray pyrolysis (SP) technique is used (Fig. 12), which has benefits such as cheap cost, simple and continuous process, production of high purity powder, chemical flexibility, and easy control. Furthermore, the photocatalytic performance of the manufactured samples was tested by degrading and decreasing methyl orange (MO) under visible-light irradiation [119].

The electrospinning method is a strategy to form diverse photocatalysts into nanofibers with some advantages of simple, adaptable, and cost-effective method for mass-producing tunable structures. Electrospun fibrous materials have widespread use in catalysis because of their specific large surface area, variable chemical composition, shape, fiber diameter, and high foaming [120]. The direct electrospinning procedure for photocatalytic fibers consists of three simple steps: photocatalyst production, continuous spinning solution preparation including photocatalyst and polymer material, and direct electrospinning. The essential influencing parameters to be regulated are the polymer type, electrospinning conditions, and spinning solution property [121].

By combining electrospinning and hydrothermal procedures, Ag_3PO_4-TiO_2 NPs were formed into carbon nanofibers (Ag_3PO_4-TiO_2-CNFs) material photocatalyst (Fig. 13). Ag_3PO_4 NPs on the TiO_2-CNFs composite enhance light absorbance under visible light and provide a low-resistance route for charge separation when

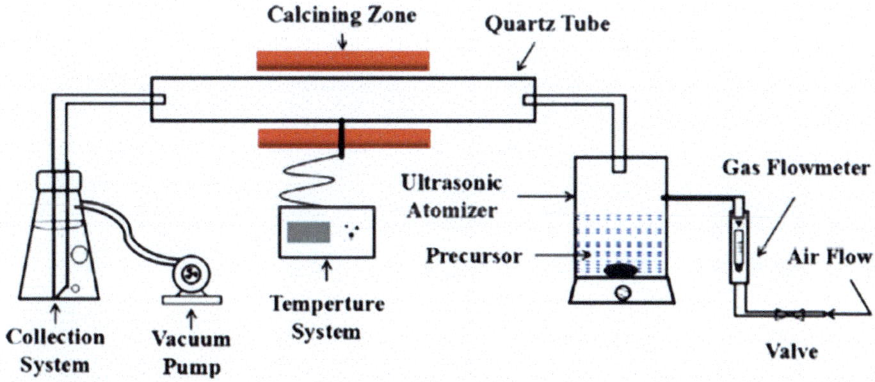

Fig. 12 Schematic of spray pyrolysis process. Reproduced with permission. [119] Copyright 2022, Elsevier (License Number: 5319690503879)

stacked up to TiO_2 NFs. Performance photocatalytic activity of the Ag_3PO_4-TiO_2-CNFs composite was investigated using degradation of a methylene blue (MB) tested under visible-light irradiation. The Ag_3PO_4-TiO_2-CNFs composite outperformed pristine Ag_3PO_4 NPs and TiO_2 NFs in photocatalytic activity. Within 10 min of being exposed to visible light, the Ag_3PO_4-TiO_2-CNFs composite completely removed MB dye. Furthermore, the reusability test revealed that the photocatalyst is relatively stable. The high specific surface area, higher visible-light sensitivity, rapid charge separation, and dye adsorption characteristic of the composite material contribute to the strong photocatalytic activity. In addition, the photocatalyst demonstrated vigorous antibacterial activity against Escherichia coli (E. coli) germs. The as-prepared Ag_3PO_4-TiO_2-CNFs photocatalyst with a remarkable dye removal capacity and high antibacterial capability shows a promising photocatalyst in wastewater treatment [122] (Table 1).

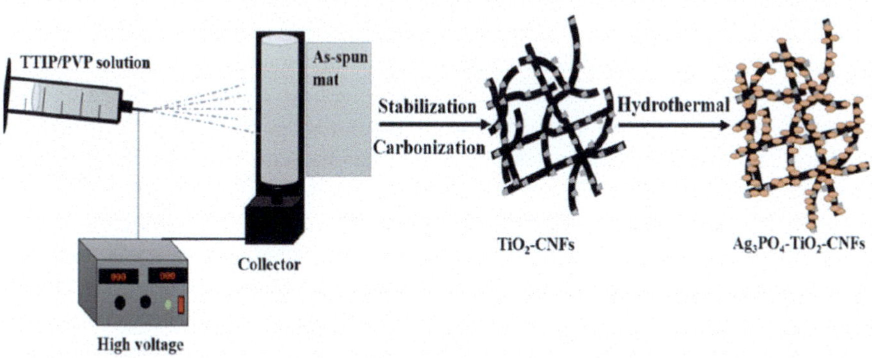

Fig. 13 The synthetic protocol of Ag_3PO_4-TiO_2-CNFs photocatalyst. Reproduced with permission. [122] Copyright 2022, Elsevier (License Number: 317530698921)

Table 1 Various synthesis methods to produce carbon-based photocatalyst materials at different temperatures and times (related to Table 2)

No	Materials	Methods	Reaction	Fig. number	References
1	CF/TiO_2	Hydrothermal	160 °C for 10 h	1	[123]
2	CQDs/BiOBr	Hydrothermal	140 °C for 24 h	2	[124]
3	CQDs/$KNbO_3$	Hydrothermal and mixed-calcination	160 °C for 12 h	3	[125]
4	Bi_2WO_6/gC_3N_4/ACF	Thermal polycondensation	500 °C for 4 h	4	[117]
5	g-C_3N_4 QDs/$BiVO_4$	Sol–gel and mixed hydrothermal	200 °C for 10 h	5	[126]
6	Ag_3PO_4-TiO_2-CNFs	Electrospinning and hydrothermal methods	60 °C for 12 h	6	[122]

2.2 *Binary and Ternary Carbon-Based Photocatalysts*

Several ways indicate that wastewater treatment techniques are available for wastewater purification, such as electrochemical, bioremediation, electrodialysis, adsorption, flotation, coagulation, and ultrafiltration [127–130]. However, some of the techniques require high costs with various limitations. Therefore, developing a novel and cheap treatment approach for enterprise waste and wastewater is predictable. There is a critical demand for a clean (eco-friendly), economic, and technological feasible way to produce green energy and effectively decompose pollutants to CO_2 and H_2O [131, 132]. Hydrogen appears to be a potential answer for overcoming the energy issue as renewable energy, clean, and sustainable source [133]. Because hydrogen is an abundant element that occurs naturally in the form of water, its generation from water using solar energy has attracted researchers' interest owing to its potential to fulfill global energy needs and address environmental issues [134, 135].

Carbon-based materials including graphene, graphitic carbon nitride (g-C_3N_4), fullerenes, carbon quantum dots, and carbon nanotubes (CNTs) have been studied for Li-ion batteries, supercapacitors, energy storage, biosensors, molecular imaging, fuel cells, and catalysis over the last two decades [136, 137]. Recently, photocatalysis has increased interest in carbon-based materials and semiconductor nanocomposites. Researchers have reported most pioneering works on nanocarbon–semiconductor interface engineering for environmental and energy applications. They often begun in binary or tertiary semiconductors by absorbing light energy equivalent to or greater than the bandgap of the semiconductor photocatalyst. Electrons are stimulated from the semiconductor valence band (VB) to the empty conduction band (CB), resulting in the formation of electron–hole pairs [138].

Table 2 The morphology of the results obtained from carbon-based photocatalyst materials

No	Figure	References
1		Copyright 2022, Elsevier (License Number: 5310080120301) [123]
2	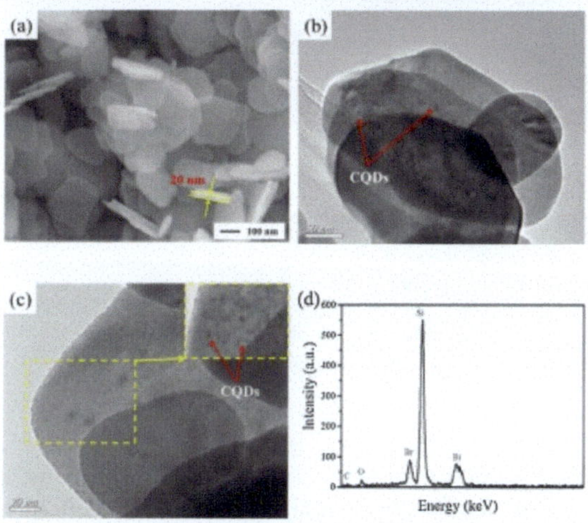	Copyright 2022, Elsevier (LicenseNumber: 5310090546516) [124]
3	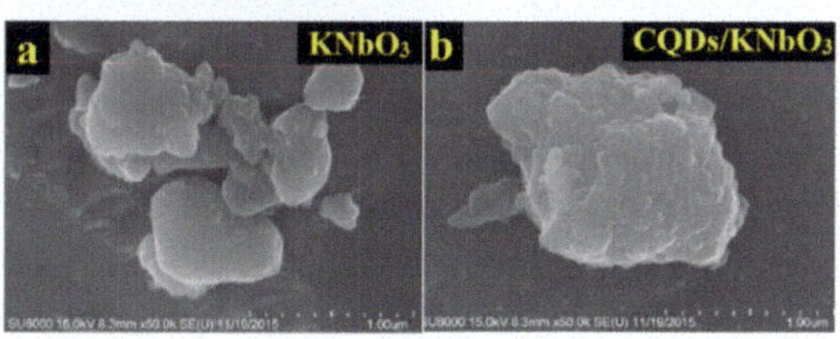	Copyright 2022, Elsevier (License Number: 5310090820453 [125]

(continued)

Table 2 (continued)

No	Figure	References
4	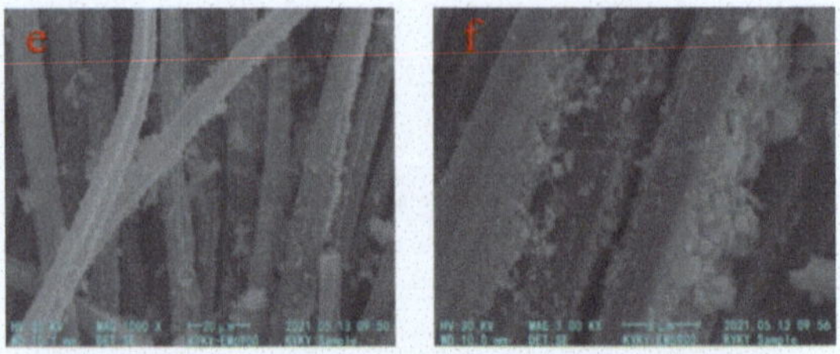	Copyright 2022, Elsevier (License Number: 5310091017897 [117]
5	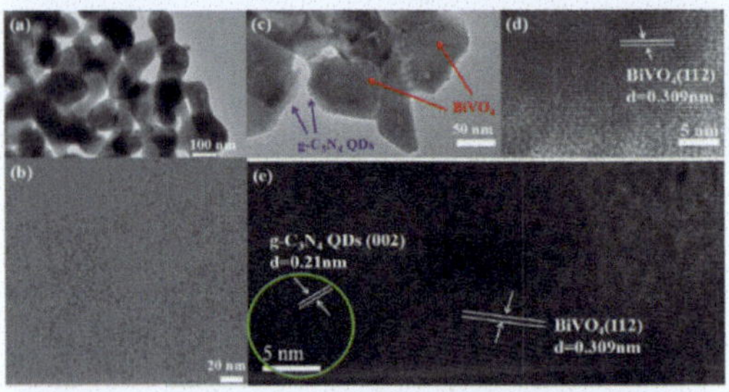	Copyright 2022, Elsevier (License Number: 5310091209541 [126]
6	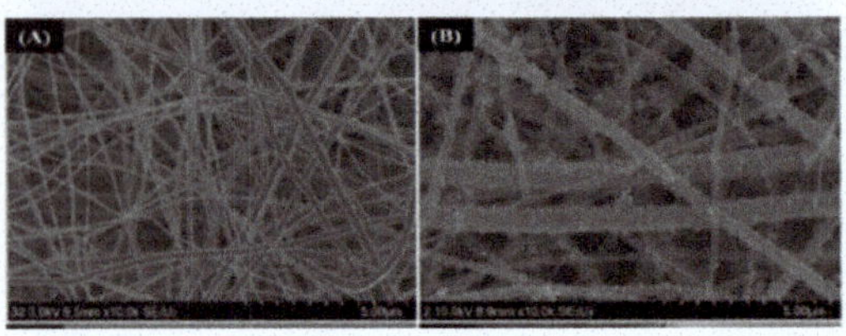	Copyright 2022, Elsevier (License Number: 5310091421188 [122]

In addition, the reason for developing carbon-based semiconductor material is abundance in nature, high performance for photocatalysis, unique 2D structure, high surface area, and easy charge-carrier mobility [139, 140]. They can transform photon energy into chemical energy, allowing them to eliminate organic contaminants from the air and liquid streams [141]. The various techniques can be applied to synthesize carbon-based photocatalysis applications, starting from the most frequently encountered methods such as hydrothermal, solvothermal, and sol–gel [142, 143]. In addition, atomic layer deposition (ALD), electrodeposition, electrophoretic deposition, chemical vapor deposition, physical vapor deposition, and electrospinning are all suitable ways of creating carbon-based photocatalysis applications [144–147]. Many researchers have previously produced carbon-based materials for the photocatalyst process with their respective details and strategies.

Graphene

The first substance identified by Oxford scientist Benjamin Brodie in 1859 was graphene, which is the exfoliation of graphite oxide [148]. The breakthrough came in atomically thin graphene oxide sheets distributed on a simple medium. Physicists are interested in graphene because of its peculiar electronic structure, which includes a linear dispersion of Dirac electrons [149].

The presence of many types of oxygen-containing functional groups on the basal plane and sheet edges enables GO to interact with a wide range of organic and inorganic materials in non-covalent, covalent, and ionic ways, making functional hybrids simple to fabricate [150]. GO is also an electrical hybrid material with conducting states from sp2 carbon sites and a substantial energy gap (carrier transport gap) between σ-states of its sp3-bonded carbons [151]. Adjusting the sp2 and sp3 fraction ratios via reduction chemistry is a potent approach to tailor the bandgap and hence influence the transition of graphene from an insulator to a semiconductor [152, 153]. When bombarded with light energy surpassing their bandgap, their sp2 domains also operate as semiconductors and display photoresponse or photoreactivity [154].

Jian Wu and colleagues used an ambient temperature in situ co-precipitation to make 2D graphene oxide with modified α-$AgVO_3$ nanorods (GO/ α-$AgVO_3$) photocatalysis, as indicated in Fig. 14. Conduction upon irradiating visible light, the surface photocatalyst GO/α-$AgVO_3$, in the valence band (VB), electrons (e^-) are excited and move toward the conductor band (CB). Also equivalently, hole (h^+) lags in VB [156]. The obtained results were excellent for decomposing rhodamine B (RhB), which was 18 times more efficient than that of pure α-$AgVO_3$ under visible-light irradiation [155]. Raja and his group also carried out other modifications using the hydrothermal method. He combined TiO_2-Cds-rGO, intending to add GO to enhance the photoefficiency by separating photogenerated electron–hole pairs [157, 158]. This ternary carbon-based was used in photoremoval methylene blue (MB) dye with 96.4% success after 50 min light irradiation [159] (Table 3).

Carbon Nanotube

CNTs have unique properties that make them an essential parameter in applied research and theory. The carbon nanotube (CNT) length is measured in microns, while its diameter is measured in nanometers, with a reasonably large aspect ratio

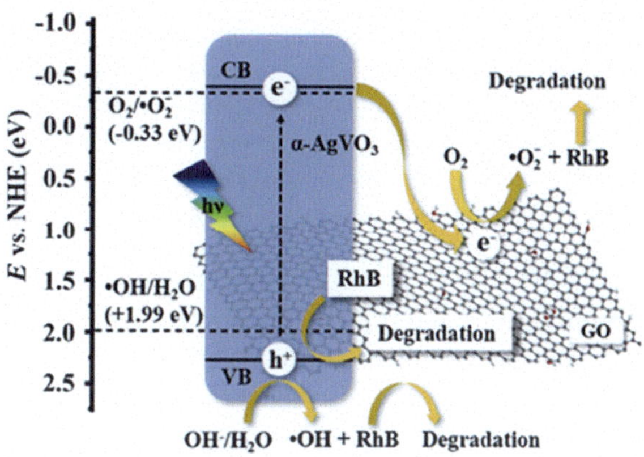

Fig. 14 Photocatalytic mechanism of GO/α-$AgVO_3$ [155]

Table 3 Several carbon materials with corresponding types of pollutant removal

Graphene-based material	Organic compounds	References
GO/α-$AgVO_3$	Rhodamine B	[155]
TiO_2-Cds-rGO	Methylene Blue	[159]
g-C_3N_4	Organic dyes, phenolic, pharmaceuticals, salt ions, heavy metals, and oils	[160]
ZnO/Au/g-C_3N_4	Methylene Blue	[161]
Ag/BiOBr/GO	Rhodamine B	[162]
BiOBr-GO-polyaniline	Phenol	[163]
g-C_3N_4/Bi_2WO_6/rGO	Ibuprofen	[164]
Bi_2WO_6/CuS/g-C_3N_4	Ciprofloxacin	[165]

[166]. The statement was discovered by the group of found research led by Iijima in 1991 [167]. It resembles a cylinder with a hexagonal arrangement of sp2 hybridized carbon atoms in terms of structure [168]. The inside is hollow, with one or more layers of graphene sheets on the surface. CNTs are formed from single-layer graphene, according to NEC and IBM, until 1993 [168].

Depending on the presence of the layer, CNTs fall into three categories, single-walled carbon nanotubes (SWNTs), double-walled (DWCNTs), and multi-walled carbon nanotubes (MWNTs) [134]. The first SWNTs introduced in 1993 were single-layer graphene sheets with a 1–2 nm diameter [169]. DWNTs, on the other hand, consist of two carbon nanotubes separated by an outer tube, which encloses the inner tube [156, 170]. The outer tube is 2–4 nm in diameter, while the inner tube is 1–3 nm. The MWNT (multiple graphene sheets) has an inner tube radius of 0.34 nm and an outer layer radius of 2 nm, respectively, reaching 20–30 nm [151].

Because of their large surface area, mesoporous characteristics, and chemical stability, MWCNTs might be a potential candidate for catalytic and photocatalytic materials. MWCNT-based nanocomposites have lately garnered much attention because of their excellent one-dimensional molecular structure, robust adsorption capacity, high surface area (>150 m^2g^{-1}), superior mechanical capabilities, chemical stability, and heat and dispersion properties. CNT-based catalyst significantly improves electronic properties and electron storage capacity in Fig. 15. They are pretty accommodating at room temperature, with hardly any resistance.

CNT has several distinctive properties in its application and is vastly used as a support material for several catalysts as well as CNTs-TiO_2 photocatalyst has been studied by several researchers (Table 4).

Carbon Quantum Dots

Carbon quantum dots (CQD) are nanostructures with a nucleus and surface of carbon atoms that may be functionally coupled with organic or biomolecules [177]. Usually, CQDs have a size below 10 nm with amorphous to nanocrystalline core types with carbon atoms predominantly having sp3 hybridized and amorphous, while GQDs are

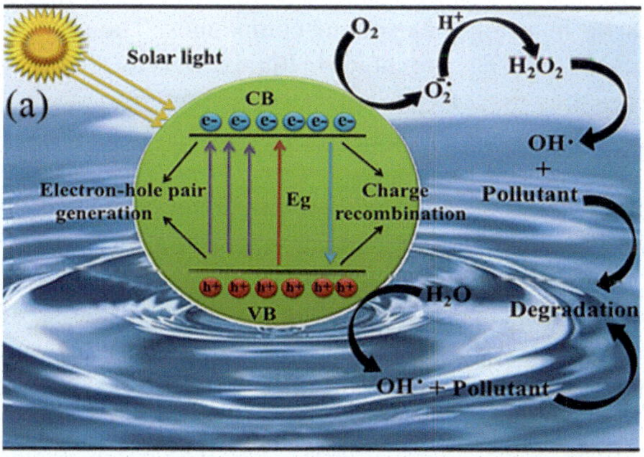

Fig. 15 Photocatalytic mechanism on the organic pollutant. Reproduced with permission. [85] Copyright 2022, Elsevier (License Number: 5323330708803)

Table 4 Several CNT materials and their corresponding application for organic pollutants removal

CNT-based material	Organic compounds	References
TiO_2@CNTs/AgNPs & TiO_2@CNTs/AgNPs	Methylene Blue	[171]
CNTs–TiO_2/Al_2O_3		[172]
CNTs-Ag-TiO_2	Organic dyes degradation	[173]
Fe-CNTs	Rhodamine B	[174]
TiO_2@CNTs nanofibers	Eriochrome Black T Dye	[175]
TiO_2/CNTs	Rhodamine B	[176]

sp2 hybridized and crystalline [178]. Based on the quantum dots family members, CQDs are interested in photocatalysis due to their abundance in nature, low toxicity, low cost, and long lifetime [179]. CQDs also have excellent aqueous solubility, chemical stability, and strong biocompatibility and are simple to fabricate and functionalize [180]. They are distinct from conventional semiconductor quantum dots. Water solubility is affected by the presence of functional groups in CQD, such as carbonyl, carboxyl, hydroxyl, and epoxy, which alters the platform for functionalization [181].

The bandgap of CQD differs from conventional semiconductors because they are subject to the quantum confinement effect [182]. Photoexcitation is based on π-plasmon absorption in the core carbon nanoparticles [183]. CQD absorption cores cover the UV/Vis spectra with an extended near-IR range [183]. CQD fluorescents are mechanically and efficiently described in electron acceptors or donor scavengers in the CQD with the most effective surface passivation.

The mechanism of e^-/h^+ couples arises in photocatalytic reactions when photogenerated electrons migrate from VB to CB after light irradiation and activation of semiconductors [184, 185]. The photogenerated e^-/h^+ couples produce reactive oxygen species (ROS) such as superoxide and hydroxyl radicals by simulating oxidation and reduction processes on the photocatalyst surface [186, 187]. The radicals are created to interact with the pollutant molecular structure, causing them to break down into minor hazardous compounds. The photocatalysis reaction refers to the reaction mechanism shown in Fig. 16. Several studies with different catalytic systems regarding these CQDs are shown in Table 5.

Fullerene

Researchers have recently produced composite nano-photocatalytic materials based on C60 fullerenes, with some impressive results. The formation of C60 fullerenes and their derivatives is shown directly in Fig. 17. Based on a C60-based carbon atomic mass spectrogram obtained in the lab, the cage structure of the C60 molecule is proposed [193]. Fullerenes were found in 1970 due to a study of the carbon atom family's spectrum in orbital space [193, 194]. The 60 carbon atoms are divided into 20 six-membered rings and 12 five-membered rings with 60 vertices each, and all carbon atoms are equal [195, 196]. The bonds in fullerenes C60 are distinct from those in graphite sp2 hybrid bonds and diamond SP^3 hybrid bonds because they are spherical [197].

Fullerenes and their derivatives in photocatalysis must have a wide range of functions and application effects. Fullerenes can play three separate roles in photocatalytic processes: electron acceptor, energy transfer mediator, and electron donor [195]. Photogenerated electrons arise in VB when light activates a semiconductor

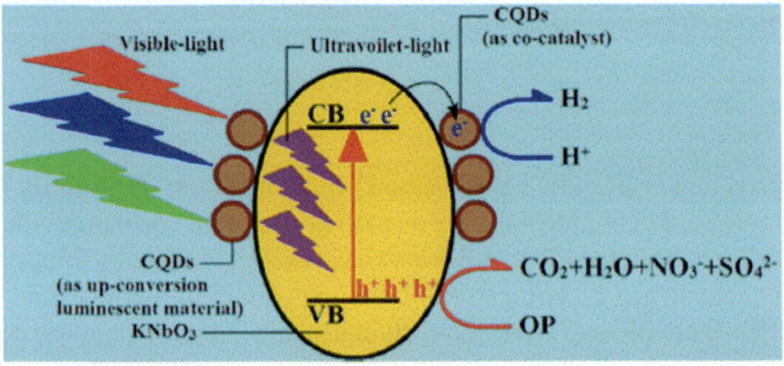

Fig. 16 Principle and mechanism on photocatalytic degradation of crystal violet with simultaneous hydrogen evolution by CQDs/$KNbO_3$ photocatalyst. Reproduced with permission. [125] Copyright 2022, Elsevier (License Number: 5323291115468)

Table 5 Several CQD materials and types of pollutants removed

CQD-based material	Organic compounds	References
CQD-ZnO composites	Phenol	[188]
CQD-g-C_3N_4	Methylene blue (MB) or Rhodamine B (RhB)	[189]
N-CQDs/SnS_2	Toxic organic pollutant	[183, 190]
(CQDs)/KNbO	Organic pollutant	[125]
CQD-PCN	Bisphenol A	[191]
GQDs, GCNQDs, CQDs	Organic pollutant	[192]

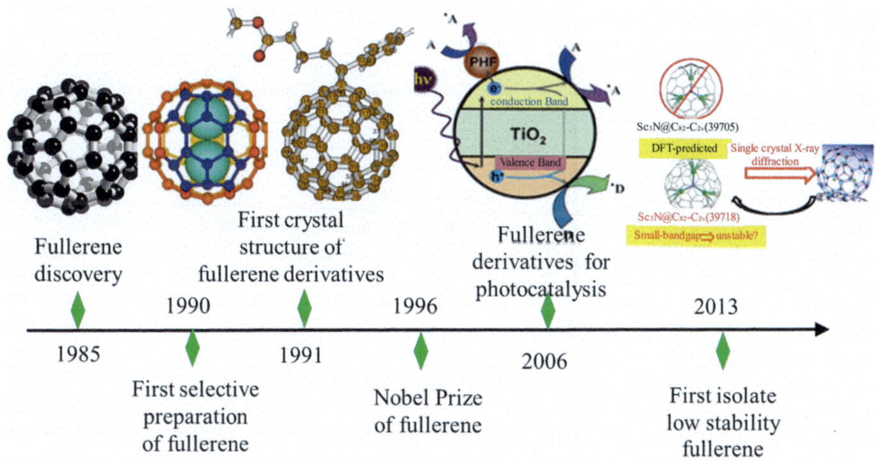

Fig. 17 Several fullerenes and their derivatives. [193] Copyright 2022, Elsevier (License Number: 5323500373424)

Table 6 Several carbon materials and types of pollutants removed

Fullerene-based material	Organic compounds	References
F-TiO_2(B)/fullerene	Organic pollutant	[199]
Zinc porphyrin functionalized fullerenes	Phenol and Methylene Blue	[200]
Fullerene C60	Rhodamine B	[201]
ZnCPP-Fullerol@TiO_2	Rhodamine B	[202]
Fullerene-Flowerlike $Cu_3(PO_4)_2$	Rhodamine B and Rhodamine 6G	[203]

material. Once the process is complete, the electrons are carried to fullerenes, which act as electron acceptors and carriers [195, 198]. Some related studies with fullerene materials are indicated in Table 6.

3 Electron Transfer Mechanism

When electrons are transported from an atom or molecule to another chemical entity, this is known as electron transfer (ET). ET is a mechanistic description of numerous forms of electron transfer redox processes. The ET reaction is an electrochemical process. Photosynthesis and respiration are also affected by ET responses. Transition metal complexes are commonly used in ET reactions. ET is a step in various commercial polymerization processes in organic chemistry and photoredox catalysis is based on it. The process of electron transfer is from a donating species to

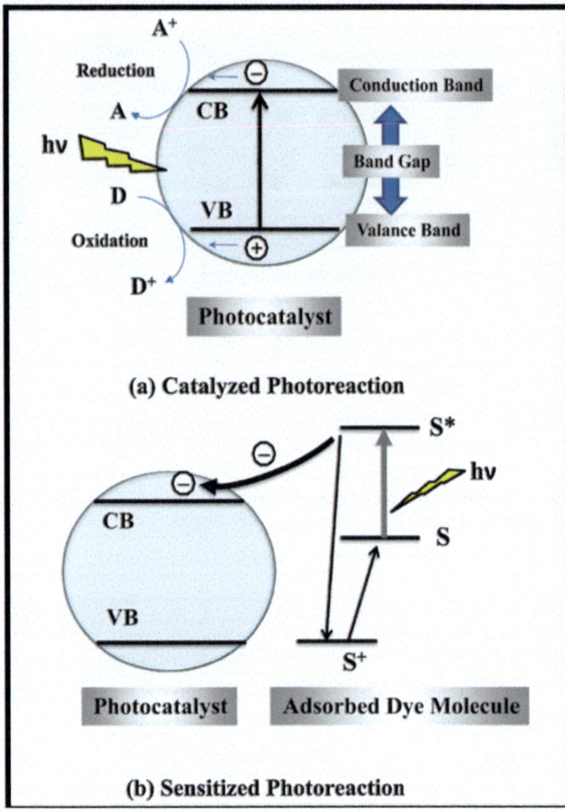

Fig. 18 Schematic of photocatalytic processes. Adapted with permission. [205] Copyright 2022, Elsevier (License Number: 5323460884776)

an accepting species which is known as light-induced electron transfer. One of the components is stimulated by light before electron transfer [204]. The whole process is illustrated as shown in Fig. 18.

As an example, this section provides a brief bio-electrochemical electron transfer. The process of extracellular electron transfer (EET) pathway can be divided into two sections based on the produced electron-carrying intermediates: (a) direct electron transfer (DET) and (b) indirect electron transfer (IET) [206]. Using a nanomaterial as carriers/hosts to immobilize redox enzymes has the potential to increase enzymatic biosensor electron transfer and stability. It is clearly revealed that a single nanomaterial will not be able to provide all of the advantages pointed to the previous development and recent use of nanocomposite materials [207, 208]. Combining two or more nanoparticles in a single unit can significantly increase mechanical, thermal, optical, and electrical properties and improve the activity of the catalytic system [209].

Because the shape of the structure is three dimensional of the redox protein/enzyme, it makes the electroactive center inaccessible, and its high adsorption to the electrode surface for further passivation and electron transfer is directly challenging to achieve [210]. At the same time, the causes for transferring electron inhibition between electrodes and proteins/enzymes could be the unique similarity and compatibility of the biomolecules on the electrode surface [211]. Specific enzyme changes are necessary to boost the electron transfer/redox rate of the protein. The redox protein hemoglobin (Hb) has been employed as a promising biomolecule to improve the performance of hydrogen peroxide (H_2O_2) biosensors based on processes that were direct electron transfer processes (DET) [212–214]. The bio-based carbon can act as a boost and help with supplement maintenance. Biochar decreases pollutants and heavy metal leaching through physical and chemical absorption of organics [215].

There are many minor problems or obstacles when using the DET process, but because DET has several advantages, many have done experiments or improved the system to do DET. Because using the DET system or method is very difficult to apply in the laboratory, many attempts have been attempted to produce DET by modifying an electrode surface with CNTs or improving precipitation with CNTs. DET has been achieved between several proteins and surface electrodes using CNT-determining electrodes [216]. The DET of glucose oxidase (GOD) and hemoglobin on the electrode surface area is measured using CNT [217, 218]. In addition, several investigators have found that similar results have been obtained with various analytes, such as catalase and bilirubin oxidase [219, 220], and CNTs at the substrate electrode, either covalently or non-covalently [221–223].

For bioremediation and bioenergy generation, high-performance microbial EET is required. The electron transfer efficiency of carbon point-modified electroactive microorganisms (CD) increased [224]. EET allows electrochemically active bacteria (EAB) to transport electrons generated by organic matter metabolism to electron acceptors, which are essential in the fields of energy product fabrication, repair, or protection of deteriorating environments, and geophysical and geochemical cycling systems [225–228]. The microbial EET process is critical for optimizing electrical pathways in biomanufacturing and has sparked much interest in the scientific community. For bioelectrical collection, EABs can use a variety of EET mechanisms, including (1) DET, which involves conductive nanowires or cytochrome c on the outer membrane of bacteria (OM); (2) mediated electron transfer (MET), which involves exogenous diffuse mediators and endogenous electron transfer [229]. Nevertheless, the slow rate of EET transfer from some microorganisms to electrodes indicates a limited power output, limiting the use of bio-electrochemical systems [230–232]. The CD with group teams containing oxygen on the surface will boost the electron transfer chain packing density and improve the CD adhesion system at the anode, promoting DET even more. Flavin secretion can be triggered by the proper quantity of CD, which will make the MET level higher or further. As a result of the synergistic impact, the EET was considerably increased. Due to biotoxicity, CD overdose had a limited promotional effect on EET performance. This approach of current augmentation from S. oneidensis MR-1 by adding a CD barrier promises to improve EET applications and understand some of the new roles of nanomaterials in mechanisms and processes [224].

3.1 Formation of Heterojunction

Various photocatalysts have been with carbon-based photocatalysts being particularly popular recently due to their thermal stability/good chemical, wide area on contact surfaces, good conductivity in electronic properties, and low manufacturing costs [233–235]. Carbon nanomaterials (CNM) are the most promising possibility for material usage in photocatalysts due to their unique characteristics [236].

Many attempts have been made to utilize photoexcited charge carriers fully and inhibit hole-electron pair recombination when photocatalytic activity is in progress. Heterojunction formation is one of the most effective methods for increasing charge separation efficiency and reducing photogenerated electron–hole pair recombination [237, 238]. The most commonly studied heterojunctions are made of two material semiconductors in a solid state, forming a sophisticated solid structure with Z-scheme-type electron transfer. The primary mechanism activity for Z-scheme heterojunction depends on the process of photosynthesis on materials. Under the light, electrons of photocatalyst I recombine with holes in photocatalyst II via electron mediators, inhibiting photo-induced charge-carrier recombination and maintaining the redox reaction characteristics of the photocatalysts [239–243].

CNMs such as fullerenes, graphene, CNTs, and their descendants have been exploited as electron mediators in various heterojunctions to enhance the conductivity of semiconductor photocatalysts [244–247]. Some of the carbon-based photocatalysts are included in Table 7. Graphene has numerous advantages over other carbon-based heterojunction photocatalysts, including low manufacturing cost, a large contact surface area, and an appropriate band structure [248, 249]. When compared to the pure compound without graphene, Jiang et al. created a solid state of Z-scheme heterojunction of Bi_2WO_6/CNTs/g-C_3N_4 composites with CNTs acting as electron mediators [250]. Gebreslassie et al. using graphene carbon based as a mediator in electron transfer could dramatically boost photocatalytic activity, resulting in a 7–15-fold higher hydrogen gas production [251]. The schematic of CNMs can be seen in Fig. 19.

CQD has recently been exploited as an electron mediator in fabricating solid-state Z-scheme heterojunctions. By connecting titanium dioxide and Cd0.5Zn0.5S with CQD, Liu et al. created a Z-scheme heterojunction CQD-based that displayed the extraordinary photocatalytic activity for hydrogen evolution [253]. Pan et al., on the other hand, create a sandwich-type structure in which CQD is sandwiched between CdS and BiOCl [254]. Compared to CdS/BiOCl, BiOCl, and CQDs/BiOCl, the CdS/CQDs/BiOCl heterojunctions had better properties on the photocatalytic activity in phenol and RhB degradation under the visible and UV irradiation [252].

Table 7 Summary of carbon-based photocatalysts. Modified from Syed et al. [252]

Material photocatalysts	Type of heterojunction	Synthesis method
PPTA/MWNTs	–	Polycondensation
$g\text{-}C_3N_4$/graphene/$NiFe_2O_4$	Solid-state Z-scheme	Hydrothermal
CN/CNT/BWO	Solid-state Z-scheme	–
Bi_2WO_6/$g\text{-}C_3N_4$	Direct Z-scheme	Hydrothermal
ZnO/$g\text{-}C_3N_4$	Direct Z-scheme	Solid-state
Cd0.5Zn0.5S/CQD/TiO_2	Solid-state Z-scheme	Hydrothermal
Cds/CQDs/BiOCl	Solid-state Z-scheme	Facile-region
Ru/$SrTiO_3$	Z-scheme	Hummers method
SnS_2/$g\text{-}C_3N_4$	Z-scheme	Hydrothermal
$SnO_2 - x$/$g\text{-}C_3N_4$	Z-scheme	Solid-state synthesis
CdS/SiC	Z-scheme	Hydrothermal
CdS/graphene	–	–
$ZnIn_2S_4$/RGO	–	Solvothermal
Bi_2WO_6/graphene	–	Sonochemical
Graphene/$g\text{-}C_3N_4$	–	Impregnation—chemical reduction
Nanoparticle/graphene	–	One-pot solution
TiO_2/graphene	–	Sol–gel method
TiO_2/carbon dots	–	Hydrothermal
CdS/graphene	–	Hydrothermal
Ta_2O_5/CNT	Schottky heterojunction	–
Ni/GO-CdS	–	Photodeposition
La-CNTs/TiO_2	–	Sol–gel method
TiO_2/CQD	–	Green synthesis

3.2 *Performance of Photocatalyst*

To date, a variety of catalysts have been utilized to photodegrade organic pollutants, including titanium dioxide (TiO_2), carbon dots (CDs), zinc oxide (ZnO), and molybdenum disulfide (MoS_2) [255–260]. Titanium dioxide has been devoted a lot to the photocatalysis mechanism because of its inexpensive cost, low toxicity, and excellent efficiency [261]. Under visible light, however, pure titanium dioxide has poor photocatalytic activity [261, 262]. To solve this challenge, combining titanium dioxide with photocatalysts or some other cocatalysts can significantly utilize sunlight while also improving catalytic mechanism or performance [263, 264]. Due to the unusual shape of the structure and strong photogenerated electron splitting ability, $g\text{-}C_3N_4$ is a suitable foundation for fabricating composite materials as photocatalysts [265]. Carbon nitride ($g\text{-}C_3N_4$) is one of the materials that has polymeric properties with outstanding performance on photocatalytic for the synthesis of organic material

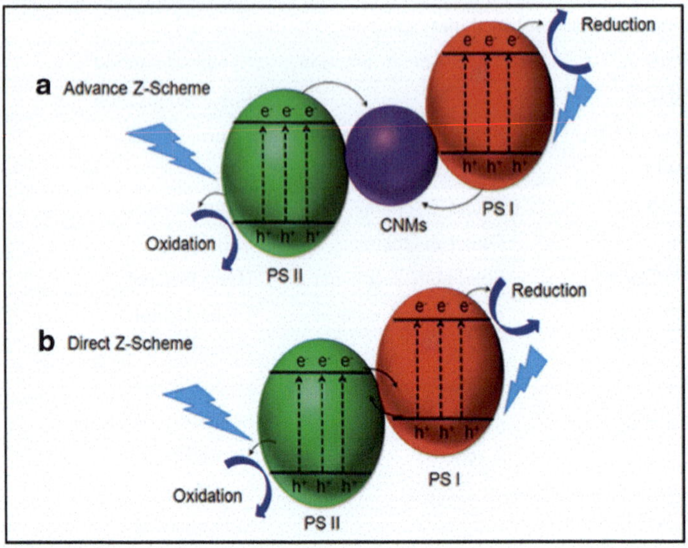

Fig. 19 Schematic conceptions with CNMs as an electron mediator with **a** advanced Z-scheme heterojunction and **b** the direct Z-scheme heterojunction. Adapted with permission [252]

[266–272], decontamination of an environmental system [273–277], and hydrogen and hydrogen peroxide production [78, 278–281], which has piqued researchers' attention as an optical semiconductor. The number bandgap is 2.7 eV, it allows visible-light excitation, and it has a significant advantage of g-C_3N_4-based materials over typical semiconductors like titanium dioxide [282, 283]. The use of g-C_3N_4 photocatalyst powder, on the other hand, revealed some drawbacks, including poor water dispersion, easy agglomeration, and recycling difficulties [284, 285]. As titanium dioxide and carbon nitride band edge positions are matched correctly, the photogenerated electron separation capability and yield, as well as the photocatalytic performance of the titanium dioxide/g-C_3N_4 nanocomposite, are increased. Several investigations [286, 287] have documented the fabrication of titanium dioxide/g-C_3N_4 nanocomposites utilizing various methods for degradation of a photocatalytic system, reduction of CO_2 compounds, and photocatalytic hydrogen element fabrication. When titanium dioxide/g-C_3N_4 nanocomposite is utilized for photocatalytic removal of antibiotics and dyes from organic compounds, it increases the ability to absorb contaminants while also providing a more active site for photocatalytic processes.

Surface modification can improve electron transport efficiency in semiconductor particles. Various processes, including semiconductor coupling [288], surface complexation [289, 290], and precious metal cluster deposition [290], can be used to select the surface of the material particles. Small particle sizes cause the band structure of semiconductor particles to be disrupted; small semiconductor particles display different properties in photophysical than bulk materials, which is named the "quantum size effect" [291]. The distance between energy levels in smaller colloidal particles is of the order EF/N, where EF is known as the Fermi energy and N is the atom number in a single particle. The charge carriers are divided by the bandgap excitation of semiconductor particles. A few of small fraction of these charge carriers can be exploited to undertake reaction in redox at the position of the interface due to their small particle size and high recombination rate [292, 293]. The proper deposition of transitions on semiconductor nanoparticles is one method that promises to overcome the limits in generating increased photoconversion. The Schottky barrier can form between the metal particles and the semiconductor when they come into contact, resulting in a rectified charge carrier [288, 294]. Attempts to employ metal and semiconductor composite nanoparticles to improve interfacial charge transfer efficiency and facilitate charge improvement in semiconductor nanomaterial structure have been made [295–297]. Another method for preventing electron–hole reassociation through the illumination and so extending the lifespan of photogenerated electrons is to use transition metal species doping. This method can also be used to harvest photons in visible light. The specific surface area of titanium dioxide, which is assured to be larger in continuous shape structures than particles that discrete, is one of the most critical elements influencing its photocatalytic activity. This can be due to the facile electron transport in its structure [298, 299]. Deposition of metal groups noble such as Au, Pt, and Ag [300–303], doping with the metal transition ions [304, 305], and nonmetals such as nitrogen and carbon [306] have all been used to change mesoporous nanostructures.

The greater specific surface area will benefit the photocatalytic degradation performance by increasing the interaction between the catalyst and small molecules. The close interaction between titanium dioxide and C_3N_4 suggests that C_3N_4 and titanium dioxide form of heterojunction structure. Between titanium dioxide and C_3N_4, a heterojunction can be formed. The photocatalytic degradation of tetracycline (TC) in several photocatalysts was investigated under with same conditions [307]. In the presence of titanium dioxide/carbon nitride (TCN), the degradation of a photocatalytic system efficiency of TC remained almost unchanged after three cycles of testing. It demonstrates that the TCN photocatalyst is more stable and reusable. IPA ($OH^{\cdot}$ scavengers), Vc (O_2^- scavengers), and EDTA (h^+ scavengers) were used to examine several types of reactive species in photocatalytic systems to investigate further the mechanism of photocatalytic degradation [308–312]. After adding IPA, Vc, and EDTA, radical trapping studies revealed that 47.28%, 85.63%, and 58.84% of TC were degraded, respectively. After adding IPA, the photodegradation efficiency was practically unaltered, demonstrating that $OH^{\cdot}$ is not the primary radical. Any radical substance that is active in the mechanism of the photodegradation process examined by assay TCN includes a hole (h^+) and superoxide radical ($O_2^{\cdot -}$). TCN photocatalyst stability was examined for three cycles of test, and no evident loss of its activity, showing that the TCN is a photocatalyst with excellent stability properties. The PXRD signal was constant before the reaction and after the photocatalytic reaction, representing that TCN structure remained stable during the reaction [307].

TCN exhibits a higher photocurrent response under visible light than titanium dioxide or C_3N_4, implying a higher ability or precision of photogenerated electron–hole pair splitting or separation and faster transfer between interfacial charges [313–315]. From above findings suggest that electron transport is aided by the intimate contact between titanium dioxide and C_3N_4. The arc radius of Nyquist on TCN clearly shows smaller yields than pure C_3N_4 or titanium dioxide, implying that the synergy between titanium dioxide and C_3N_4 improves the charge transfer efficiency. The Mott–Schottky slope of C_3N_4 and titanium dioxide exhibits n-type semiconductors. The primary active species for photodegradation are hole (h^+) and superoxide radical ($O_2^{\cdot -}$), while hydroxyl radical ($OH^{\cdot}$) is not engaged. As a Z-scheme is a heterojunction between titanium dioxide and C_3N_4 that cannot be produced, a type-II heterojunction is generated, efficiently separating electrons and holes. Electrons created by photons travel from CB of C_3N_4 to CB of titanium dioxide. Dissolved oxygen content in water is collected and stored in CB of C_3N_4, where it is converted to O_2, which aids in the photocatalytic destruction of contaminants. In the photocatalyst pits, specific organic contaminants are oxidized. As a result, the TCN hybrid structure promotes photocatalytic activity by improving photo-induced carrier separation efficiency [307, 316].

The usage of zinc oxide in photocatalytic applications is second in popularity after titanium dioxide, and numerous synthetic methods for carbon-based zinc oxide systems have been documented. Zinc oxide is a direct semiconductor (bandgap 3.4 eV) with a 60 meV exciton binding energy; it has poorer quantum efficiency and photocatalytic stability than titanium dioxide due to the photocorrosion effect during light irradiation [317]. Fabrication of zinc oxide/carbon nanocomposites can be done using various methods. Zinc oxide/graphene oxide composites can be prepared with a solvothermal method by spreading 15 nm zinc oxide nanoparticles over a wrinkled graphene layer [318]. In the other method, green hydrothermal method can be utilized to synthesize a hierarchy of graphene oxide/zinc oxide nanocomposites as an efficient photocatalyst for the photodegradation of blue dye [319]. Graphene-based zinc oxide nanocomposites were prepared using a multistep solution-based approach, which had several drawbacks, including low productivity and waste generation. Ball milling [320], ultrasonic spray [321], zinc oxide chemical precipitation on single-walled carbon nanotubes [322], and pulsed laser for deposition of zinc oxide onto multi-walled carbon nanotubes via hydrothermal synthesis [323, 324] have been used to create hybrid nanocomposites containing carbon QDs. In this scenario, acid treatment of carbon nanotube bundles is critical for establishing anchor points for zinc oxide nanoparticle development.

4 Application of Carbon-Based Material in Wastewater Treatment

4.1 Membrane Technologies for Wastewater Management

Due to the composition complexity and abundance of reactive elements, textile wastewater is a complex feed stream for membrane separation treatment [325]. The acidity of textile wastewater is considerable, with high suspended particles, nitrates, chlorides, metals such as sodium, manganese, iron, lead, copper, and chromium, as well as a high biochemical oxygen demand (BOD) and chemical oxygen demand (COD) values [326]. Textile wastewater can be treated in various ways, including physical, chemical, and biological approaches [327]. Physical treatment includes physically removing colors from textile waste, while chemical treatment involves chemical processes to remove dyes, and biological treatment involves dye biodegradation. Electrochemical technologies have also been effectively utilized in the cleaning of textile effluent. Graphite- or carbon-based material has long been employed as an anode in electrochemical therapy because it is reasonably inexpensive and produces excellent results [328]. As shown in Fig. 20 and Table 8, the four treatments are explained in detail below:

- **Physical treatment:** Adsorption, ion exchange, and irradiation, as well as filtration, are the most commonly used procedures in the treatment of textile effluent. This approach is frequently employed in the industry because of its excellent dye removal capacity and cheap operating costs [329]. In the dye waste treatment process, adsorbents such as activated carbon [330] and coal [331], fly ash [332], silica [333], wood [334], clay materials [335], agricultural waste [336], zeolite [337], and cotton waste [338] are employed. The irradiation procedure is better for neutralizing at low volumes across a wide range, but dye breakdown in textile wastewater requires a lot of dissolved oxygen [339]. For colors and other pollutants in wastewater, ion exchange is quite selective. Ion exchange has significant constraints for removing colors in textile wastewater [340].
- **Chemical treatment:** Physical techniques are ineffective for completely removing colors from textile waste since it requires further research to remove waste from textile waste, which increases processing costs. Although they are commonly utilized, chemical treatments offer disadvantages due to their ease and cost savings. Organic contaminants are often removed using chemical procedures like flocculation and coagulation [341]. The coagulation approach is particularly successful for decomposing insoluble components in textile waste but not so much for soluble substances [342]. The primary disadvantage of this procedure is the production of sludge and the added cost of removing it [343]. Total operational costs rise as a result of this.
- **Biological treatment:** Physical and chemical procedures for removing colors from textile waste are not commonly used due to their high cost, limited efficiency, and specialized features. The biological approach is a green technology

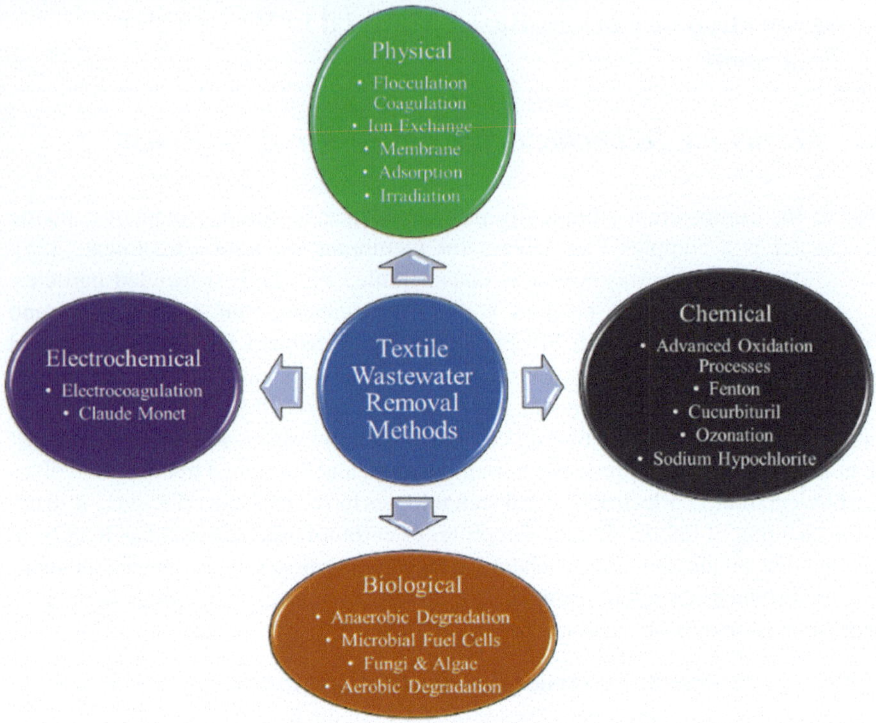

Fig. 20 Several methods for removing color from textile wastewater. Adapted with permission. [329] Copyright 2022, Elsevier (License Number: 5311831104403)

for removing colors from textile waste at a low cost and with the shortest possible turnaround time. Biological approaches can successfully remove COD and turbidity; however, they are ineffective in removing waste color from textiles. The most favorable setting for decolorization in biological treatment for textile effluent is an anaerobic series followed by aerobic treatment.

– **Electrochemical process:** In recent years, combining photocatalysis with electrochemical processes has usually been experienced. Electro Photocatalysis (EP) is a technique for reducing photogenerated electronic space recombination. After 180 min of treatment with immobilized TiO_2, the color has lost more than 90% of its intensity, resulting in methyl orange. In the EF (Electro Fenton) method, an electrochemical unit is used in addition to the standard Fenton process [344].

The whole process typically includes the following steps: first, initial treatment via a screening step; second, homogenization; third, primary treatment via coagulation; fourth, pH modification; fifth, secondary treatment via bioremediation (aerobic or anaerobic); and finally, refinement treatment via a membrane. Furthermore, many kinds of pollutants belong to textile industrial wastewater, including acid blue 25 (C.I. N. 62,055), acid orange 10 (C.I. 16,230), basic blue 9 (methylene blue), direct

Table 8 Benefits and drawbacks of the present textile wastewater treatment system

Treatment method	Advantages	Disadvantages	References
Biological	Decolorize a wide range of dyes at a low cost with minimal foam formation	All dye particles are not entirely removed	[345]
	Environmentally friendly and economically practical	Recalcitrant chemicals have a low biodegradability	
	Sludge production is reduced	The procedures need an enormous amount of land	
	Metabolites that are not harmful		
	Water use is reduced		
Physical	COD and BOD are decreased significantly	It includes using nonrecyclable chemical compounds (coagulants, flocculants, aid chemicals)	[346–348]
	A wide variety of chemicals are commercially accessible	Sludge volume output has increased	
		A residual aluminum in high level	
Chemical	Suitable for both soluble and insoluble dye removal	It takes too long	[346, 347]
	Both soluble and insoluble dyes are effectively decolored	Sludge production	
	It can develop reactive radicals in situ	For small and medium businesses, it is economically unsustainable	
	Chemical intake is minimal to non-existent	Technical constraints	
	Mineralization of pollutants	Product development	
	Effective against resistive molecules		
Electrochemical	There are no chemicals or sludge build-ups	Electricity is expensive	[329]
	It is flexible in oxidation, whether direct or indirect	Sludge of ferrous oxide	
	High-performance processing while preserving space. Environmentally friendly methods have been included		

red 28 (Congo red), basic violet 10, basic red 22 (C.I. 11,055), acid violet (C.I. 42,650), rhodamine B, acid red 14, direct red 80, acid orange 52 (methyl orange), direct red 28, brilliant green, etc. [349]. Because most textile manufacturing waste is recognized as poisonous, carcinogenic, and mutagenic to aquatic creatures, it must be eliminated [350]. The technology that has been discovered, namely, membrane

technology, has been confirmed to be an essential component in acid mine water treatment to reduce water shortages. Membrane technology satisfies a high degree of applicability in terms of simplicity of use, versatility, environmental effect, and managing water shortages [351]. Membrane technology advancements have boosted the number of water reuse projects throughout the world (United States, Singapore, Australia, Belgium, South Africa, Nambia, United Kingdom, etc.) in recent years [352]. The membrane separation technique is gaining popularity due to its benefits, including low energy consumption, ease of operation, reduced space requirements, high elimination efficiency, no need for extra chemicals, and affordable operating expenses [353]. The membrane separation technique also allows the removal and recovery of synthetic colors from wastewater, which is a potential technical route [354]. Nanofiltration (NF) and reverse osmosis (RO) have been recognized as the best approaches for eliminating different synthetic colors from wastewater using membrane separation technologies. Although NF and RO can remove dye with excellent separation efficiency, their pore sizes range from 10^{-3} to 10^{-2} μm for NF, and 10^{-3} μm for RO, resulting in low water permeate flow. As a result, more immense operating pressures are required to satisfy the increased permeate fluxes [355].

Due to their excellent removal performance, low operating costs, and space-saving processes, many membrane technologies have been used to remove synthetic colors from wastewater. The following are the advantages and disadvantages of using various types of membrane technology to remove colors from wastewater, as shown in Table 9. Regarding dye rejection, NF membrane separation technology outperforms microfiltration (MF) and ultrafiltration (UF), and it has seen less membrane fouling than RO membranes. NF will remove low molecular weight organic substances like reactive dyes at good working pressures. Several pre-treatment procedures can be used on the dye or membrane waste to minimize concentration polarization and possibly membrane fouling of the NF membrane. Finally, membrane technology, particularly NF, has made significant advances in removing synthetic colors from wastewater in recent decades. However, there is little study on using RO membranes to remove dyes from wastewater. More effort, research, and testing are required to make membrane technology more competitive in the industrial sector.

The photocatalyst is also separated from the treated effluent by the membrane. Highly effective catalyst materials, such as titanium dioxide (TiO_2), have been utilized in the membrane matrix to construct photocatalytic membranes [356]. When ultraviolet (UV) or visible-light strikes the catalyst surface and serves as a catalyst motivator, a series of reactions occurs, resulting in the generation of hydroxyl radicals [357]. As a result, the hydroxyl radicals produced can transform the organic polluters attaching to them into innocuous inorganic molecules. The photocatalyst mechanism is depicted in general in Fig. 21. UV irradiation or visible light associated with the space between the valence and conduction bands, known as the "energy band" or "bandgap," can activate photocatalysts. The electrons in the bandgap must be stimulated and jump to the conduction band to have a photocatalytic reaction. A charge carrier is generated during this activation, and the photocatalyst process may be run. Thus, the process is summarized by migrating the generated e-/h^+ pair to the semiconductor surface and participation in the redox reaction. h^+, hydroxyl

Table 9 A summary of the several experiments for carbon-based materials that are used based on different types to remove the dye

Membrane technology types	Materials	Advantages	Disadvantages	References
Microfiltration (MF)	Activated carbon as adsorbent Polyethersulfone (PES) as MF membrane, polyethyleneimine (PEI), and graphene oxide (GO) as a coating layer Alkalis as a cleaning agent, carbonates, hydroxides and phosphates as typical chemicals	1. Reusable filtrate water 2. Membrane fouling is reduced significantly when activated carbon particles are present, and the MF membrane separates the activated carbon particles from the water 3. It can eliminate 96% of actual Blue Corazol (BC) dye baths from the manufacturing process 4. Surface charge changes, pH adjustments, and a reduction in the number of bonds between the foulant and the membrane surface are all possible	1. The pore size range is restricted to 0.1–10 μm 2. It is only for the removal of colloid dye from the dye bath and the removal of suspended particles in the dye 3. Foulant can clog pores by diffusing through them, accumulating on the top, forming a cake layer, and lowering membrane permeability	[361–363]
Ultrafiltration (UF)	Carbon nanotubes/polyvinylidene fluoride/polyurethane Poly(vinyl alcohol)/carbon nanotubes/polyacrylonitrile Carbon nanotubes/ polyacrylonitrile/ polyvinylpyrrolidone	Removes particles and macromolecules from a solution in an efficient manner In a non-sensitive procedure, it can be reused It can remove oil higher than 94% for carbon-based material	1. The use of UF for color removal from wastewater is restricted to less than 90% of the time 2. The most significant disadvantage is the creation of flaws in the carbon nanotubes' sidewalls, which might cause the carbon nanotubes to shatter into smaller pieces 3. CNTs are costly, and large-scale membrane manufacturing may not be feasible 4. Furthermore, CNT toxicity can pose significant health and environmental issues when discharged during water treatment operations	[354, 364–366]

(continued)

Table 9 (continued)

Membrane technology types	Materials	Advantages	Disadvantages	References
Nanofiltration (NF)	Magnetic graphene-based (MMGO) composite-modified PES-based NF membrane	NF is the most feasible approach for removing organic molecules with low molecular weight Increased MMGO content leads to increased membrane permeability The permeate flow increased as the GO content rose	The NF process confronts a fundamental problem during the filtering phase: flux decrease owing to concentration polarization and membrane fouling Congo red, methyl blue, and sunset yellow were rejected 99.6%, 99.8%, and 97.6%, respectively, by the typical polyamide (PA) thin-film composite NF membrane	[367, 368]
	NF membrane based on flat sheet polyacrylonitrile (PAN) phosphorylated chitosan (PCS) adjusted with one layer of graphene oxide (GO)			
Reverse osmosis (RO)	Flat sheet polyethersulfone/magnetic graphene-based composite	The removal of ions and macromolecules from wastewater is simple with RO technology The percentage of removal is higher than 90%, such as specific dyes direct red 16, direct black 38, xylenol orange, etc	The RO membrane applicability is more to very high salt concentrations, and the water quality of the NF permeate is not suitable for system restoration There is little research on RO membrane ability to remove colors from wastewater	[367, 368]
	Flat sheet polyacrylonitrile/phosphorylated chitosan–graphene oxide			

Fig. 21 The photocatalytic reaction has a primary mechanism. Reproduced with permission. [358] Copyright 2022, Elsevier (Licence Number: 5312370002411)

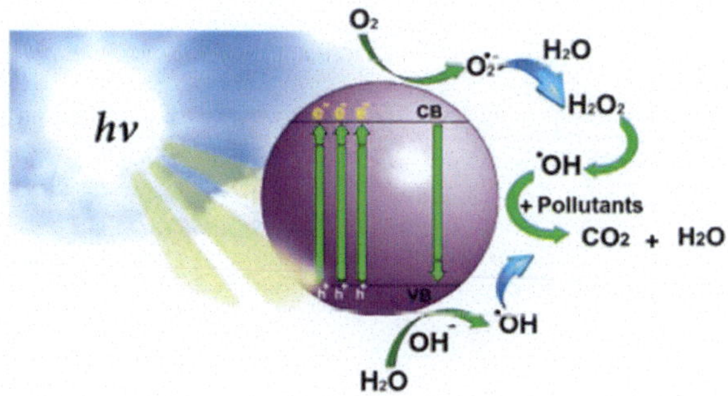

radical (•OH) as the principal oxidant in pollutant degradation in aqueous media, and superoxide radical are three separate active sites that play a vital part in the photocatalytic process ($O_2^{\bullet-}$). For OH radicals, two routes were considered: first, photogeneration of oxidized h^+ H_2O and OH in an aqueous environment to produce OH radicals. Second, photogeneration of e-reduction of O_2 in an aqueous solution to produce $O_2^{\bullet-}$ radicals, followed by •OOH radicals (the product of the reaction with h^+) and •OOH radical decomposition to produce •OH radicals. Organic contaminants must be degraded directly by h^+, and the catalyst type and oxidation conditions determine the degradation capability [358]. In addition to the main photocatalytic action, dye molecules can be directly excited under visible light due to their ability to absorb visible light (Fig. 22). In other words, direct photolysis is a photocatalytic dye degradation mechanism that includes the dye being excited from its ground state (Dye) to a triplet excited state (Dye*) under visible light (> 400 nm). The excited dye is transformed into an oxidized radical cation (Dye$^{+\bullet}$) by transferring electrons from itself to the CB of semiconductor; the excited dye is transformed into an oxidized radical cation (Dye$^{+\bullet}$). The electrons are subsequently picked up by oxygen molecules in the system, forming superoxide radical anions and hydroxyl radicals, primarily responsible for dye degradation [359].

However, there are certain drawbacks to the photocatalytic process, such as low photocatalytic efficiency, limited recycling, and low sunshine use. Meanwhile, photocatalytic membrane reactors (PMRs), in which photocatalytic processes and membranes are linked together, are one of the critical techniques for eliminating the common difficulties with the previously described photocatalytic processes [360].

4.2 The Combination of Photocatalyst and Membrane Technology

The pressure difference between the influent and permeate sides is the driving force in pressure-driven membrane processes to separate particles and solutes depending on size, shape, and charge, as shown in Fig. 23. Pressure-driven membranes such as

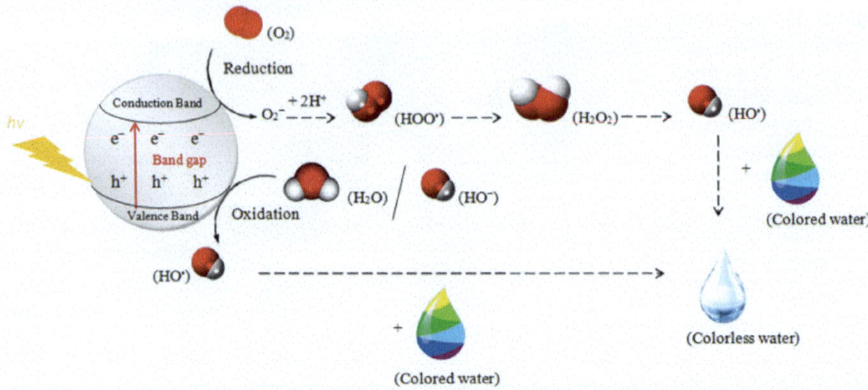

Fig. 22 Dye degradation of photocatalysis. Reproduced with permission. [339] Copyright 2022, Elsevier (Licence Number: 5312370835573)

NF, UF, and microfiltration MF membranes must serve as a separator for photocatalysts and a particular wall for contaminating compounds such as pharmaceuticals, hummus, dyes, and photodecomposition by-products in PMR. PMR has employed organic, mixed matrix membranes and inorganic membrane materials for effective wastewater treatment.

With a broad spectrum of influents, powdered activated carbon membrane has outstanding effluent quality and is suitable for plant improvement. The disadvantage is that fouling management needs extra care [358]. Functional composite membranes integrating photoelectrocatalysis and membrane filtration have been the subject of recent research. They found that putting carbon felt fibers between the polyvinylidene fluoride (PVDF) membrane and the $ZnIn_2S_4$ photocatalyst gives the organic membranes a lot of durability and electrical conductivity, which could help with photoelectrocatalytic degradation of total dissolved carbon in deionized water or laboratory-scale membrane bioreactor (MBR) effluents. The photoelectrocatalytic layer comprises $ZnIn_2S_4$ particles that are equally dispersed on the carbon fiber surface. Simultaneously, the PVDF coating layer is cast, increasing the breaking strength of carbon fiber [369].

There are several studies about photoelectrocatalytic oxidation of organics in solution with carbon-based photoanode under visible light. Wang et al. found that the phenol glucose as an organic pollutant has efficiency below 72% of flux recovery ratio (FRR) with g-C_3N_4/ CNTs/Al_2O_3 photoanode. The photogenerated electrons in the visible light illuminated by the g-C_3N_4 layer may be attracted and severed from the holes when a positive voltage is supplied to the CNT layer. Consequently, a photoelectrocatalytic membrane was created and significantly improved performance [370]. Another researcher found that the methylene blue can be degraded by about 65.9% from graphene film/TiO_2 nanotube array as photocatalytic. It was reported that high UV and visible-light absorption means more excellent transient photo-induced current [371].

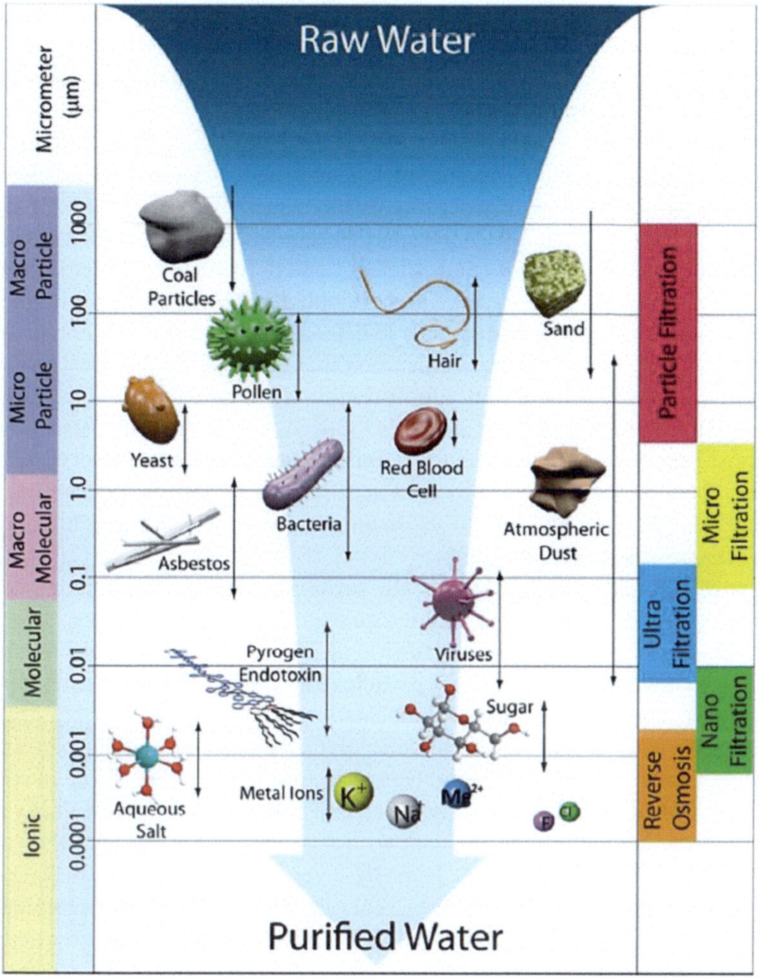

Fig. 23 The photocatalytic reaction has a primary mechanism. Reproduced with permission. [358] Copyright 2022, Elsevier (Licence Number: 5314600092500)

Another finding is that sequential deposition of a carbon graphite layer with strong electric conductivity and a layer of TiO_2 nanoparticles exhibiting photocatalytic activity on Al_2O_3 membrane support is used to design and fabricate TiO_2/carbon/Al_2O_3 membranes. When a supply voltage is applied to the membrane, the photogenerated electrons may be effectively drained from the photocatalytic layer and separated by holes, providing the membrane with a photoelectrocatalytic function. Photoelectrocatalytic membrane filtration (PECM) demonstrated higher natural organic matters (NOMs) removal and permeated flow with increasing supply voltage, according to membrane performance tests. Only UV or filtering. The NOM removal is 1.2 or 1.7 times more remarkable for the PECM process at 1.0 V than UV irradiation or filtration alone. The flow of stable permeate is 1.3 or 3 times higher for irradiated filtration. Furthermore, the PECM procedure outperforms UV irradiation or filtering alone in eliminating organic compounds (e.g., Rhodamine B), with 1.3 to 3 times greater removal rates [372]. TiO_2 has been widely used as a photocatalyst. Low cost, excellent hydrophilicity, high self-cleaning capability, and improved elimination of organic contaminants under UV light are just a few advantages of integrating TiO_2 particles into a polymer membrane matrix [373, 374] the high surface energy of TiO_2 NPs [375], in which then particle aggregation is more likely, resulting in reduced photocatalytic activity and efficiency and pore membrane blockage [376]. Several ways have been tried to solve this difficulty, including doping with metallic and nonmetallic elements or compounds [377], employing other semiconductors in the structure [378], and coupling with carbon materials [379]. GO/TiO_2 nanocomposite was employed in the PVDF ultrafiltration membrane by Xu et al. using the phase inversion method, as shown in Fig. 24 [380]. According to scientists, GO is one of the most optimistic possibilities for polymer membrane changeover. Two more attributes of GO make it one of the most acceptable alternatives for coupling with TiO_2 particles [381], in addition to its distinguishing traits such as a high surface area and charge-carrier mobility.

The novel GO/TiO_2-PVDF hybrid UF membrane enhanced photodegradation efficiency by 50–70% and 1.0–1.5 times quicker degradation kinetics. Enormous surface area and oxygen-containing functional clusters result in more significant and tighter interaction between TiO_2 particles and other components, resulting in reduced aggregation and improved photocatalytic activity [382]. Furthermore, the GO/TiO_2 nanomaterial's improved photocatalytic effectiveness might be owing to the GO's electrical characteristics and better electron transport, resulting in reduced carrier recombination [383]. Similarly, when exposed to UV, the permeability of the PVDF membrane increases by nearly two times, ensuring excellent BSA safety. Similarly, Gao et al. [384] used a layer-by-layer approach to deposit TiO_2 NPs and GO nanosheets on PES membranes. After that, ethanol/UV post-treatment was utilized to reduce GO nanosheets partially. The novel mixed membranes degraded

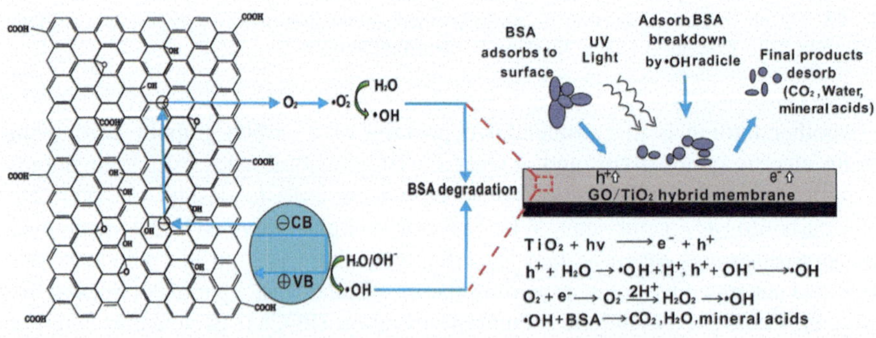

Fig. 24 Possible photocatalytic activity of GO/TiO2-PVDF membranes. Reproduced with permission. [380] Copyright 2022, Elsevier (Licence Number: 5314621393244)

methylene blue solutions 60–80% quicker under UV and 3–4 times faster in the sunshine than PES membranes guided by TiO_2 and GO. Natural organic molecules like humic acids are one of the primary causes of membrane fouling, preventing membrane technology from being widely used for water treatment. Rao et al. [385] synthesized Fe_2O_3/TiO_2/GO composites for mixing with membrane casting solutions using easy hydrothermal, colloidal, and vacuum filtering processes and reported increased performance for the changed membranes. Based on their findings, adding Fe_2O_3 to the TiO_2 photocatalyst increased its photocatalytic activity and humic acid adsorption. The photo-induced electron–hole GO sheet led to an improved split and higher photocatalytic activity. The weight ratio of Fe_2O_3:TiO_2:GO in the synthesized composite that completed the elimination of humic acid under sun irradiation in a 12-h test was 50:100:10.

Zinadini et al. [386] synthesized composites of multi-walled carbon nanotubes (MWCNTs) and zinc oxide nanoparticles (ZnO NPs), which were mixed in a PES casting solution to create mixed matrix membranes, and then examined the photocatalytic, surface, physical, and antifouling properties of the mixed membranes. They ascribed the considerable rise to the PES membrane's increased hydrophilicity and decreased surface roughness due to mixing with the nanocomposite. They employed techniques including SEM, AFM, contact angle measurement, and XRD to back up their claims. Based on their detailed description, a 0.5 weight percent ZnO/MWCNTs membrane might be the best option for permeability, antifouling capabilities, and Direct Red 16 rejection. To illustrate this performance, Fig. 25 depicts a schematic process of ZnO-coated MWCNT photocatalytic activity. The mechanical characteristics and photoactivity of the developed carbon-based materials improved significantly due to the inclusion of a photocatalyst, resulting in clean water and minimal fouling ability.

4.3 *Carbon Source for Photocatalysis and Synthesis*

Carbon dots (CDs) are novel fluorescent carbon nanomaterials with a diameter of fewer than 10 nm. They were identified as impurities in manufacturing single-walled carbon nanotubes in 2004 and purified using preparative electrophoresis, which means under the influence of an electric field, the movement of electrically charged particles in a liquid [387]. Photoluminescent CDs provide several benefits over standard semiconductor quantum dots, conversion nanoparticles, and organic dyes, including excellent photostability, high water solubility, strong chemical inertness, and ease of modification. CDs have exceptional electronic characteristics as electron donors and acceptors, giving them a wide range of applications in catalysis and optronics [388]. As shown in Fig. 26, the raw carbon sources for CDs can be artificial or natural [389].

Natural-product-derived carbon dots (NDCs) made from natural sources benefit from converting low-value biomass waste into lucrative and usable commodities. Unlike other N/S-doped CDs obtained from artificial carbon sources, which need

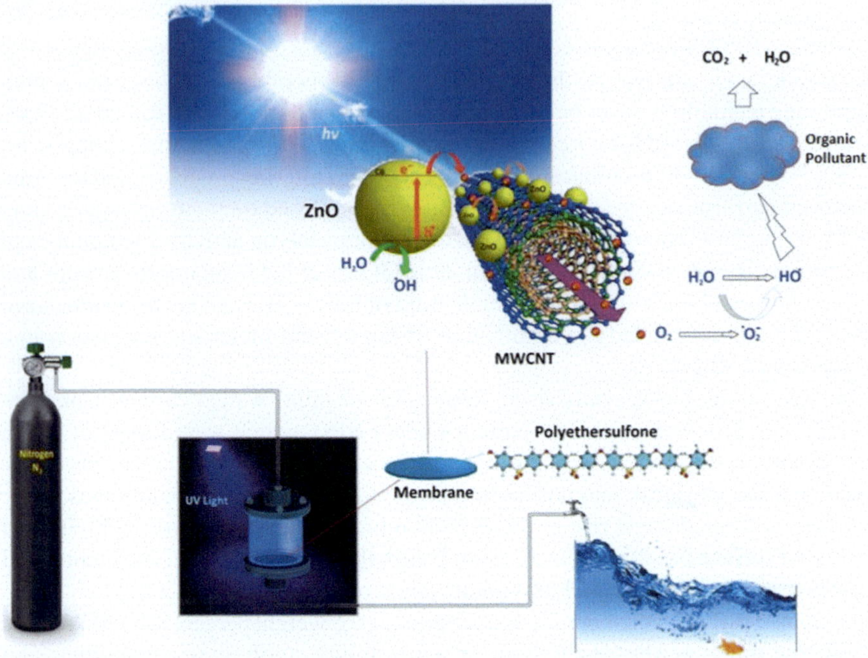

Fig. 25 The photocatalytic process using ZnO-MWCNTs/PES is depicted schematically. Reproduced with permission. [386] Copyright 2022, Elsevier (Licence Number: 5314830852832)

the inclusion of external N/S-containing compounds, natural products containing heteroatoms (N, S) are particularly ideal raw materials for the synthesis of heteroatom-doped NCDs [390, 391]. A novel method has been devised exclusively to produce NCDs from natural resources, which differs from the methods used to make CDs from artificial carbon sources [392]. Despite significant advances in this area, no study has mainly addressed NCDs.

The "top-down" and "bottom-up" techniques (Fig. 27) are the two most common synthetic procedures for preparing NCDs, while other syntheses combine the two. The "top-down" approach generates NCDs from relatively macroscopic carbon sources. It may be used to make NCDs from various natural items, including fruit, biomass materials, and garbage. Precursors are used as seeds in the "bottom-up" process. This method is suited for synthesizing NCDs from natural polymers such as chitin [393] and chitosan [394]. Both procedures were used to manufacture tiny quantities of NCDs from natural materials [395]. The discussion will concentrate on the "top-down" technique because it is the most popular way of preparing NCDs from natural sources. In addition, a new "bottom-up" approach for the preparation of NCDs developed was also discussed. The comparison of the bottom-up and top-down processes can be seen in Table 10.

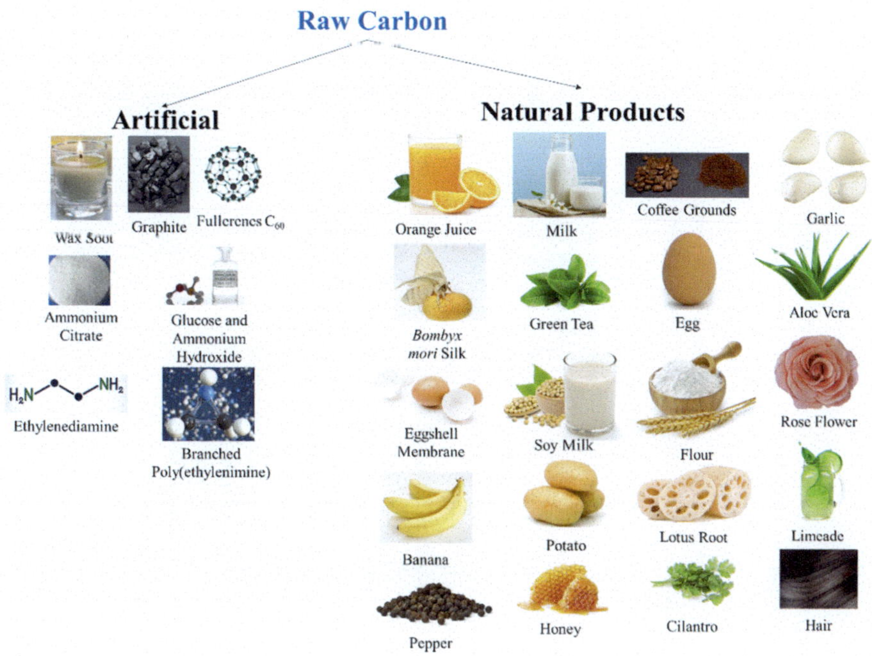

Fig. 26 Raw carbon sources for CDs

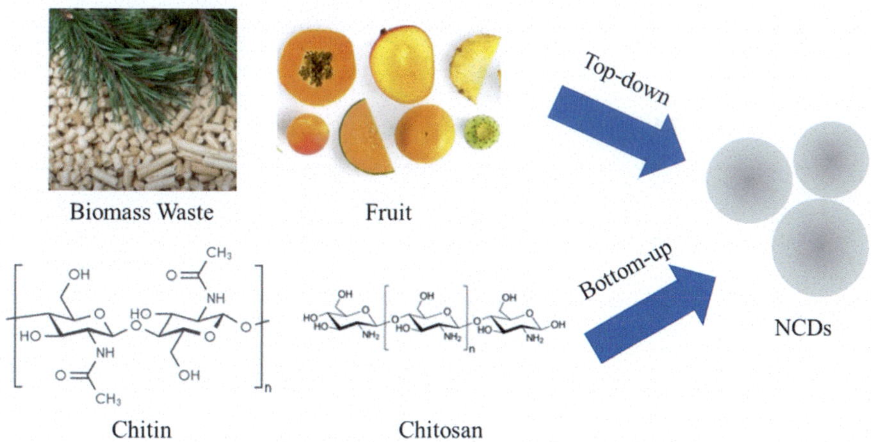

Fig. 27 Top-down and bottom-up NCDs preparation

Table 10 A summary of the common experiments for top-down and bottom-up methods to produce carbon-based materials

Method	Process	Description	Sources	Advantages	Disadvantages	References
Top-down	Hydrothermal carbonization	They are conducted in a closed reactor at a high temperature	Apple juice, hemicellulose, bamboo leaves, cabbage, and black tea	Low-cost, green, or ecologically friendly approach, non-toxic, produces high-quality NCDs, uses natural resources, and is simple to apply	NCD has a narrow size distribution (1.5–4.5 nm); therefore, it cannot employ most carbon materials, and conversion yields and particle size control are still issues. Besides, there is not a suitable technique to recycle carbon waste	[396–403]
	Extraction	NCDs can be extracted directly without the need for synthesis	Instant coffee, sugar beet molasses	NCDs are obtained in a very basic and direct treatment. The produced NCDs have high biocompatibility and may be employed in cell and fish bioimaging	This approach cannot be used since the recovered NCDs have a wide size variation, and not all natural materials contain NCDs	[404]
	Chemical oxidation	Natural goods are usually carbonized first, and then the carbonized material is processed	Dead neem leaves, starch	It can boost intrinsic state emission, resulting in a rise in fluorescence intensity. The surface features of NCDs, as well as their fluorescence emission, may easily be altered	The oxidizing reagents' residues may stay in the NCDs, increasing the biological toxicity	[405, 406]

(continued)

Table 10 (continued)

Method	Process	Description	Sources	Advantages	Disadvantages	References
	Microwave chemistry	A solvent is used to dissolve the natural product, subsequently heated in a microwave chamber	Wool, feathers, flour	In situ and transient heating may drastically enhance yields and product quality, and it is an effective and time-saving technique for the synthesis of NCDs. The NCDs could be separated and purified	Complicated equipment, low repeatability, high temperature, harsh synthesis conditions, and a hefty price tag	[407–409]
	Pyrolytic decomposition	Heat is applied to natural products, creating a carbon black substance from which the NCDs may be extracted and purified	Eggs, feathers, manure, soybean powder	The method is quick and simple to make, and the supplies are affordable	It takes place at extremely high temperatures and under carefully regulated pressure	[410, 411]
Bottom-up	Molecular aggregation	Molecular aggregation is used to prepare NCDs from biomass sources	All biomass materials	This approach does not need the use of heat or any other form of energy. Aggregates can aid fluorescence emission, and freshly generated NCDs have good bioimaging characteristics	For this approach to work, the biomass must have an aromatic structure. The chemical stability of the NCD produced by aggregation is low	[412]

5 Conclusion and Prospective Application

In photocatalysis, the electron transport process is very significant. Fundamental studies of charge transfer process dynamics are critical for developing novel photocatalysts and the fabrication of photocatalysts. Active photocatalytic electron transport plays an essential role in photoreaction. The research focuses on measuring the dynamics of the electron transfer interface using various methods, including stop flow, pulse radiolysis, EPR, photoluminescence, and time-resolved spectroscopy. The effects of many factors on electron transfer efficiency have been addressed. These results reveal that the efficacy of the electron transfer mechanism, and hence the process of photocatalytic mechanism, is significantly reliant on the surface composition, surface size, and shape of the photocatalyst. Various noble processes, including metal clustering, transition metal ion doping, semiconductor surface, and coupling, can alter the material surfaces. Many kinds of carbon-based material with Z-scheme heterojunctions with an excellent catalytic activity have been created owing to the rapid development of current technology. In a solid-state Z-scheme of material heterojunction, several forms of carbon allotropes work well as electron mediators. However, choosing suitable electron mediators with specific composites from different materials is complex and critical.

Photocatalytic degradation of organic pollutants is an appealing alternative to overcome organic pollutants because of its low cost, simplicity of operation, and lack of secondary contamination. NCDs are significant in photocatalysis and electrocatalysis. They have a variety of uses, including catalytic degradation of organic mixtures in aqueous settings and catalysis of oxygen reduction processes (ORRs) in fuel cells, and catalysis of hydrogen synthesis by separating water. Carbon-based materials with NCDs are prospective to improve photocatalytic reaction due to their adsorption capability with near-infrared absorbance and emission. Incorporating NCDs into photocatalysts will expand the light absorbance, resulting in a higher catalytic activity. As a result, the photocatalytic method is a promising alternative to overcome the world energy dilemma by harnessing solar energy for different applications.

References

1. Olisah C, Adams JB, Rubidge G (2021) The state of persistent organic pollutants in South African estuaries: a review of environmental exposure and sources. Ecotoxicol Environ Saf 219:112316
2. Almroth BC, Cartine J, Jönander C, Karlsson M, Langlois J, Lindström M et al (2021) Assessing the effects of textile leachates in fish using multiple testing methods: from gene expression to behavior. Ecotoxicol Environ Saf 207:111523
3. Ali SS, Al-Tohamy R, Sun J (2022) Performance of Meyerozyma caribbica as a novel manganese peroxidase-producing yeast inhabiting wood-feeding termite gut symbionts for azo dye decolorization and detoxification. Sci Total Environ 806:150665

4. Chandanshive V, Kadam S, Rane N, Jeon B-H, Jadhav J, Govindwar S (2020) In situ textile wastewater treatment in high rate transpiration system furrows planted with aquatic macrophytes and floating phytobeds. Chemosphere 252:126513
5. Akpomie KG, Conradie J (2020) Advances in application of cotton-based adsorbents for heavy metals trapping, surface modifications and future perspectives. Ecotoxicol Environ Saf 201:110825
6. Silva AC, Silvestre AJ, Freire CS, Vilela C (2021) Modification of textiles for functional applications. In: Fundamentals of natural fibres and textiles. Elsevier, pp 303–65
7. Kishor R, Purchase D, Saratale GD, Saratale RG, Ferreira LFR, Bilal M et al (2021) Ecotoxicological and health concerns of persistent coloring pollutants of textile industry wastewater and treatment approaches for environmental safety. J Environ Chem Eng 9(2):105012
8. Al-Tohamy R, Ali SS, Li F, Okasha KM, Mahmoud YA-G, Elsamahy T, et al (2022) A critical review on the treatment of dye-containing wastewater: ecotoxicological and health concerns of textile dyes and possible remediation approaches for environmental safety. Ecotoxicol Environ Saf 231:113160
9. Parmar S, Daki S, Bhattacharya S, Shrivastav A (2022) Microorganism: an ecofriendly tool for waste management and environmental safety. In: Development in wastewater treatment research and processes. Elsevier, pp 175–93
10. Khan S, Anas M, Malik A (2019) Mutagenicity and genotoxicity evaluation of textile industry wastewater using bacterial and plant bioassays. Toxicol Rep 6:193–201
11. Hynes NRJ, Kumar JS, Kamyab H, Sujana JAJ, Al-Khashman OA, Kuslu Y et al (2020) Modern enabling techniques and adsorbents based dye removal with sustainability concerns in textile industrial sector-A comprehensive review. J Clean Prod 272:122636
12. Lellis B, Fávaro-Polonio CZ, Pamphile JA, Polonio JC (2019) Effects of textile dyes on health and the environment and bioremediation potential of living organisms. Biotechnol Res Innov 3(2):275–290
13. Li F, Zhao K, Ng TSA, Dai Y, Wang C-H (2022) Sustainable production of bio-oil and carbonaceous materials from biowaste co-pyrolysis. Chem Eng J 427:131821
14. Berradi M, Hsissou R, Khudhair M, Assouag M, Cherkaoui O, El Bachiri A et al (2019) Textile finishing dyes and their impact on aquatic environs. Heliyon 5(11):e02711
15. Islam M, Mostafa M (2018) Textile dyeing effluents and environment concerns-a review. J Environ Sci Nat Resourc 11(1–2):131–144
16. Michael-Kordatou I, Karaolia P, Fatta-Kassinos D (2018) The role of operating parameters and oxidative damage mechanisms of advanced chemical oxidation processes in the combat against antibiotic-resistant bacteria and resistance genes present in urban wastewater. Water Res 129:208–230
17. Peng S, He X, Pan H (2018) Spectroscopic study on transformations of dissolved organic matter in coal-to-liquids wastewater under integrated chemical oxidation and biological treatment process. J Environ Sci 70:206–216
18. Huong PTL, Tu N, Lan H, Van Quy N, Tuan PA, Dinh NX et al (2018) Functional manganese ferrite/graphene oxide nanocomposites: effects of graphene oxide on the adsorption mechanisms of organic MB dye and inorganic As (V) ions from aqueous solution. RSC Adv 8(22):12376–12389
19. Beheshti A, Hashemi F, Behvandi F, Mayer P, Atzei D (2018) Selective high adsorption capacity for Congo red dye of a new 3D supramolecular complex and its magnetic hybrid. Inorgan Chem Front 5(3):694–704
20. Jilani A, Othman MHD, Ansari MO, Hussain SZ, Ismail AF, Khan IU (2018) Graphene and its derivatives: synthesis, modifications, and applications in wastewater treatment. Environ Chem Lett 16(4):1301–1323
21. Zhang G, Hu L, Zhao R, Su R, Wang Q, Wang P (2018) Microwave-assisted synthesis of ZnNiAl-layered double hydroxides with calcination treatment for enhanced PNP photodegradation under visible-light irradiation. J Photochem Photobiol A 356:633–641
22. Martínez-Huitle CA, Panizza M (2018) Electrochemical oxidation of organic pollutants for wastewater treatment. Curr Opin Electrochem 11:62–71

23. Wei H, Gao B, Ren J, Li A, Yang H (2018) Coagulation/flocculation in dewatering of sludge: a review. Water Res 143:608–631
24. Couto CF, Lange LC, Amaral MCS (2018) A critical review on membrane separation processes applied to remove pharmaceutically active compounds from water and wastewater. J Water Process Eng 26:156–175
25. Tawalbeh M, Al Mojjly A, Al-Othman A, Hilal N (2018) Membrane separation as a pre-treatment process for oily saline water. Desalination 447:182–202
26. Vilardi G, Rodríguez-Rodríguez J, Ochando-Pulido JM, Verdone N, Martinez-Ferez A, Di Palma L (2018) Large laboratory-plant application for the treatment of a tannery wastewater by Fenton oxidation: Fe (II) and nZVI catalysts comparison and kinetic modelling. Process Saf Environ Prot 117:629–638
27. Sadeghassadi M, Macnab CJ, Gopaluni B, Westwick D (2018) Application of neural networks for optimal-setpoint design and MPC control in biological wastewater treatment. Comput Chem Eng 115:150–160
28. Islam MA, Ali I, Karim SA, Firoz MSH, Chowdhury A-N, Morton DW et al (2019) Removal of dye from polluted water using novel nano manganese oxide-based materials. J Water Process Eng 32:100911
29. Natarajan S, Bajaj HC, Tayade RJ (2018) Recent advances based on the synergetic effect of adsorption for removal of dyes from waste water using photocatalytic process. J Environ Sci 65:201–222
30. Gurbuz F, Ozcan A, Çiftçi H, Acet O, Odabasi M (2019) Treatment of textile effluents through bio-composite column: decolorization and COD reduction. Int J Environ Sci Technol 16(12):8653–8662
31. Nasrabadi T, Ruegner H, Schwientek M, Bennett J, Fazel Valipour S, Grathwohl P (2018) Bulk metal concentrations versus total suspended solids in rivers: time-invariant & catchment-specific relationships. PLoS One 13(1):e0191314
32. Xiao N, Wu R, Huang JJ, Selvaganapathy PR (2020) Development of a xurographically fabricated miniaturized low-cost, high-performance microbial fuel cell and its application for sensing biological oxygen demand. Sens Actuat B Chem 304:127432
33. Meng X, Khoso SA, Jiang F, Zhang Y, Yue T, Gao J et al (2020) Removal of chemical oxygen demand and ammonia nitrogen from lead smelting wastewater with high salts content using electrochemical oxidation combined with coagulation–flocculation treatment. Sep Purif Technol 235:116233
34. Aldossary MHA, Ahmad S, Bahraq AA (2020) Effect of total dissolved solids-contaminated water on the properties of concrete. J Build Eng 32:101496
35. López Grimau V (2021) Study of a hybrid system: moving bed biofilm reactor-membrane bioreactor (MBBR-MBR) in the treatment and reuse of textile industrial effluents
36. Samsami S, Mohamadizaniani M, Sarrafzadeh M-H, Rene ER, Firoozbahr M (2020) Recent advances in the treatment of dye-containing wastewater from textile industries: overview and perspectives. Process Saf Environ Prot 143:138–163
37. Cao G, Wang R, Ju Y, Jing B, Duan X, Ao Z et al (2021) Synchronous removal of emulsions and soluble organic contaminants via a microalgae-based membrane system: performance and mechanisms. Water Res 206:117741
38. Behera M, Nayak J, Banerjee S, Chakrabortty S, Tripathy SK (2021) A review on the treatment of textile industry waste effluents towards the development of efficient mitigation strategy: an integrated system design approach. J Environ Chem Eng 9(4):105277
39. Burakov AE, Galunin EV, Burakova IV, Kucherova AE, Agarwal S, Tkachev AG et al (2018) Adsorption of heavy metals on conventional and nanostructured materials for wastewater treatment purposes: a review. Ecotoxicol Environ Saf 148:702–712
40. Mudhoo A, Ramasamy DL, Bhatnagar A, Usman M, Sillanpää M (2020) An analysis of the versatility and effectiveness of composts for sequestering heavy metal ions, dyes and xenobiotics from soils and aqueous milieus. Ecotoxicol Environ Saf 197:110587
41. Li F, Katz L, Hu Z (2019) Adsorption of major nitrogen-containing components in microalgal bio-oil by activated carbon: equilibrium, kinetics, and ideal adsorbed solution theory (IAST) model. ACS Sustain Chem Eng 7(19):16529–16538

42. Abu-Nada A, Abdala A, McKay G (2021) Removal of phenols and dyes from aqueous solutions using graphene and graphene composite adsorption: a review. J Environ Chem Eng 9(5):105858
43. Jadhav AC, Jadhav NC (2021) Treatment of textile wastewater using adsorption and adsorbents. In: Sustainable technologies for textile wastewater treatments. Elsevier, pp 235–73
44. Brião GV, Jahn SL, Foletto EL, Dotto GL (2018) Highly efficient and reusable mesoporous zeolite synthetized from a biopolymer for cationic dyes adsorption. Colloids Surf A 556:43–50
45. Madan S, Shaw R, Tiwari S, Tiwari SK (2019) Adsorption dynamics of Congo red dye removal using ZnO functionalized high silica zeolitic particles. Appl Surf Sci 487:907–917
46. Keskin B, Ersahin ME, Ozgun H, Koyuncu I (2021) Pilot and full-scale applications of membrane processes for textile wastewater treatment: a critical review. J Water Process Eng 42:102172
47. Willet J, Wetser K, Vreeburg J, Rijnaarts HH (2019) Review of methods to assess sustainability of industrial water use. Water Resourc Indus 21:100110
48. Haque MS, Nahar N, Sayem SM (2021) Industrial water management and sustainability: development of SIWP tool for textile industries of Bangladesh. Water Resourc Indus 25:100145
49. Gracia-de-Rentería P, Barberán R, Mur J (2020) The groundwater demand for industrial uses in areas with access to drinking publicly-supplied water: a microdata analysis. Water 12(1):198
50. Khattab TA, Abdelrahman MS, Rehan M (2020) Textile dyeing industry: environmental impacts and remediation. Environ Sci Pollut Res 27(4):3803–3818
51. Gomes da Silva FJ, Gouveia RM (2020) Cleaner production main concept and history. In: Cleaner Production. Springer, pp 15–31
52. Madhav S, Ahamad A, Singh P, Mishra PK (2018) A review of textile industry: wet processing, environmental impacts, and effluent treatment methods. Environ Qual Manage 27(3):31–41
53. Velusamy S, Roy A, Sundaram S, Kumar MT (2021) A review on heavy metal ions and containing dyes removal through graphene oxide-based adsorption strategies for textile wastewater treatment. Chem Rec 21(7):1570–1610
54. Yogalakshmi KN, Das A, Rani G, Jaswal V, Randhawa JS (2020) Nano-bioremediation: a new age technology for the treatment of dyes in textile effluents. In: Bioremediation of Industrial Waste for Environmental Safety. Springer, 313–347
55. Zhou Y, Lu J, Zhou Y, Liu Y (2019) Recent advances for dyes removal using novel adsorbents: a review. Environ Pollut 252:352–365
56. Singh AP, Dhadse K, Ahalawat J (2019) Managing water quality of a river using an integrated geographically weighted regression technique with fuzzy decision-making model. Environ Monit Assess 191(6):1–17
57. Srinivas R, Singh AP (2018) Impact assessment of industrial wastewater discharge in a river basin using interval-valued fuzzy group decision-making and spatial approach. Environ Dev Sustain 20(5):2373–2397
58. Mani S, Chowdhary P, Bharagava RN (2019) Textile wastewater dyes: toxicity profile and treatment approaches. Emerging and eco-friendly approaches for waste management. Springer, pp 219–244
59. Agarwal S, Singh AP (2022) Performance evaluation of textile wastewater treatment techniques using sustainability index: an integrated fuzzy approach of assessment. J Clean Prod 130384
60. Chowdhary P, Bharagava RN, Mishra S, Khan N (2020) Role of industries in water scarcity and its adverse effects on environment and human health. In: Environmental concerns and sustainable development. Springer, pp 235–256
61. Patil CS, Kadam AN, Gunjal DB, Naik VM, Lee S-W, Kolekar GB et al (2020) Sugarcane molasses derived carbon sheet@ sea sand composite for direct removal of methylene blue from textile wastewater: industrial wastewater remediation through sustainable, greener, and scalable methodology. Sep Purif Technol 247:116997
62. Oyeniran DO, Sogbanmu TO, Adesalu TA (2021) Antibiotics, algal evaluations and subacute effects of abattoir wastewater on liver function enzymes, genetic and haematologic biomarkers in the freshwater fish Clarias gariepinus. Ecotoxicol Environ Saf 212:111982

63. Bhattacharya S, Gupta AB, Gupta A, Pandey A (2018) Introduction to water remediation: importance and methods. In: Water remediation. Springer, pp 3–8
64. Singha K, Pandit P, Maity S, Sharma SR (2021) Harmful environmental effects for textile chemical dyeing practice. In: Green Chemistry for Sustainable Textiles. Elsevier, pp 153–164
65. Jiku MAS, Singha A, Faruquee M, Rahaman MA, Alam MA, Ehsanullah M (2021) Toxic wastewater status for irrigation usage at Gazipur and Savar industrial vicinity of Bangladesh. Acta Ecol Sin 41(4):358–364
66. Sojobi AO, Zayed T (2022) Impact of sewer overflow on public health: a comprehensive scientometric analysis and systematic review. Environ Res 203:111609
67. Al Sawaf MB, Karaca F (2018) Different stakeholders' opinions toward the sustainability of common textile wastewater treatment technologies in Turkey: a case study Istanbul province. Sustain Cities Soc 42:194–205
68. Chen G, An X, Li H, Lai F, Yuan E, Xia X, et al (2021) Detoxification of azo dye Direct Black G by thermophilic Anoxybacillus sp. PDR2 and its application potential in bioremediation. Ecotoxicol Environ Saf 214:112084
69. Chen Y-G, Huang J-H, Luo R, Ge H-Z, Wołowicz A, Wawrzkiewicz M et al (2021) Impacts of heavy metals and medicinal crops on ecological systems, environmental pollution, cultivation, and production processes in China. Ecotoxicol Environ Saf 219:112336
70. Soni V, Keswani K, Bhatt U, Kumar D, Singh H (2021) In vitro propagation and analysis of mixotrophic potential to improve survival rate of Dolichandra unguis-cati under ex vitro conditions. Heliyon. 7(2):e06101
71. Khalaj M, Kamali M, Khodaparast Z, Jahanshahi A (2018) Copper-based nanomaterials for environmental decontamination–an overview on technical and toxicological aspects. Ecotoxicol Environ Saf 148:813–824
72. Tounsadi H, Metarfi Y, Taleb M, El Rhazi K, Rais Z (2020) Impact of chemical substances used in textile industry on the employee's health: epidemiological study. Ecotoxicol Environ Saf 197:110594
73. Jin X, Wu C, Tian X, Wang P, Zhou Y, Zuo J (2021) A magnetic-void-porous MnFe2O4/carbon microspheres nano-catalyst for catalytic ozonation: preparation, performance and mechanism. Environ Sci Ecotechnol 7:100110
74. Wu L, Xu Y, Lv X, Chang X, Ma X, Tian X et al (2021) Impacts of an azo food dye tartrazine uptake on intestinal barrier, oxidative stress, inflammatory response and intestinal microbiome in crucian carp (Carassius auratus). Ecotoxicol Environ Saf 223:112551
75. Piątkowska M, Jedziniak P, Olejnik M, Żmudzki J, Posyniak A (2018) Absence of evidence or evidence of absence? A transfer and depletion study of Sudan I in eggs. Food Chem 239:598–602
76. Haq I, Raj A (2018) Biodegradation of Azure-B dye by Serratia liquefaciens and its validation by phytotoxicity, genotoxicity and cytotoxicity studies. Chemosphere 196:58–68
77. Ayub I, Munir A, Amjad W, Ghafoor A, Nasir MS (2018) Energy-and exergy-based thermal analyses of a solar bakery unit. J Therm Anal Calorim 133(2):1001–1013
78. Nasir MS, Yang G, Ayub I, Wang S, Wang L, Wang X et al (2019) Recent development in graphitic carbon nitride based photocatalysis for hydrogen generation. Appl Catal B 257:117855
79. Kumar P, Boukherroub R, Shankar K (2018) Sunlight-driven water-splitting using two-dimensional carbon based semiconductors. J Mater Chem A 6(27):12876–12931
80. Hasija V, Raizada P, Sudhaik A, Sharma K, Kumar A, Singh P et al (2019) Recent advances in noble metal free doped graphitic carbon nitride based nanohybrids for photocatalysis of organic contaminants in water: a review. Appl Mater Today 15:494–524
81. Hasija V, Nguyen V-H, Kumar A, Raizada P, Krishnan V, Khan AAP et al (2021) Advanced activation of persulfate by polymeric g-C3N4 based photocatalysts for environmental remediation: a review. J Hazard Mater 413:125324
82. Farhadian N, Liu S, Asadi A, Shahlaei M, Moradi S (2021) Enhanced heterogeneous Fenton oxidation of organic pollutant via Fe-containing mesoporous silica composites: a review. J Mol Liq 321:114896

83. Soni V, Raizada P, Kumar A, Hasija V, Singal S, Singh P et al (2021) Indium sulfide-based photocatalysts for hydrogen production and water cleaning: a review. Environ Chem Lett 19(2):1065–1095
84. Gong H, Jin Z, Xu H, Wang Q, Zuo J, Wu J et al (2018) Redesigning C and N mass flows for energy-neutral wastewater treatment by coagulation adsorption enhanced membrane (CAEM)-based pre-concentration process. Chem Eng J 342:304–309
85. Sharma S, Dutta V, Singh P, Raizada P, Rahmani-Sani A, Hosseini-Bandegharaei A et al (2019) Carbon quantum dot supported semiconductor photocatalysts for efficient degradation of organic pollutants in water: a review. J Clean Prod 228:755–769
86. Dutta V, Sharma S, Raizada P, Thakur VK, Khan AAP, Saini V et al (2021) An overview on WO3 based photocatalyst for environmental remediation. J Environ Chem Eng 9(1):105018
87. Sharma K, Dutta V, Sharma S, Raizada P, Hosseini-Bandegharaei A, Thakur P et al (2019) Recent advances in enhanced photocatalytic activity of bismuth oxyhalides for efficient photocatalysis of organic pollutants in water: a review. J Ind Eng Chem 78:1–20
88. Kumar A, Raizada P, Hosseini-Bandegharaei A, Thakur VK, Nguyen V-H, Singh P (2021) C-, N-Vacancy defect engineered polymeric carbon nitride towards photocatalysis: viewpoints and challenges. J Mat Chem A 9(1):111–153
89. Hasija V, Patial S, Raizada P, Khan AAP, Asiri AM, Van Le Q et al (2022) Covalent organic frameworks promoted single metal atom catalysis: strategies and applications. Coord Chem Rev 452:214298
90. Hasija V, Kumar A, Sudhaik A, Raizada P, Singh P, Van Le Q et al (2021) Step-scheme heterojunction photocatalysts for solar energy, water splitting, CO2 conversion, and bacterial inactivation: a review. Environ Chem Lett 19(4):2941–2966
91. Dutta V, Sharma S, Raizada P, Kumar R, Thakur VK, Nguyen V-H et al (2020) Recent progress on bismuth-based Z-scheme semiconductor photocatalysts for energy and environmental applications. J Environ Chem Eng 8(6):104505
92. Adegoke KA, Iqbal M, Louis H, Bello OS (2019) Synthesis, characterization and application of CdS/ZnO nanorod heterostructure for the photodegradation of Rhodamine B dye. Mater Sci Energy Technol 2(2):329–336
93. Dutta V, Sonu S, Raizada P, Thakur VK, Ahamad T, Thakur S, et al (2022) Prism-like integrated Bi2WO6 with Ag-CuBi2O4 on carbon nanotubes (CNTs) as an efficient and robust S-scheme interfacial charge transfer photocatalyst for the removal of organic pollutants from wastewater. Environ Sci Pollut Res 1–16
94. Mukherjee D, Van der Bruggen B, Mandal B (2022) Advancements in visible light responsive MOF composites for photocatalytic decontamination of textile wastewater: a review. Chemosphere 295:133835. https://doi.org/10.1016/j.chemosphere.2022.133835
95. Chen L, Caro F, Corbett CJ, Ding X (2019) Estimating the environmental and economic impacts of widespread adoption of potential technology solutions to reduce water use and pollution: application to China's textile industry. Environ Impact Assess Rev 79:106293. https://doi.org/10.1016/j.eiar.2019.106293
96. Lin H, Wu J, Zhou F, Zhao X, Lu P, Sun G et al (2023) Graphitic carbon nitride-based photocatalysts in the applications of environmental catalysis. J Environ Sci 124:570–590. https://doi.org/10.1016/j.jes.2021.11.017
97. Pirsaheb M, Asadi A, Sillanpää M, Farhadian N (2018) Application of carbon quantum dots to increase the activity of conventional photocatalysts: a systematic review. J Mol Liq 271:857–871. https://doi.org/10.1016/j.molliq.2018.09.064
98. Madima N, Mishra SB, Inamuddin I, Mishra AK (2020) Carbon-based nanomaterials for remediation of organic and inorganic pollutants from wastewater. A review. Environ Chem Lett 18(4):1169–1191. https://doi.org/10.1007/s10311-020-01001-0
99. Yao Y, Zhang H, Hu K, Nie G, Yang Y, Wang Y et al (2022) Carbon dots based photocatalysis for environmental applications. J Environ Chem Eng 10(2):107336. https://doi.org/10.1016/j.jece.2022.107336

100. Mu F, Dai B, Wu Y, Yang G, Li S, Zhang L et al (2022) 2D/3D S-scheme heterojunction of carbon nitride/iodine-deficient bismuth oxyiodide for photocatalytic hydrogen production and bisphenol A degradation. J Colloid Interface Sci 612:722–736. https://doi.org/10.1016/j.jcis.2021.12.196
101. Zhang Y, Cao W, Zhu B, Cai J, Li X, Liu J et al (2022) Fabrication of NH2-MIL-125(Ti) nanodots on carbon fiber/MoS2-based weavable photocatalysts for boosting the adsorption and photocatalytic performance. J Colloid Interface Sci 611:706–717. https://doi.org/10.1016/j.jcis.2021.12.073
102. Koe WS, Lee JW, Chong WC, Pang YL, Sim LC (2020) An overview of photocatalytic degradation: photocatalysts, mechanisms, and development of photocatalytic membrane. Environ Sci Pollut Res 27(3):2522–2565. https://doi.org/10.1007/s11356-019-07193-5
103. Lee G-J, Wu JJ (2017) Recent developments in ZnS photocatalysts from synthesis to photocatalytic applications—a review. Powder Technol 318:8–22. https://doi.org/10.1016/j.powtec.2017.05.022
104. Phin H-Y, Ong Y-T, Sin J-C (2020) Effect of carbon nanotubes loading on the photocatalytic activity of zinc oxide/carbon nanotubes photocatalyst synthesized via a modified sol-gel method. J Environ Chem Eng 8(3):103222. https://doi.org/10.1016/j.jece.2019.103222
105. Martins AC, Cazetta AL, Pezoti O, Souza JRB, Zhang T, Pilau EJ et al (2017) Sol-gel synthesis of new TiO2/activated carbon photocatalyst and its application for degradation of tetracycline. Ceram Int 43(5):4411–4418. https://doi.org/10.1016/j.ceramint.2016.12.088
106. Mkhalid IA (2022) Hydrogen evolution over sol-gel prepared visible-light-responsive Ag2O/SrAl2O4/CNT ternary photocatalyst. Ceram Int 48(2):1542–1549. https://doi.org/10.1016/j.ceramint.2021.09.233
107. Zhao Q, Feng G, Jiang F, Lan S, Chen J, Zhong F et al (2020) Nonhydrolytic sol-gel in-situ synthesis of novel recoverable amorphous Fe2TiO5/C hollow spheres as visible-light driven photocatalysts. Mater Des 194:108928. https://doi.org/10.1016/j.matdes.2020.108928
108. Dahman Y (2017) Nanotechnology and functional materials for engineers. Elsevier
109. Shen T, Wang Q, Guo Z, Kuang J, Cao W (2018) Hydrothermal synthesis of carbon quantum dots using different precursors and their combination with TiO2 for enhanced photocatalytic activity. Ceram Int 44(10):11828–11834. https://doi.org/10.1016/j.ceramint.2018.03.271
110. Perumal K, Shanavas S, Karthigeyan A, Ahamad T, Alshehri SM, Murugakoothan P (2020) Hydrothermal assisted precipitation synthesis of highly stable g-C3N4/BiOBr/CdS photocatalyst with enhanced visible light photocatalytic degradation of tetracycline. Diam Relat Mater 110:108091. https://doi.org/10.1016/j.diamond.2020.108091
111. Rani K, Gupta V (2022) Surfactant assisted solvothermal synthesis of Bi2Te3 nanostructure for thermoelectric applications. Mater Today Proc https://doi.org/10.1016/j.matpr.2022.04.109.
112. Wang C, Yang K, Wei X, Ding S, Tian F, Li F (2018) One-pot solvothermal synthesis of carbon dots/Ag nanoparticles/TiO2 nanocomposites with enhanced photocatalytic performance. Ceram Int 44(18):22481–22488. https://doi.org/10.1016/j.ceramint.2018.09.017
113. Guan Y, Wang S, Du Q, Wu M, Zheng Z, Li Z et al (2022) C-scheme electron transfer mechanism: an efficient ternary heterojunction photocatalyst carbon quantum dots/Bi/BiOBr with full ohmic contact. J Colloid Interface Sci. https://doi.org/10.1016/j.jcis.2022.05.091
114. Yu W, Wan S, Yuan D, Sun L, Wang Y, Wang M (2021) Microwave solvothermal-assisted calcined synthesis of Bi2WxMo1−XO6 solid solution photocatalysts for degradation and detoxification of bisphenol A under simulated sunlight irradiation. Sep Purif Technol 275:119175. https://doi.org/10.1016/j.seppur.2021.119175
115. Wang X, Meng J, Zhang X, Liu Y, Ren M, Yang Y et al (2021) Controllable approach to carbon-deficient and oxygen-doped graphitic carbon nitride: robust photocatalyst against recalcitrant organic pollutants and the mechanism insight. Adv Func Mater 31(20):2010763
116. Raizada P, Sudhaik A, Singh P, Hosseini-Bandegharaei A, Thakur P (2019) Converting type II AgBr/VO into ternary Z scheme photocatalyst via coupling with phosphorus doped g-C3N4 for enhanced photocatalytic activity. Sep Purif Technol 227:115692. https://doi.org/10.1016/j.seppur.2019.115692

117. Zhang R, Jiang J, Zeng K (2022) Synthesis of Bi2WO6/g-C3N4 heterojunction on activated carbon fiber membrane as a thin-film photocatalyst for treating antibiotic wastewater. Inorgan Chem Commun 2022:109418. https://doi.org/10.1016/j.inoche.2022.109418
118. Kafle BP (2019) Chemical analysis and material characterization by spectrophotometry. Elsevier
119. Liu P, Bao R, Fang D, Yi J, Li L (2018) A facile synthesis of CNTs/Cu2O-CuO heterostructure composites by spray pyrolysis and its visible light responding photocatalytic properties. Adv Powder Technol 29(9):2027–2034. https://doi.org/10.1016/j.apt.2018.05.009
120. Zhu S, Nie L (2021) Progress in fabrication of one-dimensional catalytic materials by electrospinning technology. J Ind Eng Chem 93:28–56. https://doi.org/10.1016/j.jiec.2020.09.016
121. Wang W, Yang R, Li T, Komarneni S, Liu B (2021) Advances in recyclable and superior photocatalytic fibers: material, construction, application and future perspective. Compos B Eng 205:108512. https://doi.org/10.1016/j.compositesb.2020.108512
122. Pant B, Prasad Ojha G, Acharya J, Park M (2021) Ag3PO4-TiO2-Carbon nanofiber composite: an efficient visible-light photocatalyst obtained from eelectrospinning and hydrothermal methods. Sep Purif Technol 276:119400. https://doi.org/10.1016/j.seppur.2021.119400
123. Shen X, Song L, Luo L, Zhang Y, Zhu B, Liu J et al (2018) Preparation of TiO2/C3N4 heterojunctions on carbon-fiber cloth as efficient filter-membrane-shaped photocatalyst for removing various pollutants from the flowing wastewater. J Colloid Interface Sci 532:798–807. https://doi.org/10.1016/j.jcis.2018.08.028
124. Ji M, Zhang Z, Xia J, Di J, Liu Y, Chen R et al (2018) Enhanced photocatalytic performance of carbon quantum dots/BiOBr composite and mechanism investigation. Chin Chem Lett 29(6):805–810. https://doi.org/10.1016/j.cclet.2018.05.002
125. Qu Z, Wang J, Tang J, Shu X, Liu X, Zhang Z et al (2018) Carbon quantum dots/KNbO3 hybrid composites with enhanced visible-light driven photocatalytic activity toward dye waste-water degradation and hydrogen production. Mol Catal 445:1–11. https://doi.org/10.1016/j.mcat.2017.11.002
126. Li C, Che H, Liu C, Che G, Charpentier PA, Xu WZ et al (2019) Facile fabrication of g-C3N4 QDs/BiVO4 Z-scheme heterojunction towards enhancing photodegradation activity under visible light. J Taiwan Inst Chem Eng 95:669–681. https://doi.org/10.1016/j.jtice.2018.10.011
127. Peng H, Guo J (2020) Removal of chromium from wastewater by membrane filtration, chemical precipitation, ion exchange, adsorption electrocoagulation, electrochemical reduction, electrodialysis, electrodeionization, photocatalysis and nanotechnology: a review. Environ Chem Lett 18(6):2055–2068. https://doi.org/10.1007/s10311-020-01058-x
128. Ihsanullah I, Jamal A, Ilyas M, Zubair M, Khan G, Atieh MA (2020) Bioremediation of dyes: current status and prospects. J Water Proc Eng 38:101680. https://doi.org/10.1016/j.jwpe.2020.101680
129. Hansima MACK, Makehelwala M, Jinadasa KBSN, Wei Y, Nanayakkara KGN, Herath AC, et al (2021) Fouling of ion exchange membranes used in the electrodialysis reversal advanced water treatment: a review. Chemosphere 263:127951. https://doi.org/10.1016/j.chemosphere.2020.127951
130. Tahir MB, Kiran H, Iqbal T (2019) The detoxification of heavy metals from aqueous environment using nano-photocatalysis approach: a review. Environ Sci Pollut Res 26(11):10515–10528. https://doi.org/10.1007/s11356-019-04547-x
131. Sutar S, Patil P, Jadhav J (2022) Recent advances in biochar technology for textile dyes wastewater remediation: a review. Environ Res 209:112841. https://doi.org/10.1016/j.envres.2022.112841
132. Danish MS, Estrella LL, Alemaida IMA, Lisin A, Moiseev N, Ahmadi M, et al (2021) Photocatalytic applications of metal oxides for sustainable environmental remediation. Metals 11(1). https://doi.org/10.3390/met11010080
133. Bilal Tahir M, Nadeem Riaz K, Asiri AM (2019) Boosting the performance of visible light-driven WO3/g-C3N4 anchored with BiVO4 nanoparticles for photocatalytic hydrogen evolution. Int J Energy Res 43(11):5747–5758. https://doi.org/10.1002/er.4673

134. Anzar N, Hasan R, Tyagi M, Yadav N, Narang J (2020) Carbon nanotube–a review on synthesis, properties and plethora of applications in the field of biomedical science. Sens Int 1:100003. https://doi.org/10.1016/j.sintl.2020.100003
135. Vidas L, Castro R (2021) Recent developments on hydrogen production technologies: state-of-the-art review with a focus on green-electrolysis. Appl Sci 11(23). https://doi.org/10.3390/app112311363
136. Egbedina AO, Bolade OP, Ewuzie U, Lima EC (2022) Emerging trends in the application of carbon-based materials: a review. J Environ Chem Eng 10(2):107260. https://doi.org/10.1016/j.jece.2022.107260
137. Chen T-W, Kalimuthu P, Veerakumar P, Lin K-C, Chen S-M, Ramachandran R, et al (2022) Recent developments in carbon-based nanocomposites for fuel cell applications: a review. Molecules 27(3). https://doi.org/10.3390/molecules27030761
138. Hao L, Huang H, Zhang Y, Ma T (2021) Oxygen vacant semiconductor photocatalysts. Adv Funct Mater 31(25):2100919. https://doi.org/10.1002/adfm.202100919
139. Hossain S, Chu W-S, Lee CS, Ahn S-H, Chun D-M (2019) Photocatalytic performance of few-layer Graphene/WO3 thin films prepared by a nano-particle deposition system. Mater Chem Phys 226:141–150. https://doi.org/10.1016/j.matchemphys.2019.01.026
140. Zhang C, Bai J, Ma L, Lv Y, Wang F, Zhang X et al (2018) Synthesis of halogen doped graphite carbon nitride nanorods with outstanding photocatalytic H2O2 production ability via saturated NH4X (X = Cl, Br) solution-hydrothermal post-treatment. Diam Relat Mater 87:215–222. https://doi.org/10.1016/j.diamond.2018.06.013
141. Vaiano V, Sacco O, Matarangolo M (2018) Photocatalytic degradation of paracetamol under UV irradiation using TiO2-graphite composites. Catal Today 315:230–236. https://doi.org/10.1016/j.cattod.2018.02.002
142. Lopes JL, Martins MJ, Nogueira HIS, Estrada AC, Trindade T (2021) Carbon-based heterogeneous photocatalysts for water cleaning technologies: a review. Environ Chem Lett 19(1):643–668. https://doi.org/10.1007/s10311-020-01092-9
143. Liu Z, Ling Q, Cai Y, Xu L, Su J, Yu K et al (2022) Synthesis of carbon-based nanomaterials and their application in pollution management. Nanoscale Adv 4(5):1246–1262. https://doi.org/10.1039/D1NA00843A
144. Filik H, Avan AA (2020) Review on applications of carbon nanomaterials for simultaneous electrochemical sensing of environmental contaminant dihydroxybenzene isomers. Arab J Chem 13(7):6092–6105. https://doi.org/10.1016/j.arabjc.2020.05.009
145. Mao H, Zhang F, Du M, Dai L, Qian Y, Pang H (2021) Review on synthesis of porous TiO2-based catalysts for energy conversion systems. Ceram Int 47(18):25177–25200. https://doi.org/10.1016/j.ceramint.2021.06.039
146. Nguyen D-B, Dong TMT, Nguyen TMN, Nguyen T-T, Vuong V-D, Thanh Phong M, et al (2021) Multi-layered thin film nanocomposite MoS2@MoO2/MWCNP/ITO-PET: electrochemical approaches for synthesis and structural characterizations. Appl Surf Sci 565:150508. https://doi.org/10.1016/j.apsusc.2021.150508
147. Shi Y, Huang J, Zeng G, Cheng W, Hu J (2019) Photocatalytic membrane in water purification: is it stepping closer to be driven by visible light? J Membr Sci 584:364–392. https://doi.org/10.1016/j.memsci.2019.04.078
148. Jain VP, Chaudhary S, Sharma D, Dabas N, Lalji RSK, Singh BK, et al (2021) Advanced functionalized nanographene oxide as a biomedical agent for drug delivery and anti-cancerous therapy: a review. Eur Polymer J 142:110124. https://doi.org/10.1016/j.eurpolymj.2020.110124
149. Davydov VN (2018) The recurrent relations for the electronic band structure of the multilayer graphene. Proc R Soc A Math Phys Eng Sci 474(2220):20180439. https://doi.org/10.1098/rspa.2018.0439
150. Lv M, Yan L, Liu C, Su C, Zhou Q, Zhang X et al (2018) Non-covalent functionalized graphene oxide (GO) adsorbent with an organic gelator for co-adsorption of dye, endocrine-disruptor, pharmaceutical and metal ion. Chem Eng J 349:791–799. https://doi.org/10.1016/j.cej.2018.04.153

151. Patnaik S, Behera A, Parida K (2021) A review on g-C3N4/graphene nanocomposites: multifunctional roles of graphene in the nanohybrid photocatalyst toward photocatalytic applications. Catal Sci Technol 11(18):6018–6040. https://doi.org/10.1039/D1CY00784J
152. Stathis A, Bouza Z, Papadakis I, Couris S (2022) Tailoring the nonlinear optical response of some graphene derivatives by ultraviolet (UV) irradiation. Nanomaterials 12(1). https://doi.org/10.3390/nano12010152
153. Van Khai T, Viet Hai L, Thi Thu Ha N, Thi Thom N, Van Trang N, Thi Nam P, et al (2021) Combined experimental and theoretical studies on enlarged bandgap and improved photoelectrochemical properties of reduced graphene oxide film by hydrogen annealing. J Electroanalyt Chem 900:115722. https://doi.org/10.1016/j.jelechem.2021.115722
154. Heng ZW, Chong WC, Pang YL, Koo CH (2021) An overview of the recent advances of carbon quantum dots/metal oxides in the application of heterogeneous photocatalysis in photodegradation of pollutants towards visible-light and solar energy exploitation. J Environ Chem Eng 9(3):105199. https://doi.org/10.1016/j.jece.2021.105199
155. Wu J, Li L, Li X-A, Min X, Xing Y (2022) A novel 2D graphene oxide modified α-AgVO3 nanorods: design, fabrication, and enhanced visible-light photocatalytic performance. J Adv Ceram 11(2):308–320. https://doi.org/10.1007/s40145-021-0534-6
156. Long X, Feng C, Yang S, Ding D, Feng J, Liu M, et al (2022) Oxygen doped graphitic carbon nitride with regulatable local electron density and band structure for improved photocatalytic degradation of bisphenol A. Chem Eng J 435:134835. https://doi.org/10.1016/j.cej.2022.134835
157. Raja A, Rajasekaran P, Selvakumar K, Arunpandian M, Kaviyarasu K, Asath Bahadur S, et al (2020) Visible active reduced graphene oxide-BiVO4-ZnO ternary photocatalyst for efficient removal of ciprofloxacin. Separat Purif Technol 233:115996. https://doi.org/10.1016/j.seppur.2019.115996
158. Selvakumar K, Raja A, Arunpandian M, Stalindurai K, Rajasekaran P, Sami P et al (2019) Efficient photocatalytic degradation of ciprofloxacin and bisphenol A under visible light using Gd2WO6 loaded ZnO/bentonite nanocomposite. Appl Surf Sci 481:1109–1119. https://doi.org/10.1016/j.apsusc.2019.03.178
159. Raja A, Rajasekaran P, Selvakumar K, Ganapathi Raman R, Swaminathan M (2020) Visible active TiO2-CdS-rGO ternary nanocomposite for enhanced photodecomposition of methylene blue. Mater Today Proc 29:1125–1128. https://doi.org/10.1016/j.matpr.2020.05.272
160. Li X, Huang G, Chen X, Huang J, Li M, Yin J, et al (2021) A review on graphitic carbon nitride (g-C3N4) based hybrid membranes for water and wastewater treatment. Sci Total Environ 792:148462. https://doi.org/10.1016/j.scitotenv.2021.148462
161. Lee SJ, Begildayeva T, Jung HJ, Koutavarapu R, Yu Y, Choi M, et al (2021) Plasmonic ZnO/Au/g-C3N4 nanocomposites as solar light active photocatalysts for degradation of organic contaminants in wastewater. Chemosphere 263:128262. https://doi.org/10.1016/j.chemosphere.2020.128262
162. Li C, Wang B, Zhang F, Song N, Liu G, Wang C et al (2020) Performance of Ag/BiOBr/GO composite photocatalyst for visible-light-driven dye pollutants degradation. J Market Res 9(1):610–621. https://doi.org/10.1016/j.jmrt.2019.11.001
163. Liu X, Cai L (2019) A novel double Z-scheme BiOBr-GO-polyaniline photocatalyst: study on the excellent photocatalytic performance and photocatalytic mechanism. Appl Surf Sci 483:875–887. https://doi.org/10.1016/j.apsusc.2019.03.273
164. Zaaba NI, Foo KL, Hashim U, Tan SJ, Liu W-W, Voon CH (2017) Synthesis of graphene oxide using modified hummers method: solvent influence. Proc Eng 184:469–477. https://doi.org/10.1016/j.proeng.2017.04.118
165. Bai Y, Mao W, Wu Y, Gao Y, Wang T, Liu S (2021) Synthesis of novel ternary heterojunctions via Bi2WO6 coupling with CuS and g-C3N4 for the highly efficient visible-light photodegradation of ciprofloxacin in wastewater. Coll Surf A Physicochem Eng Aspects 610:125481. https://doi.org/10.1016/j.colsurfa.2020.125481
166. Aqel A, El-Nour KMMA, Ammar RAA, Al-Warthan A (2012) Carbon nanotubes, science and technology part (I) structure, synthesis and characterisation. Arab J Chem 5(1):1–23. https://doi.org/10.1016/j.arabjc.2010.08.022

167. Masugi Y, Kawai H, Ejiri M, Hirano H, Fujiwara Y, Tanaka T, et al (2022) Early strong predictors of decline in instrumental activities of daily living in community-dwelling older Japanese people. PLOS One 17(4):e0266614-e
168. Rathinavel S, Priyadharshini K, Panda D (2021) A review on carbon nanotube: an overview of synthesis, properties, functionalization, characterization, and the application. Mater Sci Eng B 268:115095. https://doi.org/10.1016/j.mseb.2021.115095
169. Jauhari H, Grover R, Gupta N, Nanda O, Mehta DS, Saxena K (2018) Solid state dye sensitized solar cells with polyaniline-thiourea based polymer electrolyte composition. J Renew Sustain Energy 10(3):033502. https://doi.org/10.1063/1.5019293
170. Poudel YR, Li W (2018) Synthesis, properties, and applications of carbon nanotubes filled with foreign materials: a review. Mater Today Phys 7:7–34. https://doi.org/10.1016/j.mtphys.2018.10.002
171. Azzam EMS, Fathy NA, El-Khouly SM, Sami RM (2019) Enhancement the photocatalytic degradation of methylene blue dye using fabricated CNTs/TiO2/AgNPs/Surfactant nanocomposites. J Water Proc Eng 28:311–321. https://doi.org/10.1016/j.jwpe.2019.02.016
172. Zhao H, Li H, Yu H, Chang H, Quan X, Chen S (2013) CNTs–TiO2/Al2O3 composite membrane with a photocatalytic function: fabrication and energetic performance in water treatment. Sep Purif Technol 116:360–365. https://doi.org/10.1016/j.seppur.2013.06.007
173. Yang Y, Liu K, Sun F, Liu Y, Chen J (2022) Enhanced performance of photocatalytic treatment of Congo red wastewater by CNTs-Ag-modified TiO2 under visible light. Environ Sci Pollut Res 29(11):15516–15525. https://doi.org/10.1007/s11356-021-16734-w
174. Zhang K, Meng Z, Oh W (2010) Degradation of Rhodamine B by Fe-Carbon Nanotubes/TiO2 composites under UV light in aerated solution. Chin J Catal 31(7):751–758. https://doi.org/10.1016/S1872-2067(09)60084-X
175. Zhu J, Jiang Z (2021) Electrochemical photocatalytic degradation of eriochrome black T dye using synthesized TiO2@ CNTs nanofibers. Int J Electrochem Sci 16:210318. https://doi.org/10.20964/2021.03.55
176. Huang Y, Chen D, Hu X, Qian Y, Li D (2018) Preparation of TiO2/carbon nanotubes/reduced graphene oxide composites with enhanced photocatalytic activity for the degradation of Rhodamine B. Nanomaterials 8(6). https://doi.org/10.3390/nano8060431
177. Gagic M, Kociova S, Smerkova K, Michalkova H, Setka M, Svec P et al (2020) One-pot synthesis of natural amine-modified biocompatible carbon quantum dots with antibacterial activity. J Colloid Interface Sci 580:30–48. https://doi.org/10.1016/j.jcis.2020.06.125
178. Vieira KO, Bettini J, de Oliveira LFC, Ferrari JL, Schiavon MA (2017) Synthesis of multicolor photoluminescent carbon quantum dots functionalized with hydrocarbons of different chain lengths. New Carbon Mater 32(4):327–337. https://doi.org/10.1016/S1872-5805(17)60126-4
179. Liang Q, Ma W, Shi Y, Li Z, Yang X (2013) Easy synthesis of highly fluorescent carbon quantum dots from gelatin and their luminescent properties and applications. Carbon 60:421–428. https://doi.org/10.1016/j.carbon.2013.04.055
180. Pan M, Xie X, Liu K, Yang J, Hong L, Wang S (2020) Fluorescent carbon quantum dots—synthesis, functionalization and sensing application in food analysis. Nanomaterials 10(5). https://doi.org/10.3390/nano10050930
181. Molaei MJ (2020) Principles, mechanisms, and application of carbon quantum dots in sensors: a review. Anal Methods 12(10):1266–1287. https://doi.org/10.1039/C9AY02696G
182. Chen H, Pina JM, Hou Y, Sargent EH (2022) Synthesis, applications, and prospects of quantum-dot-in-perovskite solids. Adv Energy Mater 12(4):2100774. https://doi.org/10.1002/aenm.202100774
183. Fernando KAS, Sahu S, Liu Y, Lewis WK, Guliants EA, Jafariyan A et al (2015) Carbon quantum dots and applications in photocatalytic energy conversion. ACS Appl Mater Interfaces 7(16):8363–8376. https://doi.org/10.1021/acsami.5b00448
184. Murillo-Sierra JC, Hernández-Ramírez A, Zhao Z-Y, Martínez-Hernández A, Gracia-Pinilla MA (2021) Construction of direct Z-scheme WO3/ZnS heterojunction to enhance the photocatalytic degradation of tetracycline antibiotic. J Environ Chem Eng 9(2):105111. https://doi.org/10.1016/j.jece.2021.105111

185. Xu X, Lin H, Xiao P, Zhu J, Bi H, Carabineiro SAC (2022) Construction of Ag-bridged Z-Scheme LaFe0.5Co0.5O3/Ag10/graphitic carbon nitride heterojunctions for photo-fenton degradation of tetracycline hydrochloride: interfacial electron effect and reaction mechanism. Adv Mater Interfaces 9(5):2101902. https://doi.org/10.1002/admi.202101902
186. Kokilavani S, Syed A, Elgorban AM, Bahkali AH, Al-Shwaiman HA, Varma RS, et al (2022) Designing Z-scheme AgIO4 nanorod embedded with Bi2S3 nanoflakes for expeditious visible light photodegradation of congo red and rhodamine B. Chemosphere 294:133755. https://doi.org/10.1016/j.chemosphere.2022.133755
187. Bokare A, Chinnusamy S, Erogbogbo F (2021) TiO2-graphene quantum dots nanocomposites for photocatalysis in energy and biomedical applications. Catalysts 11(3). https://doi.org/10.3390/catal11030319
188. Liang H, Tai X, Du Z, Yin Y (2020) Enhanced photocatalytic activity of ZnO sensitized by carbon quantum dots and application in phenol wastewater. Opt Mater 100:109674. https://doi.org/10.1016/j.optmat.2020.109674
189. Zhang L, Zhang J, Xia Y, Xun M, Chen H, Liu X, et al (2020) Metal-free carbon quantum dots implant graphitic carbon nitride: enhanced photocatalytic dye wastewater purification with simultaneous hydrogen production. Int J Mol Sci 21(3). https://doi.org/10.3390/ijms21031052
190. Wang S, Li L, Zhu Z, Zhao M, Zhang L, Zhang N, et al (2019) Remarkable improvement in photocatalytic performance for tannery wastewater processing via SnS2 modified with N-doped carbon quantum dots: synthesis, characterization, and 4-Nitrophenol-Aided Cr(VI) photoreduction. Small 15(29):1804515. https://doi.org/10.1002/smll.201804515
191. Ming H, Wei D, Yang Y, Chen B, Yang C, Zhang J, et al (2021) Photocatalytic activation of peroxymonosulfate by carbon quantum dots functionalized carbon nitride for efficient degradation of bisphenol A under visible-light irradiation. Chem Eng J 424:130296. https://doi.org/10.1016/j.cej.2021.130296
192. Mishra A, Basu S, Shetti NP, Reddy KR, Aminabhavi TM (2019) Chapter 27–Photocatalysis of graphene and carbon nitride-based functional carbon quantum dots. In: Thomas S, Pasquini D, Leu S-Y, Gopakumar D (eds) ABTNMiWP. Elsevier, pp.759–81
193. Hamblin MR (2018) Fullerenes as photosensitizers in photodynamic therapy: pros and cons. Photochem Photobiol Sci 17(11):1515–1533. https://doi.org/10.1039/C8PP00195B
194. Cortés-Arriagada D, Ortega DE (2021) Fullerene–phosphorene–nanoflake nanostructures: modulation of their interaction mechanisms and electronic properties through the size of carbon fullerenes. Carbon 182:354–365. https://doi.org/10.1016/j.carbon.2021.06.036
195. Pan Y, Liu X, Zhang W, Liu Z, Zeng G, Shao B, et al (2020) Advances in photocatalysis based on fullerene C60 and its derivatives: properties, mechanism, synthesis, and applications. Appl Catal B Environ 265:118579. https://doi.org/10.1016/j.apcatb.2019.118579
196. Ghosh A, Banerjee S, Debnath T, Das AK (2022) Dehydrogenation of ammonia–borane to functionalize neutral and Li+-encapsulated C60, C70 and C36 fullerene cages: a DFT approach. Phys Chem Chem Phys 24(6):4022–4041. https://doi.org/10.1039/D1CP05770G
197. Gergeroglu H, Yildirim S, Ebeoglugil MF (2020) Nano-carbons in biosensor applications: an overview of carbon nanotubes (CNTs) and fullerenes (C60). SN Appl Sci 2(4):603. https://doi.org/10.1007/s42452-020-2404-1
198. Baskar AV, Benzigar MR, Talapaneni SN, Singh G, Karakoti AS, Yi J, et al (2022) Self-assembled fullerene nanostructures: synthesis and applications. Adv Funct Mater 32(6):2106924. https://doi.org/10.1002/adfm.202106924
199. Panahian Y, Arsalani N, Nasiri R (2018) Enhanced photo and sono-photo degradation of crystal violet dye in aqueous solution by 3D flower like F-TiO2(B)/fullerene under visible light. J Photochem Photobiol, A 365:45–51. https://doi.org/10.1016/j.jphotochem.2018.07.035
200. Regulska E, Rivera-Nazario DM, Karpinska J, Plonska-Brzezinska ME, Echegoyen L (2019) Zinc porphyrin-functionalized fullerenes for the sensitization of titania as a visible-light active photocatalyst: river waters and wastewaters remediation. Molecules 24(6). https://doi.org/10.3390/molecules24061118

201. Ajiboye TO, Oyewo OA, Onwudiwe DC (2021) Adsorption and photocatalytic removal of Rhodamine B from wastewater using carbon-based materials. FlatChem 29:100277. https://doi.org/10.1016/j.flatc.2021.100277
202. Wu Z-Y, Xu Y-J, Huang L-J, Zhang Q-X, Tang D-L (2021) Fullerene-cored star-shaped polyporphyrin-incorporated TiO2 as photocatalysts for the enhanced degradation of rhodamine B. J Environ Chem Eng 9(5):106142. https://doi.org/10.1016/j.jece.2021.106142
203. Chen M, Zhang G, Jiang Y, Yin K, Zhang L, Li H et al (2019) Fullerene-directed synthesis of flowerlike Cu3(PO4)2 crystals for efficient photocatalytic degradation of dyes. Langmuir 35(26):8806–8815. https://doi.org/10.1021/acs.langmuir.9b00193
204. Williams R (2007) Introduction to electron transfer
205. Mohamed HH, Bahnemann DW (2012) The role of electron transfer in photocatalysis: fact and fictions. Appl Catal B 128:91–104. https://doi.org/10.1016/j.apcatb.2012.05.045
206. Zhen G, Zheng S, Lu X, Zhu X, Mei J, Kobayashi T et al (2018) A comprehensive comparison of five different carbon-based cathode materials in CO2 electromethanogenesis: long-term performance, cell-electrode contact behaviors and extracellular electron transfer pathways. Biores Technol 266:382–388. https://doi.org/10.1016/j.biortech.2018.06.101
207. Shi J, Claussen JC, McLamore ES, ul Haque A, Jaroch D, Diggs AR, et al (2011) A comparative study of enzyme immobilization strategies for multi-walled carbon nanotube glucose biosensors. Nanotechnology 22(35):355502. https://doi.org/10.1088/0957-4484/22/35/355502
208. Hwa K-Y, Subramani B (2014) Synthesis of zinc oxide nanoparticles on graphene–carbon nanotube hybrid for glucose biosensor applications. Biosens Bioelectron 62:127–133. https://doi.org/10.1016/j.bios.2014.06.023
209. Liu X, Yan R, Zhang J, Zhu J, Wong DKY (2015) Evaluation of a carbon nanotube-titanate nanotube nanocomposite as an electrochemical biosensor scaffold. Biosens Bioelectron 66:208–215. https://doi.org/10.1016/j.bios.2014.11.028
210. Gu H-Y, Yu A-M, Chen H-Y (2001) Direct electron transfer and characterization of hemoglobin immobilized on a Au colloid–cysteamine-modified gold electrode. J Electroanal Chem 516(1):119–126. https://doi.org/10.1016/S0022-0728(01)00669-6
211. Marcus RA, Sutin N (1985) Electron transfers in chemistry and biology. Biochim Biophys Acta (BBA) Rev Bioenerget 811(3):265–322. https://doi.org/10.1016/0304-4173(85)90014-X
212. Das P, Das M, Chinnadayyala SR, Singha IM, Goswami P (2016) Recent advances on developing 3rd generation enzyme electrode for biosensor applications. Biosens Bioelectron 79:386–397. https://doi.org/10.1016/j.bios.2015.12.055
213. Palanisamy S, Cheemalapati S, Chen S-M (2012) Highly sensitive and selective hydrogen peroxide biosensor based on hemoglobin immobilized at multiwalled carbon nanotubes–zinc oxide composite electrode. Anal Biochem 429(2):108–115. https://doi.org/10.1016/j.ab.2012.07.001
214. Kafi AKM, Crossley MJ (2013) Synthesis of a conductive network of crosslinked carbon nanotube/hemoglobin on a thiol-modified Au surface and its application to biosensing. Biosens Bioelectron 42:273–279. https://doi.org/10.1016/j.bios.2012.10.040
215. Bagreev A, Bandosz TJ, Locke DC (2001) Pore structure and surface chemistry of adsorbents obtained by pyrolysis of sewage sludge-derived fertilizer. Carbon 39(13):1971–1979. https://doi.org/10.1016/S0008-6223(01)00026-4
216. Rivas GA, Rubianes MD, Rodríguez MC, Ferreyra NF, Luque GL, Pedano ML et al (2007) Carbon nanotubes for electrochemical biosensing. Talanta 74(3):291–307. https://doi.org/10.1016/j.talanta.2007.10.013
217. Cai C, Chen J (2004) Direct electron transfer and bioelectrocatalysis of hemoglobin at a carbon nanotube electrode. Anal Biochem 325(2):285–292. https://doi.org/10.1016/j.ab.2003.10.040

218. Cai C, Chen J (2004) Direct electron transfer of glucose oxidase promoted by carbon nanotubes. Anal Biochem 332(1):75–83. https://doi.org/10.1016/j.ab.2004.05.057
219. Tominaga M, Nomura S, Taniguchi I (2008) Bioelectrocatalytic current based on direct heterogeneous electron transfer reaction of glucose oxidase adsorbed onto multi-walled carbon nanotubes synthesized on platinum electrode surfaces. Electrochem Commun 10(6):888–890. https://doi.org/10.1016/j.elecom.2008.04.011
220. Salimi A, Noorbakhsh A, Ghadermarz M (2005) Direct electrochemistry and electrocatalytic activity of catalase incorporated onto multiwall carbon nanotubes-modified glassy carbon electrode. Anal Biochem 344(1):16–24. https://doi.org/10.1016/j.ab.2005.05.035
221. Wang J, Li M, Shi Z, Li N, Gu Z (2002) Direct electrochemistry of cytochrome c at a glassy carbon electrode modified with single-wall carbon nanotubes. Anal Chem 74(9):1993–1997. https://doi.org/10.1021/ac010978u
222. Li J, Wang Y-B, Qiu J-D, Sun D-C, Xia X-H (2005) Biocomposites of covalently linked glucose oxidase on carbon nanotubes for glucose biosensor. Anal Bioanaly Chem 383(6):918–922. https://doi.org/10.1007/s00216-005-0106-6
223. Manesh KM, Kim HT, Santhosh P, Gopalan AI, Lee K-P (2008) A novel glucose biosensor based on immobilization of glucose oxidase into multiwall carbon nanotubes–polyelectrolyte-loaded electrospun nanofibrous membrane. Biosens Bioelectron 23(6):771–779. https://doi.org/10.1016/j.bios.2007.08.016
224. Zhang P, Yang C, Li Z, Liu J, Xiao X, Li D et al (2022) Accelerating the extracellular electron transfer of Shewanella oneidensis MR-1 by carbon dots: the role of carbon dots concentration. Electrochim Acta 421:140490. https://doi.org/10.1016/j.electacta.2022.140490
225. Fredrickson JK, Romine MF, Beliaev AS, Auchtung JM, Driscoll ME, Gardner TS et al (2008) Towards environmental systems biology of Shewanella. Nat Rev Microbiol 6(8):592–603. https://doi.org/10.1038/nrmicro1947
226. Ter Heijne A, Schaetzle O, Gimenez S, Fabregat-Santiago F, Bisquert J, Strik DPBTB, et al (2011) Identifying charge and mass transfer resistances of an oxygen reducing biocathode. Energy Environ Sci 4(12):5035–5043. https://doi.org/10.1039/C1EE02131A
227. Li W-W, Yu H-Q, He Z (2014) Towards sustainable wastewater treatment by using microbial fuel cells-centered technologies. Energy Environ Sci 7(3):911–924. https://doi.org/10.1039/C3EE43106A
228. Lovley DR (2011) Live wires: direct extracellular electron exchange for bioenergy and the bioremediation of energy-related contamination. Energy Environ Sci 4(12):4896–4906. https://doi.org/10.1039/C1EE02229F
229. Schröder U (2007) Anodic electron transfer mechanisms in microbial fuel cells and their energy efficiency. Phys Chem Chem Phys 9(21):2619–2629. https://doi.org/10.1039/B703627M
230. Lovley DR (2006) Bug juice: harvesting electricity with microorganisms. Nat Rev Microbiol 4(7):497–508. https://doi.org/10.1038/nrmicro1442
231. Zhang T, Cui C, Chen S, Yang H, Shen P (2008) The direct electrocatalysis of Escherichia coli through electroactivated excretion in microbial fuel cell. Electrochem Commun 10(2):293–297. https://doi.org/10.1016/j.elecom.2007.12.009
232. Lovley DR (2011) Powering microbes with electricity: direct electron transfer from electrodes to microbes. Environ Microbiol Rep 3(1):27–35. https://doi.org/10.1111/j.1758-2229.2010.00211.x
233. Yang M-Q, Zhang N, Pagliaro M, Xu Y-J (2014) Artificial photosynthesis over graphene–semiconductor composites. Are we getting better? Chem Soc Rev 43(24):8240–8254. https://doi.org/10.1039/C4CS00213J
234. Xia Y, Li Q, Lv K, Tang D, Li M (2017) Superiority of graphene over carbon analogs for enhanced photocatalytic H2-production activity of ZnIn2S4. Appl Catal B 206:344–352. https://doi.org/10.1016/j.apcatb.2017.01.060
235. Ma X, Xiang Q, Liao Y, Wen T, Zhang H (2018) Visible-light-driven CdSe quantum dots/graphene/TiO2 nanosheets composite with excellent photocatalytic activity for E. coli disinfection and organic pollutant degradation. Appl Surf Sci 457:846–855. https://doi.org/10.1016/j.apsusc.2018.07.003

236. Yu H, Zhao Y, Zhou C, Shang L, Peng Y, Cao Y et al (2014) Carbon quantum dots/TiO2 composites for efficient photocatalytic hydrogen evolution. J Mater Chem A 2(10):3344–3351. https://doi.org/10.1039/C3TA14108J
237. Moniz SJA, Shevlin SA, Martin DJ, Guo Z-X, Tang J (2015) Visible-light driven heterojunction photocatalysts for water splitting—a critical review. Energy Environ Sci 8(3):731–759. https://doi.org/10.1039/C4EE03271C
238. Wang J, Wang G, Wei X, Liu G, Li J (2018) ZnO nanoparticles implanted in TiO2 macrochannels as an effective direct Z-scheme heterojunction photocatalyst for degradation of RhB. Appl Surf Sci 456:666–675. https://doi.org/10.1016/j.apsusc.2018.06.182
239. Yuan Y, Guo R-T, Hong L-F, Ji X-Y, Lin Z-D, Li Z-S, et al (2021) A review of metal oxide-based Z-scheme heterojunction photocatalysts: actualities and developments. Mater Today Energy 21:100829. https://doi.org/10.1016/j.mtener.2021.100829
240. Cao W, Jiang C, Chen C, Zhou H, Wang Y (2021) A novel Z-scheme CdS/Bi4O5Br 2 heterostructure with mechanism analysis: enhanced photocatalytic performance. J Alloy Compd 861:158554. https://doi.org/10.1016/j.jallcom.2020.158554
241. Wong KT, Kim SC, Yun K, Choong CE, Nah IW, Jeon B-H et al (2020) Understanding the potential band position and e–/h+ separation lifetime for Z-scheme and type-II heterojunction mechanisms for effective micropollutant mineralization: comparative experimental and DFT studies. Appl Catal B 273:119034. https://doi.org/10.1016/j.apcatb.2020.119034
242. He S, Yan C, Chen X-Z, Wang Z, Ouyang T, Guo M-L et al (2020) Construction of core-shell heterojunction regulating α-Fe2O3 layer on CeO2 nanotube arrays enables highly efficient Z-scheme photoelectrocatalysis. Appl Catal B 276:119138. https://doi.org/10.1016/j.apcatb.2020.119138
243. Zhang M, Lai C, Li B, Huang D, Zeng G, Xu P et al (2019) Rational design 2D/2D BiOBr/CDs/g-C3N4 Z-scheme heterojunction photocatalyst with carbon dots as solid-state electron mediators for enhanced visible and NIR photocatalytic activity: kinetics, intermediates, and mechanism insight. J Catal 369:469–481. https://doi.org/10.1016/j.jcat.2018.11.029
244. Natarajan TS, Thampi KR, Tayade RJ (2018) Visible light driven redox-mediator-free dual semiconductor photocatalytic systems for pollutant degradation and the ambiguity in applying Z-scheme concept. Appl Catal B 227:296–311. https://doi.org/10.1016/j.apcatb.2018.01.015
245. Zhang P, Wang T, Chang X, Gong J (2016) Effective charge carrier utilization in photocatalytic conversions. Acc Chem Res 49(5):911–921. https://doi.org/10.1021/acs.accounts.6b00036
246. Lee SL, Chang C-J (2019) Recent developments about conductive polymer based composite photocatalysts. Polymers 11(2). https://doi.org/10.3390/polym11020206
247. Guo Y, Li H, Ma W, Shi W, Zhu Y, Choi W (2020) Photocatalytic activity enhanced via surface hybridization. Carbon Energy 2(3):308–349. https://doi.org/10.1002/cey2.66
248. Kandy MM (2020) Carbon-based photocatalysts for enhanced photocatalytic reduction of CO2 to solar fuels. Sustain Energy Fuels 4(2):469–484. https://doi.org/10.1039/C9SE00827F
249. Li X, Shen R, Ma S, Chen X, Xie J (2018) Graphene-based heterojunction photocatalysts. Appl Surf Sci 430:53–107. https://doi.org/10.1016/j.apsusc.2017.08.194
250. Jiang D, Ma W, Xiao P, Shao L, Li D, Chen M (2018) Enhanced photocatalytic activity of graphitic carbon nitride/carbon nanotube/Bi2WO6 ternary Z-scheme heterojunction with carbon nanotube as efficient electron mediator. J Colloid Interface Sci 512:693–700. https://doi.org/10.1016/j.jcis.2017.10.074
251. Gebreslassie G, Bharali P, Chandra U, Sergawie A, Boruah PK, Das MR et al (2019) Novel g-C3N4/graphene/NiFe2O4 nanocomposites as magnetically separable visible light driven photocatalysts. J Photochem Photobiol A 382:111960. https://doi.org/10.1016/j.jphotochem.2019.111960
252. Syed N, Huang J, Feng Y, Wang X, Cao L. Carbon-based nanomaterials via heterojunction serving as photocatalyst. Front Chem 7
253. Liu E, Xu C, Jin C, Fan J, Hu X (2019) Carbon quantum dots bridged TiO2 and Cd0.5Zn0.5S film as solid-state Z-scheme photocatalyst with enhanced H2 evolution activity. J Taiwan Inst Chem Eng 97:316–325. https://doi.org/10.1016/j.jtice.2019.02.027

254. Pan J, Liu J, Zuo S, Khan UA, Yu Y, Li B (2018) Structure of Z-scheme CdS/CQDs/BiOCl heterojunction with enhanced photocatalytic activity for environmental pollutant elimination. Appl Surf Sci 444:177–186. https://doi.org/10.1016/j.apsusc.2018.01.189
255. Xia T, Lin Y, Li W, Ju M (2021) Photocatalytic degradation of organic pollutants by MOFs based materials: a review. Chin Chem Lett 32(10):2975–2984. https://doi.org/10.1016/j.cclet.2021.02.058
256. Cao H-L, Cai F-Y, Yu K, Zhang Y-Q, Lü J, Cao R (2019) Photocatalytic degradation of tetracycline antibiotics over CdS/nitrogen-doped–carbon composites derived from in situ carbonization of metal-organic frameworks. ACS Sustain Chem Eng 7(12):10847–10854. https://doi.org/10.1021/acssuschemeng.9b01685
257. Miao R, Luo Z, Zhong W, Chen S-Y, Jiang T, Dutta B et al (2016) Mesoporous TiO2 modified with carbon quantum dots as a high-performance visible light photocatalyst. Appl Catal B 189:26–38. https://doi.org/10.1016/j.apcatb.2016.01.070
258. Yu H, Shi R, Zhao Y, Waterhouse GIN, Wu L-Z, Tung C-H et al (2016) Smart utilization of carbon dots in semiconductor photocatalysis. Adv Mater 28(43):9454–9477. https://doi.org/10.1002/adma.201602581
259. Wang Z, Wang J, Li L, Zheng J, Jia S, Chen J et al (2017) Fabricating efficient CdSe–CdS photocatalyst systems by spatially resetting water splitting sites. J Mater Chem A 5(38):20131–20135. https://doi.org/10.1039/C7TA06085H
260. Lian Z, Sakamoto M, Kobayashi Y, Tamai N, Ma J, Sakurai T et al (2018) Durian-shaped CdS@ZnSe Core@Mesoporous-shell nanoparticles for enhanced and sustainable photocatalytic hydrogen evolution. J Phys Chem Lett 9(9):2212–2217. https://doi.org/10.1021/acs.jpclett.8b00789
261. Tan Y, Shu Z, Zhou J, Li T, Wang W, Zhao Z (2018) One-step synthesis of nanostructured g-C3N4/TiO2 composite for highly enhanced visible-light photocatalytic H2 evolution. Appl Catal B 230:260–268. https://doi.org/10.1016/j.apcatb.2018.02.056
262. Wang W, Zhu S, Cao Y, Tao Y, Li X, Pan D et al (2019) Edge-enriched ultrathin MoS2 embedded yolk-shell TiO2 with boosted charge transfer for superior photocatalytic H2 evolution. Adv Func Mater 29(36):1901958. https://doi.org/10.1002/adfm.201901958
263. Meng A, Zhang L, Cheng B, Yu J (2019) Dual cocatalysts in TiO2 photocatalysis. Adv Mater 31(30):1807660. https://doi.org/10.1002/adma.201807660
264. Guo X, Li X, Qin L, Kang S-Z, Li G (2019) A highly active nano-micro hybrid derived from Cu-bridged TiO2/porphyrin for enhanced photocatalytic hydrogen production. Appl Catal B 243:1–9. https://doi.org/10.1016/j.apcatb.2018.10.030
265. Xiao X, Gao Y, Zhang L, Zhang J, Zhang Q, Li Q et al (2020) A promoted charge separation/transfer system from Cu single atoms and C3N4 layers for efficient photocatalysis. Adv Mater 32(33):2003082. https://doi.org/10.1002/adma.202003082
266. Lopes JC, Sampaio MJ, Fernandes RA, Lima MJ, Faria JL, Silva CG (2020) Outstanding response of carbon nitride photocatalysts for selective synthesis of aldehydes under UV-LED irradiation. Catal Today 357:32–38. https://doi.org/10.1016/j.cattod.2019.03.050
267. Lima MJ, Pastrana-Martínez LM, Sampaio MJ, Dražić G, Silva AMT, Faria JL et al (2018) Selective production of benzaldehyde using metal-free reduced graphene oxide/carbon nitride hybrid photocatalysts. Chem Select 3(28):8070–8081. https://doi.org/10.1002/slct.201800962
268. Bellardita M, García-López EI, Marcì G, Krivtsov I, García JR, Palmisano L (2018) Selective photocatalytic oxidation of aromatic alcohols in water by using P-doped g-C3N4. Appl Catal B 220:222–233. https://doi.org/10.1016/j.apcatb.2017.08.033
269. Wang C, Dai Y, Fu X, Lu H, Zhang J (2021) A novel layer-layer crossed structure of bentonite/g-C3N4 for enhanced photocatalytic oxidation of arsenic(III) in a wide pH range. Surf Interf 26:101365. https://doi.org/10.1016/j.surfin.2021.101365
270. Zhou M, Yang P, Wang S, Luo Z, Huang C, Wang X (2018) Structure-mediated charge separation in boron carbon nitride for enhanced photocatalytic oxidation of alcohol. Chemsuschem 11(22):3949–3955. https://doi.org/10.1002/cssc.201801827

271. Long J, Wang S, Ding Z, Wang S, Zhou Y, Huang L et al (2012) Amine-functionalized zirconium metal–organic framework as efficient visible-light photocatalyst for aerobic organic transformations. Chem Commun 48(95):11656–11658. https://doi.org/10.1039/C2CC34620F
272. Chen Y, Zhang J, Zhang M, Wang X (2013) Molecular and textural engineering of conjugated carbon nitride catalysts for selective oxidation of alcohols with visible light. Chem Sci 4(8):3244–3248. https://doi.org/10.1039/C3SC51203G
273. Svoboda L, Praus P, Lima MJ, Sampaio MJ, Matýsek D, Ritz M et al (2018) Graphitic carbon nitride nanosheets as highly efficient photocatalysts for phenol degradation under high-power visible LED irradiation. Mater Res Bull 100:322–332. https://doi.org/10.1016/j.materresbull.2017.12.049
274. Torres-Pinto A, Sampaio MJ, Silva CG, Faria JL, Silva AMT (2019) Metal-free carbon nitride photocatalysis with in situ hydrogen peroxide generation for the degradation of aromatic compounds. Appl Catal B 252:128–137. https://doi.org/10.1016/j.apcatb.2019.03.040
275. Moreira NFF, Sampaio MJ, Ribeiro AR, Silva CG, Faria JL, Silva AMT (2019) Metal-free g-C3N4 photocatalysis of organic micropollutants in urban wastewater under visible light. Appl Catal B 248:184–192. https://doi.org/10.1016/j.apcatb.2019.02.001
276. Panimalar S, Uthrakumar R, Selvi ET, Gomathy P, Inmozhi C, Kaviyarasu K et al (2020) Studies of MnO2/g-C3N4 hetrostructure efficient of visible light photocatalyst for pollutants degradation by sol-gel technique. Surf Interf 20:100512. https://doi.org/10.1016/j.surfin.2020.100512
277. Liu W, Li Y, Liu F, Jiang W, Zhang D, Liang J (2019) WITHDRAWN: Visible-light-driven photocatalytic degradation of diclofenac by carbon quantum dots modified porous g-C3N4: mechanisms, degradation pathway and DFT calculation. Water Res 150:431–441. https://doi.org/10.1016/j.watres.2018.12.001
278. fiqar Z, Tao J, Yang T, Liu Q, Hu J, Tang H (2021) Designing 0D/2D CdS nanoparticles/g-C3N4 nanosheets heterojunction as efficient photocatalyst for improved H2-evolution. Surf Interf 26:101312. https://doi.org/10.1016/j.surfin.2021.101312
279. Pawar RC, Kang S, Park JH, Kim J-H, Ahn S, Lee CS (2016) Room-temperature synthesis of nanoporous 1D microrods of graphitic carbon nitride (g-C3N4) with highly enhanced photocatalytic activity and stability. Sci Rep 6(1):31147. https://doi.org/10.1038/srep31147
280. Torres-Pinto A, Sampaio MJ, Teixo J, Silva CG, Faria JL, Silva AMT (2020) Photo-Fenton degradation assisted by in situ generation of hydrogen peroxide using a carbon nitride photocatalyst. J Water Proc Eng 37:101467. https://doi.org/10.1016/j.jwpe.2020.101467
281. Han Q, Wang B, Gao J, Cheng Z, Zhao Y, Zhang Z et al (2016) Atomically thin mesoporous nanomesh of graphitic C3N4 for high-efficiency photocatalytic hydrogen evolution. ACS Nano 10(2):2745–2751. https://doi.org/10.1021/acsnano.5b07831
282. Ong W-J, Tan L-L, Ng YH, Yong S-T, Chai S-P (2016) Graphitic carbon nitride (g-C3N4)-based photocatalysts for artificial photosynthesis and environmental remediation: are we a step closer to achieving sustainability? Chem Rev 116(12):7159–7329. https://doi.org/10.1021/acs.chemrev.6b00075
283. Wen J, Xie J, Chen X, Li X (2017) A review on g-C3N4-based photocatalysts. Appl Surf Sci 391:72–123. https://doi.org/10.1016/j.apsusc.2016.07.030
284. Chen S, Lu W, Han J, Zhong H, Xu T, Wang G et al (2019) Robust three-dimensional g-C3N4@cellulose aerogel enhanced by cross-linked polyester fibers for simultaneous removal of hexavalent chromium and antibiotics. Chem Eng J 359:119–129. https://doi.org/10.1016/j.cej.2018.11.110
285. Chen Y, Lu W, Shen H, Gu Y, Xu T, Zhu Z et al (2019) Solar-driven efficient degradation of emerging contaminants by g-C3N4-shielding polyester fiber/TiO2 composites. Appl Catal B 258:117960. https://doi.org/10.1016/j.apcatb.2019.117960
286. Marchal C, Cottineau T, Méndez-Medrano MG, Colbeau-Justin C, Caps V, Keller V (2018) Au/TiO2–gC3N4 nanocomposites for enhanced photocatalytic H2 production from water under visible light irradiation with very low quantities of sacrificial agents. Adv Energy Mater 8(14):1702142. https://doi.org/10.1002/aenm.201702142

287. Ma L, Wang G, Jiang C, Bao H, Xu Q (2018) Synthesis of core-shell TiO2@g-C3N4 hollow microspheres for efficient photocatalytic degradation of rhodamine B under visible light. Appl Surf Sci 430:263–272. https://doi.org/10.1016/j.apsusc.2017.07.282
288. Vinodgopal K, Kamat PV (1995) Enhanced rates of photocatalytic degradation of an azo dye using SnO2/TiO2 coupled semiconductor thin films. Environ Sci Technol 29(3):841–845
289. Brus L (1986) Electronic wave functions in semiconductor clusters: experiment and theory. J Phys Chem 90(12):2555–2560
290. Wells RL, Gladfelter WL (1997) Pathways to nanocrystalline III-V (13–15) compound semiconductors. J Cluster Sci 8(2):217–238
291. Henglein A (1989) Small-particle research: physicochemical properties of extremely small colloidal metal and semiconductor particles. Chem Rev 89(8):1861–1873
292. Vogel R, Hoyer P, Weller H (2002) Quantum-sized PbS, CdS, Ag2S, Sb2S3, and Bi2S3 particles as sensitizers for various nanoporous wide-bandgap semiconductors. J Phys Chem 98(12):3183–3188
293. Hotchandani S, Kamat PV (1992) Charge-transfer processes in coupled semiconductor systems. Photochemistry and photoelectrochemistry of the colloidal cadmium sulfide-zinc oxide system. J Phys Chem 96(16):6834–6839
294. Resch U, Weller H, Henglein A (1989) Photochemistry and radiation chemistry of colloidal semiconductors. 33. Chemical changes and fluorescence in CdTe and ZnTe. Langmuir 5(4):1015–1020
295. Zaban A, Mićić OI, Gregg BA, Nozik AJ (1998) Photosensitization of nanoporous TiO2 electrodes with InP quantum dots. Langmuir 14(12):3153–3156
296. Hirakawa T, Kamat PV (2005) Charge separation and catalytic activity of Ag@ TiO2 core−shell composite clusters under UV−irradiation. J Am Chem Soc 127(11):3928–3934
297. Nasr C, Hotchandani S, Kim WY, Schmehl RH, Kamat PV (1997) Photoelectrochemistry of composite semiconductor thin films. Photosensitization of SnO2/CdS coupled nanocrystallites with a ruthenium polypyridyl complex. J Phys Chem B 101(38):7480–7487
298. Ismail AA, Bahnemann DW, Bannat I, Wark M (2009) Gold nanoparticles on mesoporous interparticle networks of titanium dioxide nanocrystals for enhanced photonic efficiencies. J Phys Chem C 113(17):7429–7435
299. Ismail AA, Bahnemann DW (2011) Mesoporous titania photocatalysts: preparation, characterization and reaction mechanisms. J Mater Chem 21(32):11686–11707
300. Tayade RJ, Kulkarni RG, Jasra RV (2006) Transition metal ion impregnated mesoporous TiO2 for photocatalytic degradation of organic contaminants in water. Ind Eng Chem Res 45(15):5231–5238
301. Schattka JH, Shchukin DG, Jia J, Antonietti M, Caruso RA (2002) Photocatalytic activities of porous titania and titania/zirconia structures formed by using a polymer gel templating technique. Chem Mater 14(12):5103–5108
302. Antonelli DM, Ying JY (1995) Synthesis of hexagonally packed mesoporous TiO2 by a modified sol–gel method. Angew Chem Int Ed Engl 34(18):2014–2017
303. Wang X, Caruso RA (2011) Enhancing photocatalytic activity of titania materials by using porous structures and the addition of gold nanoparticles. J Mater Chem 21(1):20–28
304. Jing D, Zhang Y, Guo L (2005) Study on the synthesis of Ni doped mesoporous TiO2 and its photocatalytic activity for hydrogen evolution in aqueous methanol solution. Chem Phys Lett 415(1–3):74–78
305. Huang Y, Ho W, Lee S, Zhang L, Li G, Yu JC (2008) Effect of carbon doping on the mesoporous structure of nanocrystalline titanium dioxide and its solar-light-driven photocatalytic degradation of NO x. Langmuir 24(7):3510–3516
306. Choi H, Antoniou MG, Pelaez M, De la Cruz AA, Shoemaker JA, Dionysiou DD (2007) Mesoporous nitrogen-doped TiO2 for the photocatalytic destruction of the cyanobacterial toxin microcystin-LR under visible light irradiation. Environ Sci Technol 41(21):7530–7535
307. Liu S, Zou Q, Ma Y, Chi D, Chen R, Fang H et al (2022) Metal-organic frameworks derived TiO2/carbon nitride heterojunction photocatalyst with efficient catalytic performance under visible light. Inorg Chim Acta 536:120918. https://doi.org/10.1016/j.ica.2022.120918

308. Jin M, Qian X, Gao J, Chen J, Hensley DK, Ho HC et al (2019) Solvent-free synthesis of CuO/HKUST-1 composite and its photocatalytic application. Inorg Chem 58(13):8332–8338. https://doi.org/10.1021/acs.inorgchem.9b00362
309. Liang Q, Gao W, Liu C, Xu S, Li Z (2020) A novel 2D/1D core-shell heterostructures coupling MOF-derived iron oxides with ZnIn2S4 for enhanced photocatalytic activity. J Hazard Mater 392:122500. https://doi.org/10.1016/j.jhazmat.2020.122500
310. Pi Y, Jin S, Li X, Tu S, Li Z, Xiao J (2019) Encapsulated MWCNT@MOF-derived In2S3 tubular heterostructures for boosted visible-light-driven degradation of tetracycline. Appl Catal B 256:117882. https://doi.org/10.1016/j.apcatb.2019.117882
311. Wang Q, Wang W, Zhong L, Liu D, Cao X, Cui F (2018) Oxygen vacancy-rich 2D/2D BiOCl-g-C3N4 ultrathin heterostructure nanosheets for enhanced visible-light-driven photocatalytic activity in environmental remediation. Appl Catal B 220:290–302. https://doi.org/10.1016/j.apcatb.2017.08.049
312. Liu D, Zhang J, Li C, Zhang X, Chen X, Wang F et al (2019) In-situ fabrication of atomic charge transferring path for constructing heterojunction photocatalysts with hierarchical structure. Appl Catal B 248:459–465. https://doi.org/10.1016/j.apcatb.2019.02.050
313. Liu H, Zhang J, Ao D (2018) Construction of heterostructured ZnIn2S4@NH2-MIL-125(Ti) nanocomposites for visible-light-driven H2 production. Appl Catal B 221:433–442. https://doi.org/10.1016/j.apcatb.2017.09.043
314. Zhang H, Zuo S, Qiu M, Wang S, Zhang Y, Zhang J, et al Direct probing of atomically dispersed Ru species over multi-edged TiO2 for highly efficient photocatalytic hydrogen evolution. Sci Adv 6(39):eabb9823. https://doi.org/10.1126/sciadv.abb9823
315. Li C, Liu X, Yan Y, Song X, Yan Y, Liu C, et al (2021) Synergy between Cu doping and catalytic platform in 2D Ni-MOFs/Cu-Zn0.5Cd0.5S for efficient water-to-hydrogen conversion. Chem Eng J 410:128316. https://doi.org/10.1016/j.cej.2020.128316
316. Dai D, Qiu J, Li M, Xu J, Zhang L, Yao J (2021) Construction of two-dimensional BiOI on carboxyl-rich MIL-121 for visible-light photocatalytic degradation of tetracycline. J Alloy Compd 872:159711. https://doi.org/10.1016/j.jallcom.2021.159711
317. Kumar S, Sharma V, Bhattacharyya K, Krishnan V (2016) Synergetic effect of MoS 2–RGO doping to enhance the photocatalytic performance of ZnO nanoparticles. New J Chem 40(6):5185–5197
318. Atchudan R, Edison TNJI, Perumal S, Karthikeyan D, Lee YR (2016) Facile synthesis of zinc oxide nanoparticles decorated graphene oxide composite via simple solvothermal route and their photocatalytic activity on methylene blue degradation. J Photochem Photobiol B 162:500–510
319. Rabieh S, Nassimi K, Bagheri M (2016) Synthesis of hierarchical ZnO–reduced graphene oxide nanocomposites with enhanced adsorption–photocatalytic performance. Mater Lett 162:28–31
320. Guler Ö, Guler SH, Yo F, Aydin H, Aydin C, El-Tantawy F et al (2015) Electrical and optical properties of carbon nanotube hybrid zinc oxide nanocomposites prepared by ball mill technique. Fullerenes Nanotubes Carbon Nanostruct 23(10):865–869
321. Zhang Y, Sun X, Pan L, Li H, Sun Z, Sun C et al (2009) Carbon nanotube–ZnO nanocomposite electrodes for supercapacitors. Solid State Ionics 180(32–35):1525–1528
322. Paul R, Kumbhakar P, Mitra AK (2010) Blue–green luminescence by SWCNT/ZnO hybrid nanostructure synthesized by a simple chemical route. Phys E 43(1):279–284
323. Aïssa B, Fauteux C, El Khakani MA, Therriault D (2009) Structural and photoluminescence properties of laser processed ZnO/carbon nanotube nanohybrids. J Mater Res 24(11):3313–3320
324. Saleh TA, Gondal MA, Drmosh QA (2010) Preparation of a MWCNT/ZnO nanocomposite and its photocatalytic activity for the removal of cyanide from water using a laser. Nanotechnology 21(49):495705
325. Leaper S, Abdel-Karim A, Gad-Allah TA, Gorgojo P (2019) Air-gap membrane distillation as a one-step process for textile wastewater treatment. Chem Eng J 360:1330–1340. https://doi.org/10.1016/j.cej.2018.10.209

326. Pal P (2017) Industrial water treatment process technology. Butterworth-Heinemann
327. Al-Mamun MR, Kader S, Islam MS, Khan MZH (2019) Photocatalytic activity improvement and application of UV-TiO2 photocatalysis in textile wastewater treatment: a review. J Environ Chem Eng 7(5):103248. https://doi.org/10.1016/j.jece.2019.103248
328. Vlyssides AG, Papaioannou D, Loizidoy M, Karlis PK, Zorpas AA (2000) Testing an electrochemical method for treatment of textile dye wastewater. Waste Manage 20(7):569–574. https://doi.org/10.1016/S0956-053X(00)00028-3
329. Al-Gheethi AA, Azhar QM, Senthil Kumar P, Yusuf AA, Al-Buriahi AK, Radin Mohamed RMS et al (2022) Sustainable approaches for removing Rhodamine B dye using agricultural waste adsorbents: a review. Chemosphere 287:132080. https://doi.org/10.1016/j.chemosphere.2021.132080
330. Kumari M, Chaudhary GR, Chaudhary S, Umar A (2022) Transformation of solid plastic waste to activated carbon fibres for wastewater treatment. Chemosphere 294:133692. https://doi.org/10.1016/j.chemosphere.2022.133692
331. Masengo JL, Mulopo J (2022) Synthesis and performance evaluation of adsorbents derived from sewage sludge blended with waste coal for nitrate and methyl red removal. Sci Rep 12(1):1670. https://doi.org/10.1038/s41598-022-05662-5
332. Li CJ, Zhang YJ, Chen H, He PY, Meng Q (2022) Development of porous and reusable geopolymer adsorbents for dye wastewater treatment. J Clean Prod 348:131278. https://doi.org/10.1016/j.jclepro.2022.131278
333. Qiang T, Zhu R (2022) Bio-templated synthesis of porous silica nano adsorbents to wastewater treatment inspired by a circular economy. Sci Total Environ 819:152929. https://doi.org/10.1016/j.scitotenv.2022.152929
334. Alam S, Khan MS, Bibi W, Zekker I, Burlakovs J, Ghangrekar MM et al (2021) Preparation of activated carbon from the wood of Paulownia tomentosa as an efficient adsorbent for the removal of acid red 4 and methylene blue present in wastewater. Water 13(11):1453
335. Shahadat M, Isamil S (2018) Regeneration performance of clay-based adsorbents for the removal of industrial dyes: a review. RSC Adv 8(43):24571–24587
336. Kadhom M, Albayati N, Alalwan H, Al-Furaiji M (2020) Removal of dyes by agricultural waste. Sustain Chem Pharm 16:100259. https://doi.org/10.1016/j.scp.2020.100259
337. Kamali M, Esmaeili H, Tamjidi S (2022) Synthesis of Zeolite Clay/Fe-Al hydrotalcite composite as a reusable adsorbent for adsorption/desorption of cationic dyes. Arab J Sci Eng 47(5):6651–6665. https://doi.org/10.1007/s13369-022-06580-4
338. Sajednia G, Rahimi E, Alvand N, Karbassi A, Baghdadi M (2019) Fibrous adsorbent derived from sulfonation of cotton waste: application for removal of cadmium sulfide nanoparticles from aquatic media. SN Appl Sci 1(12):1525. https://doi.org/10.1007/s42452-019-1525-x
339. Pirsaheb M, Moradi N (2021) A systematic review of the sonophotocatalytic process for the decolorization of dyes in aqueous solution: synergistic mechanisms, degradation pathways, and process optimization. J Water Proc Eng 44:102314. https://doi.org/10.1016/j.jwpe.2021.102314
340. Sarode S, Upadhyay P, Khosa MA, Mak T, Shakir A, Song S et al (2019) Overview of wastewater treatment methods with special focus on biopolymer chitin-chitosan. Int J Biol Macromol 121:1086–1100. https://doi.org/10.1016/j.ijbiomac.2018.10.089
341. Bruno P, Campo R, Giustra MG, De Marchis M, Di Bella G (2020) Bench scale continuous coagulation-flocculation of saline industrial wastewater contaminated by hydrocarbons. J Water Proc Eng 34:101156. https://doi.org/10.1016/j.jwpe.2020.101156
342. Shabir M, Yasin M, Hussain M, Shafiq I, Akhter P, Nizami A-S et al (2022) A review on recent advances in the treatment of dye-polluted wastewater. J Ind Eng Chem. https://doi.org/10.1016/j.jiec.2022.05.013
343. Tabatabaei M, Kazemzadeh F, Sabah M, Wood DA (2022) Chapter Ten–Sustainability in natural gas reservoir drilling: a review on environmentally and economically friendly fluids and optimal waste management. In: Wood DA, Cai J (eds) Sustainable natural gas reservoir and production engineering. Gulf Professional Publishing, pp 269–304

344. Collivignarelli MC, Abbà A, Carnevale Miino M, Damiani S (2019) Treatments for color removal from wastewater: state of the art. J Environ Manage 236:727–745. https://doi.org/10.1016/j.jenvman.2018.11.094
345. Salleh MAM, Mahmoud DK, Karim WAWA, Idris A (2011) Cationic and anionic dye adsorption by agricultural solid wastes: a comprehensive review. Desalination 280(1):1–13. https://doi.org/10.1016/j.desal.2011.07.019
346. Foo KY, Hameed BH (2010) Decontamination of textile wastewater via TiO2/activated carbon composite materials. Adv Coll Interface Sci 159(2):130–143. https://doi.org/10.1016/j.cis.2010.06.002
347. Crini G, Lichtfouse E (2019) Advantages and disadvantages of techniques used for wastewater treatment. Environ Chem Lett 17(1):145–155. https://doi.org/10.1007/s10311-018-0785-9
348. Sivarajasekar N, Baskar R (2015) Agriculture waste biomass valorization for cationic dyes sequestration: a concise review. J Chem Pharm Res 7(9):737–748
349. De Gisi S, Notarnicola M (2017) Industrial wastewater treatment. In: Abraham MA (ed) Encyclopedia of sustainable technologies. Elsevier, Oxford, pp 23–42
350. Kul AR, Benek V, Selçuk A, Onursal N (2017) Using natural stone pumice in van region on adsorption of some textile dyes. J Turkish Chem Soc Sect A Chem 4(2):525–536
351. Agboola O (2019) The role of membrane technology in acid mine water treatment: a review. Korean J Chem Eng 36(9):1389–1400. https://doi.org/10.1007/s11814-019-0302-2
352. Tang CY, Yang Z, Guo H, Wen JJ, Nghiem LD, Cornelissen E (2018) Potable water reuse through advanced membrane technology. Environ Sci Technol 52(18):10215–10223. https://doi.org/10.1021/acs.est.8b00562
353. Hebbar RS, Isloor AM, Zulhairun AK, Sohaimi Abdullah M, Ismail AF (2017) Efficient treatment of hazardous reactive dye effluents through antifouling polyetherimide hollow fiber membrane embedded with functionalized halloysite nanotubes. J Taiwan Inst Chem Eng 72:244–252. https://doi.org/10.1016/j.jtice.2017.01.022
354. Moradihamedani P (2022) Recent advances in dye removal from wastewater by membrane technology: a review. Polym Bull 79(4):2603–2631. https://doi.org/10.1007/s00289-021-03603-2
355. Aouni A, Fersi C, Ben Sik Ali M, Dhahbi M (2009) Treatment of textile wastewater by a hybrid electrocoagulation/nanofiltration process. J Hazard Mater 168(2):868–874. https://doi.org/10.1016/j.jhazmat.2009.02.112
356. Hir ZAM, Moradihamedani P, Abdullah AH, Mohamed MA (2017) Immobilization of TiO2 into polyethersulfone matrix as hybrid film photocatalyst for effective degradation of methyl orange dye. Mater Sci Semicond Proc 57:157–165. https://doi.org/10.1016/j.mssp.2016.10.009
357. Hadnadjev-Kostic M, Vulic T, Marinkovic-Neducin R, Lončarević D, Dostanić J, Markov S et al (2017) Photo-induced properties of photocatalysts: a study on the modified structural, optical and textural properties of TiO2–ZnAl layered double hydroxide based materials. J Clean Prod 164:1–18. https://doi.org/10.1016/j.jclepro.2017.06.091
358. Nasrollahi N, Ghalamchi L, Vatanpour V, Khataee A (2021) Photocatalytic-membrane technology: a critical review for membrane fouling mitigation. J Ind Eng Chem 93:101–116. https://doi.org/10.1016/j.jiec.2020.09.031
359. Chiu Y-H, Chang T-FM, Chen C-Y, Sone M, Hsu Y-J (2019) Mechanistic insights into photodegradation of organic dyes using heterostructure photocatalysts. Catalysts 9(5):430
360. Zhang F, Wang X, Liu H, Liu C, Wan Y, Long Y et al (2019) Recent advances and applications of semiconductor photocatalytic technology. Appl Sci 9(12):2489
361. Lee J-W, Choi S-P, Thiruvenkatachari R, Shim W-G, Moon H (2006) Submerged microfiltration membrane coupled with alum coagulation/powdered activated carbon adsorption for complete decolorization of reactive dyes. Water Res 40(3):435–444. https://doi.org/10.1016/j.watres.2005.11.034
362. Homem NC, de Camargo Lima Beluci N, Amorim S, Reis R, Vieira AMS, Vieira MF, et al (2019) Surface modification of a polyethersulfone microfiltration membrane with graphene oxide for reactive dyes removal. Appl Surf Sci 486:499–507. https://doi.org/10.1016/j.apsusc.2019.04.276

363. Gul A, Hruza J, Yalcinkaya F (2021) Fouling and chemical cleaning of microfiltration membranes: a mini-review. Polymers 13(6):846
364. Gu J, Gu H, Zhang Q, Zhao Y, Li N, Xiong J (2018) Sandwich-structured composite fibrous membranes with tunable porous structure for waterproof, breathable, and oil-water separation applications. J Coll Interf Sci 514:386–395. https://doi.org/10.1016/j.jcis.2017.12.032
365. Santosh V, Gopinath J, Babu PV, Sainath AVS, Reddy AVR (2018) Acetyl-d-glucopyranoside functionalized carbon nanotubes for the development of high performance ultrafiltration membranes. Sep Purif Technol 191:134–143. https://doi.org/10.1016/j.seppur.2017.09.018
366. Noamani S, Niroomand S, Rastgar M, Sadrzadeh M (2019) Carbon-based polymer nanocomposite membranes for oily wastewater treatment. npj Clean Water 2(1):20. https://doi.org/10.1038/s41545-019-0044-z
367. Abdi G, Alizadeh A, Zinadini S, Moradi G (2018) Removal of dye and heavy metal ion using a novel synthetic polyethersulfone nanofiltration membrane modified by magnetic graphene oxide/metformin hybrid. J Membr Sci 552:326–335. https://doi.org/10.1016/j.memsci.2018.02.018
368. Song Y, Sun Y, Chen M, Huang P, Li T, Zhang X et al (2020) Efficient removal and fouling-resistant of anionic dyes by nanofiltration membrane with phosphorylated chitosan modified graphene oxide nanosheets incorporated selective layer. J Water Proc Eng 34:101086. https://doi.org/10.1016/j.jwpe.2019.101086
369. Gao B, An J, Wang Y, Liu J, Wang L, Sillanpää M (2020) Functional photoelectrocatalytic membrane fabricated from ZnIn2S4, PVDF and carbon fibre for continuous removal of tetracycline. J Solid State Chem 290:121525. https://doi.org/10.1016/j.jssc.2020.121525
370. Wang X, Wang G, Chen S, Fan X, Quan X, Yu H (2017) Integration of membrane filtration and photoelectrocatalysis on g-C3N4/CNTs/Al2O3 membrane with visible-light response for enhanced water treatment. J Membr Sci 541:153–161. https://doi.org/10.1016/j.memsci.2017.06.046
371. Cheng X, Liu H, Chen Q, Li J, Wang P (2014) Preparation of graphene film decorated TiO2 nano-tube array photoelectrode and its enhanced visible light photocatalytic mechanism. Carbon 66:450–458. https://doi.org/10.1016/j.carbon.2013.09.021
372. Wang G, Chen S, Yu H, Quan X (2015) Integration of membrane filtration and photoelectrocatalysis using a TiO2/carbon/Al2O3 membrane for enhanced water treatment. J Hazard Mater 299:27–34. https://doi.org/10.1016/j.jhazmat.2015.06.005
373. Abdelraheem WHM, Patil MK, Nadagouda MN, Dionysiou DD (2019) Hydrothermal synthesis of photoactive nitrogen-and boron-codoped TiO2 nanoparticles for the treatment of bisphenol A in wastewater: synthesis, photocatalytic activity, degradation byproducts and reaction pathways. Appl Catal B 241:598–611. https://doi.org/10.1016/j.apcatb.2018.09.039
374. Khan JA, Sayed M, Shah NS, Khan S, Zhang Y, Boczkaj G et al (2020) Synthesis of eosin modified TiO2 film with co-exposed 001 and 101 facets for photocatalytic degradation of para-aminobenzoic acid and solar H2 production. Appl Catal B 265:118557. https://doi.org/10.1016/j.apcatb.2019.118557
375. Teow YH, Ahmad AL, Lim JK, Ooi BS (2012) Preparation and characterization of PVDF/TiO2 mixed matrix membrane via in situ colloidal precipitation method. Desalination 295:61–69. https://doi.org/10.1016/j.desal.2012.03.019
376. Anandan S, Narasinga Rao T, Sathish M, Rangappa D, Honma I, Miyauchi M (2013) Superhydrophilic graphene-loaded TiO2 thin film for self-cleaning applications. ACS Appl Mater Interfaces 5(1):207–212. https://doi.org/10.1021/am302557z
377. Chen Y, Huang W, He D, Situ Y, Huang H (2014) Construction of heterostructured g-C3N4/Ag/TiO2 microspheres with enhanced photocatalysis performance under visible-light irradiation. ACS Appl Mater Interfaces 6(16):14405–14414. https://doi.org/10.1021/am503674e
378. Li Y, Zhu L, Guo Y, Song H, Lou Z, Ye Z (2014) A new type of hybrid nanostructure: complete photo-generated carrier separation and ultrahigh photocatalytic activity. J Mater Chem A 2(34):14245–14250

379. Linley S, Liu Y, Ptacek CJ, Blowes DW, Gu FX (2014) Recyclable graphene oxide-supported titanium dioxide photocatalysts with tunable properties. ACS Appl Mater Interfaces 6(7):4658–4668. https://doi.org/10.1021/am4039272
380. Xu Z, Wu T, Shi J, Teng K, Wang W, Ma M et al (2016) Photocatalytic antifouling PVDF ultrafiltration membranes based on synergy of graphene oxide and TiO2 for water treatment. J Membr Sci 520:281–293. https://doi.org/10.1016/j.memsci.2016.07.060
381. Lv N, Li Y, Huang Z, Li T, Ye S, Dionysiou DD et al (2019) Synthesis of GO/TiO2/Bi2WO6 nanocomposites with enhanced visible light photocatalytic degradation of ethylene. Appl Catal B 246:303–311. https://doi.org/10.1016/j.apcatb.2019.01.068
382. Safarpour M, Khataee A, Vatanpour V (2015) Effect of reduced graphene oxide/TiO2 nanocomposite with different molar ratios on the performance of PVDF ultrafiltration membranes. Sep Purif Technol 140:32–42. https://doi.org/10.1016/j.seppur.2014.11.010
383. Gao P, Liu Z, Tai M, Sun DD, Ng W (2013) Multifunctional graphene oxide–TiO2 microsphere hierarchical membrane for clean water production. Appl Catal B 138–139:17–25. https://doi.org/10.1016/j.apcatb.2013.01.014
384. Gao Y, Hu M, Mi B (2014) Membrane surface modification with TiO2–graphene oxide for enhanced photocatalytic performance. J Membr Sci 455:349–356. https://doi.org/10.1016/j.memsci.2014.01.011
385. Rao G, Zhang Q, Zhao H, Chen J, Li Y (2016) Novel titanium dioxide/iron (III) oxide/graphene oxide photocatalytic membrane for enhanced humic acid removal from water. Chem Eng J 302:633–640. https://doi.org/10.1016/j.cej.2016.05.095
386. Zinadini S, Rostami S, Vatanpour V, Jalilian E (2017) Preparation of antibiofouling polyethersulfone mixed matrix NF membrane using photocatalytic activity of ZnO/MWCNTs nanocomposite. J Membr Sci 529:133–141. https://doi.org/10.1016/j.memsci.2017.01.047
387. Xu X, Ray R, Gu Y, Ploehn HJ, Gearheart L, Raker K et al (2004) Electrophoretic analysis and purification of fluorescent single-walled carbon nanotube fragments. J Am Chem Soc 126(40):12736–12737. https://doi.org/10.1021/ja040082h
388. Fang S, Xia Y, Lv K, Li Q, Sun J, Li M (2016) Effect of carbon-dots modification on the structure and photocatalytic activity of g-C3N4. Appl Catal B 185:225–232. https://doi.org/10.1016/j.apcatb.2015.12.025
389. Zhang X, Jiang M, Niu N, Chen Z, Li S, Liu S et al (2018) Natural-product-derived carbon dots: from natural products to functional materials. Chemsuschem 11(1):11–24. https://doi.org/10.1002/cssc.201701847
390. Zhao S, Lan M, Zhu X, Xue H, Ng T-W, Meng X et al (2015) Green synthesis of bifunctional fluorescent carbon dots from garlic for cellular imaging and free radical scavenging. ACS Appl Mater Interfaces 7(31):17054–17060. https://doi.org/10.1021/acsami.5b03228
391. Sun D, Ban R, Zhang P-H, Wu G-H, Zhang J-R, Zhu J-J (2013) Hair fiber as a precursor for synthesizing of sulfur- and nitrogen-co-doped carbon dots with tunable luminescence properties. Carbon 64:424–434. https://doi.org/10.1016/j.carbon.2013.07.095
392. Wang Y, Hu A (2014) Carbon quantum dots: synthesis, properties and applications. J Mater Chem C 2(34):6921–6939
393. Shchipunov YA, Khlebnikov ON, Silant'ev VE (2015) Carbon quantum dots hydrothermally synthesized from chitin. Polym Sci Ser B 57(1):16–22. https://doi.org/10.1134/S1560090415010121
394. Gogoi N, Chowdhury D (2014) Novel carbon dot coated alginate beads with superior stability, swelling and pH responsive drug delivery. J Mater Chem B 2(26):4089–4099
395. Li W, Zhang Z, Kong B, Feng S, Wang J, Wang L et al (2013) Simple and green synthesis of nitrogen-doped photoluminescent carbonaceous nanospheres for bioimaging. Angew Chem Int Ed 52(31):8151–8155
396. Liu Y, Zhao Y, Zhang Y (2014) One-step green synthesized fluorescent carbon nanodots from bamboo leaves for copper(II) ion detection. Sens Actuat B Chem 196:647–652. https://doi.org/10.1016/j.snb.2014.02.053
397. Mehta VN, Jha S, Basu H, Singhal RK, Kailasa SK (2015) One-step hydrothermal approach to fabricate carbon dots from apple juice for imaging of mycobacterium and fungal cells. Sens Actuat B Chem 213:434–443. https://doi.org/10.1016/j.snb.2015.02.104

398. Song P, Zhang L, Long H, Meng M, Liu T, Yin Y et al (2017) A multianalyte fluorescent carbon dots sensing system constructed based on specific recognition of Fe (III) ions. RSC Adv 7(46):28637–28646
399. Liang Z, Zeng L, Cao X, Wang Q, Wang X, Sun R (2014) Sustainable carbon quantum dots from forestry and agricultural biomass with amplified photoluminescence by simple NH 4 OH passivation. J Mater Chem C 2(45):9760–9766
400. Alam A-M, Park B-Y, Ghouri ZK, Park M, Kim H-Y (2015) Synthesis of carbon quantum dots from cabbage with down-and up-conversion photoluminescence properties: excellent imaging agent for biomedical applications. Green Chem 17(7):3791–3797
401. Briscoe J, Marinovic A, Sevilla M, Dunn S, Titirici M (2015) Biomass-derived carbon quantum dot sensitizers for solid-state nanostructured solar cells. Angew Chem Int Ed 54(15):4463–4468. https://doi.org/10.1002/anie.201409290
402. Sahu S, Behera B, Maiti TK, Mohapatra S (2012) Simple one-step synthesis of highly luminescent carbon dots from orange juice: application as excellent bio-imaging agents. Chem Commun 48(70):8835–8837
403. Wang Z, Yu J, Zhang X, Li N, Liu B, Li Y et al (2016) Large-scale and controllable synthesis of graphene quantum dots from rice husk biomass: a comprehensive utilization strategy. ACS Appl Mater Interfaces 8(2):1434–1439. https://doi.org/10.1021/acsami.5b10660
404. Jiang C, Wu H, Song X, Ma X, Wang J, Tan M (2014) Presence of photoluminescent carbon dots in Nescafe® original instant coffee: applications to bioimaging. Talanta 127:68–74. https://doi.org/10.1016/j.talanta.2014.01.046
405. Suryawanshi A, Biswal M, Mhamane D, Gokhale R, Patil S, Guin D et al (2014) Large scale synthesis of graphene quantum dots (GQDs) from waste biomass and their use as an efficient and selective photoluminescence on–off–on probe for Ag+ ions. Nanoscale 6(20):11664–11670
406. Yan Z, Zhang Z, Chen J (2016) Biomass-based carbon dots: synthesis and application in imatinib determination. Sens Actuat B Chem 225:469–473. https://doi.org/10.1016/j.snb.2015.10.107
407. Wang L, Bi Y, Hou J, Li H, Xu Y, Wang B et al (2016) Facile, green and clean one-step synthesis of carbon dots from wool: application as a sensor for glyphosate detection based on the inner filter effect. Talanta 160:268–275. https://doi.org/10.1016/j.talanta.2016.07.020
408. Qin X, Lu W, Asiri AM, Al-Youbi AO, Sun X (2013) Microwave-assisted rapid green synthesis of photoluminescent carbon nanodots from flour and their applications for sensitive and selective detection of mercury(II) ions. Sens Actuat B Chem 184:156–162. https://doi.org/10.1016/j.snb.2013.04.079
409. Li J-Y, Liu Y, Shu Q-W, Liang J-M, Zhang F, Chen X-P et al (2017) One-pot hydrothermal synthesis of carbon dots with efficient up- and down-converted photoluminescence for the sensitive detection of morin in a dual-readout assay. Langmuir 33(4):1043–1050. https://doi.org/10.1021/acs.langmuir.6b04225
410. Ye Q, Yan F, Luo Y, Wang Y, Zhou X, Chen L (2017) Formation of N, S-codoped fluorescent carbon dots from biomass and their application for the selective detection of mercury and iron ion. Spectrochim Acta Part A Mol Biomol Spectrosc 173:854–862. https://doi.org/10.1016/j.saa.2016.10.039
411. Sharma A, Das J (2019) Small molecules derived carbon dots: synthesis and applications in sensing, catalysis, imaging, and biomedicine. J Nanobiotechnol 17(1):1–24
412. Niu N, Ma Z, He F, Li S, Li J, Liu S et al (2017) Preparation of carbon dots for cellular imaging by the molecular aggregation of cellulolytic enzyme lignin. Langmuir 33(23):5786–5795

Syntheses and Applications of Nanomaterials-Based Photocatalysts for Air Purification

Ridhwan Haliq, Olivia Christy Tarigan, Muhammad Razaki Aprilio, Francisca Tania Deviani Wijaya, Galang Dhaifullah Abdul Aziz, Stefani Catherine, and Hairus Abdullah

Abstract Air pollution induces a variety of human ailments, including respiratory, circulatory, neurological, digestive, and urinary organ problems, as well as cancer. Volatile organic compounds (VOCs) have been primary air pollutants in indoor and urban environments, adverse to human health and the environment. The air purifier was developed to reduce indoor airborne pollutants, including particles. The discussion includes properties of air purified, photocatalytic properties on different airborne contaminants, the effect of greenhouse gases (CO_x, NO_x, SO_x) on the air purifier, classifying the effect of photocatalyst materials on ground level smoke, ozone, smogs, viruses, and bacteria. Research on air filters to overcome the Covid 19 virus outbreak is increasingly popular. Current possible materials and performance for photocatalytic air purifiers using bismuth (Bi), WO_3, silver (Ag), C_3N_4, and oxide materials. Some materials with different variations are found with the best product and removal efficiency results. The discussion is to elucidate the possible material for an air filter (air purifier). In addition, HEPA Filter with an excellent filter for dealing with mold, dust, germs, viruses, and other gases in indoor and outdoor systems is also discussed. These filters are made up of multiple layers of filter material and can eliminate at least 99.97% of particles larger than 3 microns. Finally, TiO_2 as a famous photocatalytic material is also discussed as a filter precursor in several investigations which is ideal for germs, smoke, and smells.

R. Haliq · M. R. Aprilio · F. T. D. Wijaya · G. D. A. Aziz · S. Catherine
Department of Materials Science and Engineering, National Taiwan University of Science and Technology, Taipei 10672, Taiwan, ROC

R. Haliq
Department of Mechanical Engineering, Institut Teknologi Kalimantan, Balikpapan 76127, Indonesia

O. C. Tarigan
Department of Mechanical Engineering, National Taiwan University of Science and Technology, Taipei 10672, Taiwan, ROC

H. Abdullah (✉)
Department of Industrial Engineering, Universitas Prima Indonesia, Medan, Indonesia
e-mail: hairus@unprimdn.ac.id

H. Abdullah (ed.), *Photocatalytic Activities for Environmental Remediation and Energy Conversion*, Green Energy and Technology,
https://doi.org/10.1007/978-981-19-6748-1_2

Keywords Air purification · Photocatalysis · Pollutant · Nanomaterials · VOCs

1 Introduction

Air pollution induces a variety of human ailments, including respiratory, circulatory, neurological, digestive, and urinary organ problems, as well as cancer [1, 2]. Indoor air can be compromised with microbiological and chemical toxins, creating a serious risk to human health [3]. Volatile organic compounds (VOCs) have been primary air pollutants in indoor air and urban environments, damaging human health, and the environment. Poor indoor air quality threatens millions of citizens as indoor air may include more than 900 VOCs. Better air quality is vital for health. In particular, VOCs released from building and interior materials have been linked to allergy disorders and sick building syndrome [4]. In many developing countries, the natural environment has been damaged by air pollution generated by car exhaust. Elevated levels of harmful air pollutants, such as NO_x and VOC, are associated with heavy traffic, which causes damage to the environment and causes public health issues [5]. One of the groups of contaminants known as hydrocarbons (HC) is critical to be overcome. VOCs such as benzene, toluene, ethylbenzene, o-xylene, styrene, and the terpenes -pinene and limonene were also identified in the facilities such as homes, offices, schools, and shopping malls. Human activities such as coal/fuels combustion in industrial furnaces and cars are some of the most significant producers of NO_x in large urban settings [6]. Nitrogen oxides (NO_x), such as nitric oxide (NO), and nitrogen dioxide (NO_2) are significant pollution resulted from natural nitrogen cycles and human impacts such as inefficient fossil fuel burning, engine exhaust emissions, and coal ash gases [7, 8]. Reactive pollutants make up the majority of vehicle emissions. Secondary pollutants include ozone, nitrogen oxides ($NO_x = NO + NO_2$), and O_3. The $NO–NO_2–O_3$ photochemistry has also been used to study responsive pollutants dispersion in modeled 2D street canyons [9].

Circulation with ambient air can minimize the VOCs. However, this requires a lot of energy and generates more NO_x emissions and outdoor air conditioning. According to Sidheswaran et al., a heating, ventilation, and air conditioning (HVAC) system with 15–20% outdoor air and 80–85% indoor air is practicable, providing 15–20% of VOCs are eliminated from the supply airflow [10]. The air heated by the Trombe wall can be utilized for heat, breath, and air condition of the building. It is an excellent strategy to reduce heating and ventilation expenditures [11]. The cleaning of air has a wide range of applications. NO_x, NO, and NO_2 are found in the outside air. They are substantial air pollutants created by the combustion of fuels in vehicles and power plants. Unsurprisingly, NO_x emissions, significant interior air pollutants, and VOCs like acetaldehyde and formaldehyde contribute to sick building syndrome [12]. Integrating the adsorption process by filters into AC systems is one way to reduce VOC concentration. Filters using a powdered heterogeneous catalyst (AC), bag filters, and filter cassettes comprised of flat sheet filter media. Compared to cartridges, they offer a reduced pressure drop, which might be a significant design and selection consideration in terms of ventilation system energy efficiency [13].

High-efficiency particulate absorbing (HEPA) filters, non-thermal plasma air filters, and polytetrafluoroethylene (PTFE) multi-tube high-efficiency membrane air filters have traditionally been used in buildings with heating, ventilation, and air conditioning (HVAC) systems to reach desirable suspended particulate matter levels [14]. Titanium dioxide (titania or TiO_2), a semiconductor photocatalyst, is a multifunctional material with various potential uses in various fields, including air and wastewater treatment, solar energy conversion, energy storage, and catalysis [15]. Since their potential to digest and mineralize hazardous organic pollutants into innocuous chemicals such as H_2O, CO_2, and NO_3, photocatalytic facilities have been quickly developed in recent decades to ameliorate urban air quality [16]. Most of these studies concentrate on pollutant ratios such as [NO]/[NO_2], which are critical influencing parameters for reactive pollutant dispersion [9].

In numerous studies, HEPA filters in hospitals have been seen to lessen the risks of aspergillus and other fungal infections [17]. According to research, HEPA filters can help reduce airborne levels of the severe acute respiratory syndrome coronavirus-2 (SARS-CoV-2) [18]. Health care workers (HCWs) require high-level respiratory protection during the pandemic. However, those who cannot endure prolonged usage of regular respiratory protection can employ powered air-purifying respirator (PAPR) assemblies, commonly known as 'power hoods' [19]. The filters inside the respiratory protection can be modified with particular antimicrobial materials.

To combat the COVID-19 pandemic, researchers investigate SARS-CoV-2 pathophysiology, structural mutations, virology, and biological function [20]. Viruses like Ebola, Zika, influenza (H1N1), and coronavirus have always been a source of concern worldwide. Human health and the global economy are impacted by their linked pandemics, endemics [21]. Human health is at risk from harmful microbes suspended in the air or adhered to surfaces. There have been some reported cases, such as severe acute respiratory syndrome (SARS) in 2003 and H1N1 in 2010 before the deadly coronavirus (COVID-19) outbreak in 2020 [22]. Expiratory events are the general spreading pathway, including coughing and sneezing, which produce aerosols. It can commonly discharge hundreds of tiny drops containing millions of virus particles into the air, which causes severe illness to others. Several environmental conditions can possibly influence the virus-containing aerosol-based spread. The aerosols can travel up to 30 feet in length under specific scenarios. Furthermore, environmental temperature and humidity might boost the survivability of aerosolized viruses, allowing them to persist for longer [23]. In addition, multidrug-resistant (MDR) bacteria and viruses can infect people via aerosol transmission, posing a serious threat to public health. According to the studies, large-scale epidemics of severe acute respiratory syndrome coronavirus 2 (SARS-CoV-2) are primarily transmitted through the air [24].

Various photocatalysts and composites have already been extensively used for effective NO oxidation, including TiO_2, ZnO, g-C_3N_4, and others [6]. Titanium dioxide (TiO_2) is a popular photocatalyst that has been studied extensively by scientists [25]. Because of its high melting point, hydrothermal stability, and alkaline resistance, TiO_2 could survive in challenging environments. Similarly, mesoporous materials, such as TiO_2, are generally utilized as catalyst supports since their structure facilitates mass and heat transport throughout the catalytic reaction [26]. Because of its beneficial properties of endurance, stability, and low cost, titanium dioxide (TiO_2) is the most extensively utilized photocatalyst in the environment, whether in commercial applications or scientific studies [27].

Because of its unique and vital intrinsic qualities, titanium dioxide (TiO_2) is a commonly utilized photocatalyst. It is inert and somewhat stable under extreme heat and chemical conditions, allowing it to be used again. Operations. The photocatalytic activity of titanium dioxide is strongly influenced by its crystalline phase, crystallite size, specific surface area, and porosity [28]. Photocatalysis technology has been applied to remove pollutants from the air and water by creating different photocatalyst systems with various methods. Doping, whether with metal or nonmetal, has some drawbacks. Nonmetal-doped TiO_2 has difficulty in reaction time to achieve high doping content and stability. In addition, the thermal stability of metal-doped TiO_2 declines, and the metal dopant becomes the photogenerated carrier recombination center [29]. Nano-TiO_2 has received significant attention because of its high specific surface area ratio and excellent photocatalytic action, particularly anatase TiO_2, which has already been widely used for decontamination [30]. Titanium dioxide (TiO_2)-based photocatalysts have been intensively explored to remove pollutants in heterogeneous photocatalysis due to their stability, high catalytic efficiency, non-toxic, and relatively inexpensive [31]. However, the broad bandgap energy of TiO_2 (3.0–3.2 eV), which requires UV radiations to excite the semiconductor, has exhibited numerous limits in photocatalytic activity, despite its distinctive features. Moreover, the quantum efficiency is still low due to the high recombination rate of e^-/h^+ pairings [32].

Due to its high photocatalytic activity, long-term consistency, non-toxicity, and affordable price, titanium dioxide (TiO_2) has been the most popular substance [33, 34]. Photocatalysts based on TiO_2 produce highly oxidizing free radicals that are bactericidal and antiviral against various bacteria and viruses, including influenza, rotavirus, and SARS-CoV-2 [35]. When TiO_2 is combined with other photocatalysts or cocatalysts, sunlight energy might be effectively used as the catalytic performance is improved [36]. In TiO_2, absorbed photons form electron–hole pairs that travel a path that allows some interactions with foreign molecules which physically adsorbed on its surface. The generated holes can oxidize contaminants, producing high-reactivity species such as O_2, OH, and H_2O_2, which dissolve pollutants and kill microorganisms [37]. Photocatalytic degradation (PCD) is an excellent strategy for eliminating low VOC levels in indoor environments. PCD converts VOCs to CO_2 and H_2O at standard temperature and pressure. TiO_2 has been the most extensively utilized photocatalyst in purifying the air with different photocatalysts [4].

Different visible-light stimulated photocatalysts can be made from bismuth (Bi)-based compounds. Bi semi-metallic nature permits it to be used straight as semiconductor photocatalysts or co-catalysts to optimize the photocatalytic behavior. On the other side, Bi-based semiconductors have a smaller bandgap and they may harvest light in the visible spectrum [38]. Over the years, only a few reviews of stacked Bi-based photocatalysts have already been published, concentrating on tailoring methodologies and photocatalytic applications [39]. Bi-based photocatalysts are growing in popularity, and excellent chemical stability under visible light irradiation, specifically their distinctive electronic properties, structure and regulated morphology, would benefit them [40]. They can be modified for their electrical structure and performance by utilizing ultrathin 2D nanosheets morphology. Various photocatalytic domains, such as CO_2 and N_2 reduction, can benefit from ultrathin 2D Bi-based materials [39]. Due to their great stability, resistance to deactivation, low cost, and outstanding activity, Bi-based nanomaterials are suited for photocatalytic applications [41]. Bi-based compounds have been reported to be useful in a variety of photocatalytic applications, including organic pollutant degradation [42]. The photocatalytic microstructure (i.e., size, shape, and surface to volume ratio), redox capacity, and process conditions can all affect the useful application of Bi-based photocatalysts in water purification. In this way, morphological management, surface and crystal facet engineering, heterojunction constructions, doping, and substrate immobilization can all help to improve Bi-based photocatalysts [41]. Therefore, the Bi-based photocatalytic application needs improvement. In addition, a Bi-based photocatalyst can accelerate the photocatalytic activity of organic pollutant degradation (phenolic chemicals) [43].

On the other hand, researchers have been interested in employing zinc oxide (ZnO) and WO_3 as antimicrobial semiconductor photocatalysts for decades. It is ecologically friendly, safe, and stable and contains powerful bactericidal and virucidal capabilities [44]. In addition, with some exciting properties, tungsten oxide (WO_3) is a semiconductor with a broad bandgap (Eg., 3.3 eV) and possesses a hydrophilic nature of WO_3 for membrane application. Some interesting applications are surface acoustic wave (SAW) devices, ultrasonic transducer arrays, chemisorption gas sensors, and optical waveguides [45]. Due to its environmentally favorable usage of non-toxic ingredients, the green production of silver nanoparticles (AgNPs) is widely acknowledged. AgNPs are frequently utilized as antioxidants, antibacterial, and antifungal agents in agriculture and medicine [46, 47], which are possible for photocatalytic air purifiers. Numerous discoveries have been made to manufacture the air purifier using different technologies [48]. Particulate matter (PM), sulfur dioxide, carbon monoxide, nitrogen oxides, and volatile organic compounds are the most common air contaminants that need remediation [49].

2 Properties of Air Purified

The air purifier was developed to reduce airborne pollutants indoors, mainly including particles. In China, the earliest air purifier developed was an electrostatic cleaner. It was originally invented by the author's group from the air cleaning studio of the former Institute of Air Conditioning at China Academy of Building Research as auxiliary equipment in 1971 in Tianjin to reduce the particle concentration of the vortex area in the cleanroom with the mixed flow. It was termed as the electrostatic air purifier at that time. Air pollution has been one of the biggest causes of death in the world [50]. By definition, air pollutants can be any hazardous material in the air that can harm any living thing and its supporting environment [51]. The World Health Organization (WHO) divided the types of pollutants into two general categories: ambient air pollution and household air pollution [52]. Outdoor air pollution in cities and rural regions that consists primarily of particulate matter, ground-level ozone, NO_X, and SO_X is known as ambient air pollution. In contrast, household air pollution can be described as air pollution caused by the inefficiently combusted solid fuels and kerosene that are commonly used for cooking purposes [53, 54]. According to the WHO, ambient air pollution took a toll on 4.2 million people per year in 2016, while household air pollution contributed to over 3.8 million deaths [53, 54]. These air pollutants can spread quickly throughout the world, making it easier for them to change the climate directly and indirectly [55]. Therefore, the air quality parameter is crucial to the environment and human health. It can potentially harm the well-being of humans, including increased susceptibility to respiratory diseases and even mortality [51].

There is a specific guideline that a pollutant is allowed in the air and counted as a relatively healthy amount. Table 1 shows the recommended amount of several air pollutants according to WHO 2021 Air Quality Guidelines [52]. This parameter is considered a worldwide standard; while some changes are made to fit each country's situation, it is still the leading guideline that most countries refer to. As a comparison, the air quality parameters in Taiwan and Indonesia are displayed in Table 2 and Table 3, respectively [56, 57].

The air quality index (AQI) of Taiwan presented by IQ Air shows that in 2019, an average of "moderate" air pollution was attained (12.1–35.4 $\mu g/m^3$), while, in the same year, Indonesia was ranked 6th as the most polluted country worldwide [58, 59]. As the COVID-19 pandemic started, many countries have reported a significant decrease in air pollution over the lockdown period [60–62]. The decreased air pollution was related to the movement restriction of workers from the work-from-home initiative implemented during and post-lockdown period [60–62]. Some research also indicates that more polluted cities have higher COVID-19 mortality rates [63, 64].

Table 1 WHO's 2021 air quality guidelines [52]

Pollutant	Averaging time	2021 AQG level ($\mu g/m^3$)
$PM_{2.5}$	Annual	5
	24-h	15
PM_{10}	Annual	15
	24-h	45
O_3	Peak season	60
	8-h	100
NO_2	Annual	10
	24-h	25
SO_2	24-h	40
CO_2	24-h	4000

Table 2 Taiwan's 2020 environmental protection administration air quality standard [56]

Pollutant	Averaging time	Standard value (μg/m^3)
$PM_{2.5}$	Annual	15
	24-h	35
PM_{10}	Annual	50
	24-h	100
O_3	8-h moving	60
	Hourly	120
NO_2	Annual	30
	Hourly	100
SO_2	Annual	20
	Hourly	75
CO	8-h moving	9000
	Hourly	35,000
Pb	3-month	0.15

Table 3 Indonesian 2021 government regulation on air quality parameter [57]

Pollutant	Averaging time	Standard value (μg/m^3)
$PM_{2.5}$	Annual	15
	24-h	55
PM_{10}	Annual	40
	24-h	75
O_3	Annual	35
	8-h moving	100
	Hourly	150
NO_2	Annual	50
	24-h	65
	Hourly	200
SO_2	Annual	45
	24-h	75
	Hourly	150
CO	8-h moving	4000
	Hourly	10,000
Pb	24-h	2
Non-methane HC	3-h	160

This phenomenon suggests that pre-existing respiratory diseases from high exposure to air pollutants significantly impact the mortality rate from COVID-19 and potentially future pandemics [63, 64]. However, a study by Agarwal et al. shows that the indoor air pollutant has also decreased, linking it to several measurements that households use to improve air quality in their home [65]. This condition includes non-pharmaceutical measures, such as the usage of facemasks, social distancing, and engineering control, i.e., the installation of air purifiers and the addition of indoor ventilation [65].

3 Photocatalytic Properties on Different Airborne Contaminants

3.1 Volatile Organic Compounds (VOCs)

VOCs are organic compounds that have vapor pressure >10 Pa at 25 °C with a boiling point of up to 260 °C at atmospheric pressure and up to 15 carbon atoms [66]. VOCs consist of carbonyls, alkanes, alkenes, alcohols, esters, ethers, aromatics, and amides [66]. Most VOCs are produced through human activities, whether small or large, i.e., driving vehicles, painting, and combustion of fuels. In ambient conditions, even humans emit VOCs from their breath and skin, most of them being acetone, isoprene, methanol, and acetic acid [66, 67]. However, the earth also naturally emits some VOCs from activities such as reduced carbon dioxide from photosynthesis and chemical mimicry from some animals. Listed below are some VOCs commonly found in households and their risks to human health.

1. **Benzene, Toluene, Ethylbenzene, and Xylenes (BTEX)**

BTEX consists of four types of organic compounds: benzene, toluene, ethylbenzene, and xylenes [68]. These materials are commonly found in petroleum products. Listed below are the properties and dangers of each organic compound to human health.

2. **Benzene**

Benzene is a colorless liquid with a sweet odor and is highly flammable [69]. It is heavier than air in its vapor form, with a density of 2.77, making it sink closer to ground level [69, 70]. The solubility of benzene, at only 1.79 g/L, is considered small in water even though it is still larger compared to alkanes of similar size [71, 72]. This property, as well as the fact that benzene is lighter than water, makes it float on top of the water [72]. Benzene acts by causing cells to malfunction. Long-term exposure to benzene has a profound effect on the blood, e.g., causing the bone marrow to stop producing enough red blood cells, resulting in anemia [69].

3. **Toluene**

Similar to benzene in structure, toluene consists of an aromatic hydrocarbon ring and a methyl group attached to one of its carbons [73] with a vapor density of 3.2 compared to air and solubility of 0.5 g/L, it tends to lay low on the ground and float on the surface of the water [73, 74]. Long-term exposure to toluene affects the human's central nervous system [75]. The symptoms of toluene exposures are headache, drowsiness, dizziness, disorientation, nausea, slowed judgment, slowed stride, and blurred vision [76].

4. **Ethylbenzene**

Ethylbenzene is a highly flammable liquid that has no color and a similar odor to gasoline [77]. Same as the previous materials, ethylbenzene has a higher vapor density compared to air at 3.7, making it lay on the lower ground level [77, 78]. It has a very low solubility of 0.00206 g/L, which makes it tend to float on water [77]. Ethylbenzene is a potential human carcinogen, according to the international agency for research on cancer (IARC) [79]. In animals, exposed to relatively low quantities of ethylbenzene for several days to weeks, irreversible damage to the inner ear and hearing has been seen, while the same amount of exposure for several months to years leads them to suffer kidney injury [79].

5. **Xylene**

Xylene is a disubstituted molecule benzene ring with two methyl groups that is colorless, highly flammable and has a sweet odor [80, 81]. Its vapor form has a higher density compared to air at 3.7 and low solubility in water with <0.1 g/L, making it float on the water surface and sink to the lower grounds [81]. There is no known mechanism through which xylene causes toxicity; though the most serious areas that are affected by xylene exposures are the central nervous system, respiratory system, and cardiovascular system [82].

The breakdown process starts with the attack of the BTEX molecule by ·OH radicals, SO^{4-} and other aggressive chloro-species, such as $ClO_3/ClO_2/ClO^-$ [83]. This may be termed a degradation pathway because active species caused benzene rings to open and complete mineralization [83]. The addition of HO· to benzene started the oxidation of benzene in the breakdown pathway, resulting in radicals [83, 84]. The radical disproportionation produced phenol, which quickly interacted with hydroquinone and p-benzoquinone to yield hydroquinone [83, 85]. HO· might join the benzene ring of toluene, yielding o-, m-, and p-cresol respectively [83]. HO· would also attack the methyl group, forming $C_6H_5{\cdot}CH_2$, which would then be disproportioned to generate benzyl alcohol or benzaldehyde due to the subsequent disproportionation [83, 85]. HO could be attached to the aromatic ring of ethylene, resulting in the formation of ethyl phenol isomers [83]. Ethyl carbons were easily attacked by HO, resulting in the formation of various intermediates such as -phenyl ethanol, and phenylacetaldehyde [83, 86]. m-tolylalcohol, m-tolualdehyde, 1,3-benzenedimethanol, and 1,3-benzenedialdehyde were also found after xylene breakdown, which was the result of attaching HO·. The methyl carbons were easily attacked by HO·, resulting in m-tolylalcohol and 1,3-benzenedimethanol in the system, which yielded m-tolualdehyde and 1,3-benzenedidialdehyde respectively [83, 87]. Finally, the rings were eliminated, and intermediates such as glycerol, maleic acid, succinic acid, and propionaldehyde were transformed into linear compounds [83]. With the interaction of Cl and aromatics during the breakdown of PEC, Due to an increase in the production of electrons by the electrolysis process, photoelectrocatalytic (PEC) can boost removal efficiency by increasing current density. various intermediates were generated [83]. During the breakdown of 2,4-Dichlorophenol (2,4-DCP), intermediate products such as chlorophenols, hydroquinone, benzoquinone, and phenol were discovered [83]. The aromatic ring of 2,4-DCP can be hydroxylated by adding ·OH to it, yielding 3,5-dichlorocatechol, 4,6-dichlororesorcinol, and chlorophenols. Meanwhile, ·OH was used to replace the chlorine atom at the para-position on the aromatic ring, yielding chlorohydroquinone and phenol. By oxidation with another hydroxyl radical followed by Cl· abstraction, chlorohydroquinone is transformed into hydroxyhydroquinone [83]. The hydroxybenzoquinone is generated by the oxidation of hydroxyhydroquinone followed by reduction, or by the hydroxyl radical attack on benzoquinone. When hydroxyl radicals attack hydroxybenzoquinone and benzoquinone, the ring opens, forming aliphatics [83, 88].

6. **Chloroform**

Chloroform or trichloromethane ($CHCl_3$) is a volatile liquid with an ether-like odor and no color [89]. It is denser than water and air, with its density of 1.492 mg/L and a vapor density of 4.2 [90, 91]. It was used as an inhalation anesthetic in the 1800s and was replaced due to its toxicity to human health [91]. Currently, chloroform is common in refrigerants and solvents, as well as used in chemical manufacturing [92]. According to the US EPA, chronic effects of long-term exposure to chloroform can affect the liver and central nervous system [93].

The reaction mechanism of chloroform degradation can be described in the equations below [94]. The complete mineralization of chloroform is hydrochloric acid and carbon dioxide [94].

$$Cl\cdot + CHCl_3 \rightarrow HCl + \cdot CCl_3 \tag{1}$$

$$\cdot CHCl_2 + CHCl_3 \rightarrow CH_2Cl_2 + \cdot CCl_3 \tag{2}$$

$$\cdot CCl_3 + O_2 \rightarrow CCl_3OO \tag{3}$$

$$CCl_3OO\cdot \rightarrow 2\,CCl_3O\cdot + O_2 \tag{4}$$

$$CCl_3O\cdot \rightarrow COCl_2 + Cl \tag{5}$$

$$COCl_2 + H_2O \rightarrow 2\,HCl + CO_2 \tag{6}$$

7. **Formaldehyde**

Formaldehyde of methanal (CH_2O) is a colorless, flammable gas with a pungent odor [95]. It has a slightly larger vapor density at 1.03 compared to air, and is soluble in water [96]. Due to its solubility, formaldehyde is widely used as a preservative when mixed with water, also known as formalin [95]. It is found in emissions from unvented fuel-burning appliances, cigarette smoke, as well as a result of combustion and some other natural processes [97]. Formaldehyde might enter into the household from the existence of smokers in the room, or new products or constructions, since it occurs in fabrics and newly manufactured woods [98]. Formaldehyde causes myeloid leukemia and uncommon malignancies, including tumors of the paranasal sinuses, nasal cavity, and nasopharynx, according to studies of employees exposed to high quantities of formaldehydes [99].

Under UV light, formaldehyde can be photocatalytically degraded on the surface of nanoscale TiO_2, where the photo-production holes and electronics react with the water molecules that adsorb on the catalyst surface to produce hydroxyl [100]. The hydroxyl generated is crucial in the photocatalytic elimination of formaldehyde [100]. The reaction mechanism can be seen as follows.

$$TiO_2 + hv \rightarrow h^+ + e^- \tag{7}$$

$$H_2O + h^+ \rightarrow \cdot OH + H^+ \tag{8}$$

$$O_2 + e^- \rightarrow \cdot O_2 \tag{9}$$

$$CH_2O + \cdot OH \rightarrow \cdot CHO + H_2O \tag{10}$$

$$\cdot CHO + \cdot OH \rightarrow HCOOH \tag{11}$$

$$\cdot CHO + \cdot O_2 \rightarrow HCO_3 - \xrightarrow{+H^+} HCOOOH \xrightarrow{+CH_2O} CHOOH \tag{12}$$

$$HCOOH \xrightarrow{-H^+} HCOO^- \xrightarrow{h^+} H^+ + \cdot CO_2^- \tag{13}$$

$$\cdot CO_2^- \xrightarrow{[O][\cdot OH],h^+} CO_2 \tag{14}$$

4 Methyl Ethyl Ketone (MEK)

MEK or 2-butanone (C_4H_8O) is a flammable liquid with no color and a sharp odor. It is mainly used as a solvent in a lot of industries, such as synthetic rubber, adhesives, and protective coating materials [101, 102]. It has a half-life of 14–15 h in air while it lasts longer in the soil and water, with 3–12 days through evaporation and more

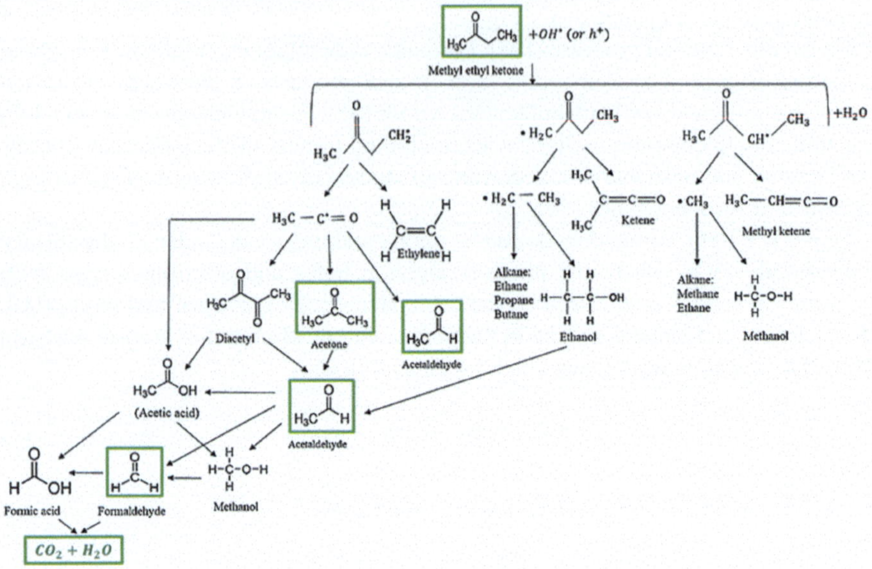

Fig. 1 Possible reaction pathway of the breakdown of methyl ethyl ketone. Reproduced with permission. [105] Copyright 2022, Elsevier 5326540985929

than 3 days by biodegradation [102]. MEK might leech out to the environment through hazardous waste sites, as well as from cigarette smoke and releases from building materials [103]. It is an irritant to the eyes and nose of humans and can cause drowsiness or dizziness and skin dryness or cracking from prolonged exposure [104]. Figure 1 shows all of the possible breakdown processes of MEK [105].

In short, the photocatalytic reaction will generate electrons and holes under UV irradiation [105]. The product of the reaction, which is electrons and holes, will then react with oxygen from the surrounding air and available water to form OH-radical groups [105]. These OH-radicals will further react with the MEK molecule, resulting in the formation of three different alkyl radicals and water, which will then decompose into smaller radicals (ethylene, ketone, and methyl ketene) [105]. These smaller radicals will proceed to react with the TiOH on the surface and/or mix with the other radicals to generate alkane, alcohol, ketone, aldehyde, and acetic acid, among other organic chemicals [105]. The formed acetaldehyde can also be a result of the oxidation of the smaller by-products listed before—diacetyl, acetone, and ethanol [105]. Following that, the acetaldehyde can then be degraded into acetic acid, formaldehyde, and methanol. MEK photocatalytic process will then be completed with the oxidation of acetic acid and formaldehyde to formic acid followed by another oxidation process which results in CO_2 and H_2O, leading to complete mineralization [105].

4.1 *Greenhouse Gases (CO_x, NO_x, SO_x)*

People's lifestyles and cultures had changed due to industrialization. Energy produced from fossil fuels was used to operate newly invented machines. Fossil fuels are widely used in power generation and transportation technology. When a fossil fuel burns, it produces several types of gases such as carbon dioxide (CO_2), methane (CH_4), nitrous oxide (N_2O), and fluorinated. These gases in the Earth atmosphere act as a greenhouse barrier and trap the solar radiation. These gases that absorb and trap heat in the atmosphere are known as greenhouse gases (GHG). They play a crucial role in maintaining the earth temperature by absorbing infrared radiation. Global warming caused by greenhouse gases is the trapping of heat in the atmosphere [106]. Global warming has been caused by human activities, which have increased GHG beyond their limit, altering the living conditions on Earth. Many studies have predicted that CO_2 and other greenhouse gases will increase substantially. Overproduction of greenhouse gases causes changes in the climate and ecosystem that, in turn, lead to environmental and health issues. Due to the changes in rainfall patterns, a decline in agricultural production has been observed, leading to food shortage. As reported by the WHO in 2018, global warming will lead to infectious diseases like malaria, dengue fever, and diarrhea [107]. GHG also significantly pollutes the air. Each year, 6.5 million people die from air pollution, including 3 million from open-air pollution. Approximately 4.5 million deaths will be caused by outdoor pollution by the end of 2040 [108].

Radiative forcing refers to the influence of greenhouse gases on the Earth atmosphere by altering incoming and outgoing energy states. This is a measure (expressed in Wm^{-2}) of the importance and contribution of GHGs to current climate change [109, 110]. Table 4 shows GHGs characteristics, the global radiative forcing in Wm^{-2}, global warming potential GWP100 (time horizon of 100 years), atmospheric lifetimes in years and relative importance. Recently, many studies have been conducted to reduce or remove greenhouse gases from the atmosphere. One process that has been reported for efficient GHG capture and utilization on a commercial scale is photocatalyst utilization for the removal of GHGs.

4.1.1 Carbon Dioxide (CO_2)

CO_2 emissions in the atmosphere have grown tremendously due to heavy dependence on fossil fuels. Since the industrial revolution in 1750, atmospheric CO_2 levels have rapidly increased from 280 to 400 ppm in 2014 [108]. CO_2 is one of the main greenhouse gases that are the leading cause of global warming. Reducing emissions is the easiest way to deal with this problem, but it might not be sufficient or possible in all cases, especially in cases where fossil fuels cannot simply be avoided.

Currently, there are three approaches being taken to reduce the amount of CO_2 in the atmosphere, (1) direct reduction of CO_2 emissions; (2) CO_2 capture and storage (CCS); and (3) CO_2 utilization and conversion. As the need for energy for

Table 4 Principal GHGs and their characteristics [110]

GHG	Global radiative forcing in 2008 ($W\ m^{-2}$)	Relative importance	GWP_{100}	Atmospheric lifetime (years)
CO_2	1.739	63.5%	1	>100
CH_4	0.500	18.2%	25	12
N_2O	0.171	6.2%	298	114
CFC-12 or CCl_2F_2	0.169	6.2%	10,900	100
CFC-11 CCl_3F	0.061	2.2%	4800	55
15 other minor GHGs	0.100	3.7%	–	–
HCFC-22 or $CHClF_2$	0.0330 (in 2005)	1.2%	1700	12
CFC-113 or CCl_2FCClF_2	0.0240 (in 2005)	0.9%	4800	85
HFC-134a or CFH_2CF_3	0.0055 (in 2005)	<0.2%	1430	14
PFC-14 or CF_4	0.0034 (in 2005)	<0.2%	5700	50,000
HCFC-142b or CH_3CF_2Cl	0.0031 (in 2005)	<0.2%	1800	18
SF_6	0.0029 (in 2005)	<0.2%	22,800	3200
NF_3	–	–	17,200	550
Total RF in 2008	2.740	100%	–	–

life increases over time, direct reductions in CO_2 emissions are considered unrealistic. CCS is risky for leakage and compression, uneconomical, and limited to wide implementation. Therefore, utilization and conversion of CO_2 are more attractive in ameliorating energy shortages and environmental problems. A good solution is to convert CO_2 to high-value carbonaceous fuels and chemical building blocks by using solar energy. A highly efficient photocatalyst is needed to realize such CO_2 conversion practically.

The conversion of CO_2 requires the absorption of a suitable amount of light because it is difficult to break the C=O double bond. This C=O double bond has a higher bond energy (750 kJ mol^{-1}) compared to C—H(411 kJ mol^{-1}), C—O (327 kJ mol^{-1}), and C—C (336 kJ mol^{-1}) single bonds. Various products can be produced, including C1 compounds such as CO, CH_4, HCOOH, CH_3OH, HCHO, and C_2 molecules such as CH_2CH_2, C_2H_5OH, and CH_3COOH. Table 5 shows the common products of photocatalytic CO_2 reduction with their corresponding reduction potentials (E^o) (see Fig. 2).

There are three main phases of photocatalytic CO_2 conversion: (1) Photocatalysts absorb photons at or near their bandgap energies and generate electron–hole pairs; (2) Separation of the photogenerated electron–hole pairs, electrons migrate to the active sites, CO_2 molecules are absorbed; (3) Redox reaction among photogenerated charge carriers and the adsorbed CO_2 molecules. In Table 6, we list recent research about photocatalysts for the reduction of CO_2 and the obtained products.

Table 5 The reduction potentials versus NHE at pH 7 in aqueous solution, 25 °C and 1 atm gas pressure [111]. Some typical semiconductors and their redox potentials

No.	Equation	Product	*E* (V)
1	$CO_2 + e^- \rightarrow CO_2^-$	Carbonate anion radical	−1.85
2	$CO_2 + 2\,H^+ + 2e^- \rightarrow HCOOH$	Formic acid	−0.61
3	$CO_2 + 2\,H^+ + 2e^- \rightarrow CO + H_2O$	Carbon monoxide	−0.53
4	$CO_2 + 4\,H^+ + 4e^- \rightarrow HCHO + H_2O$	Formaldehyde	−0.48
5	$CO_2 + 4\,H^+ + 4e^- \rightarrow C + 2H_2O$	Carbon	−0.20
6	$CO_2 + 6\,H^+ + 6e^- \rightarrow CH_3OH + H_2O$	Methanol	−0.38
7	$CO_2 + 8\,H^+ + 8e^- \rightarrow CH_4 + 2H_2O$	Methane	−0.24
8	$2CO_2 + 12\,H^+ + 12e^- \rightarrow C_2H_4 + 4H_2O$	Ethylene	−0.34
9	$2CO_2 + 12\,H^+ + 12e^- \rightarrow C_2H_5OH + 3H_2O$	Ethanol	−0.33
10	$2CO_2 + 14\,H^+ + 14e^- \rightarrow C_2H_6 + 4H_2O$	Ethane	−0.27
11	$3CO_2 + 18\,H^+ + 18e^- \rightarrow C_3H_7OH + 5H_2O$	Propanol	−0.32
12	$2\,H^+ + 2e^- \rightarrow H_2$	Hydrogen	−0.42

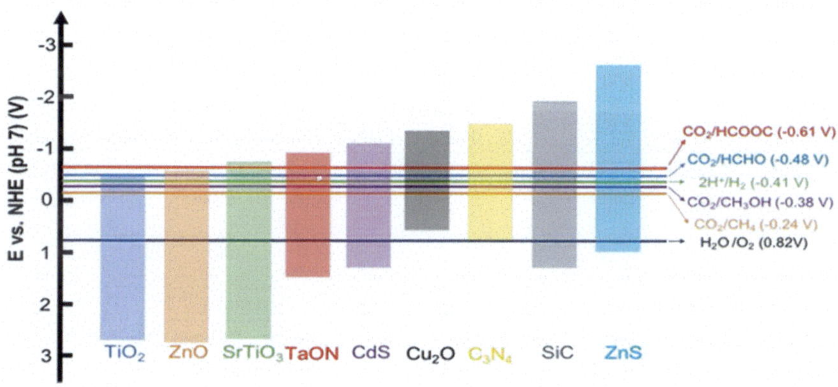

NB.: In order to proceed, the reduction potential at the catalyst must be lower than that of the desired product.

Fig. 2 CB and VB Energy levels of different semiconductor photocatalysts, electrochemical potentials of CO_2 reduction reactions to derived compounds, and H_2O reduction and oxidation potentials [111]

4.1.2 Photocatalytic Reduction of Nitrous Oxide (N_2O)

Nitrous Oxide (N_2O) is one of the greenhouse gases that play a role in global warming, accounting for 6.2% of the total global radiative forcing. The global warming potential (GWP100) of N_2O (global warming potential) is nearly 298 times greater than that of CO_2 [120]. N_2O is a very harmful and poisonous gas emitted primarily from

Table 6 Several photocatalyst systems for the reduction of CO_2

Photocatalyst	Light source	Analyzer	Main product	References
$NiMoO_4$	LED lamp 20 W (400-800 nm)	Gas chromatograph	Product: CH_3OH Production rate: 2436 μmol/(g-h)	[112]
Fe, Ni-doped CeO_2	LED lamp 20 W (λ > 400 nm)	CP-3800 gas chromatograph	Product: CH_3OH Production rate: 7039 μmol-g^{-1}	[113]
$AgVO_3/Ag/TiO_2$-NWs	UV light reactor 8 W (λ > 254 nm)	FID-equipped gas chromatograph	Product: CH_3OH Production rate: 9561 $\mu mol\ g^{-1}\ h^{-1}$	[114]
CuS/g-C_3N_4	300 W xenon lamp	Gas chromatograph	Product: CO Production rate: 13.24 $\mu mol\ g^{-1}\ h^{-1}$	[109]
In_2O_3@InP/Cu_2O	300 W xenon lamp	Techcomp GC 7900	Product: CO and CH4 Production rate CH4: 38.8 $\mu mol\ g^{-1}\ h^{-1}$ CO: 13.7 $\mu mol\ g^{-1}\ h^{-1}$	[115]
Bi_2WO_6/Si	300 W xenon lamp (λ > 400 nm)	QP2010 ultra	Product: C_2H_5OH Production rate: 402.76 μmol/g	[116]
$SnTa_2O_6$	300 W xenon lamp	Flame ionization detector (FID)	Product: CO Production rate: 28 $\mu mol\ g^{-1}\ h^{-1}$	[117]
NF@ZnO/Au@ZIF-8	300 W xenon lamp	GC-9860-5 C	Product: CO and CH_4 Production rate CH_4: 270.12 $\mu mol\ g^{-1}\ h^{-1}$ CO: 21.36 $\mu mol\ g^{-1}\ h^{-1}$	[118]
$ZnIn_2S_4$ nanorods	300 W xenon lamp (λ > 800 nm)	GC, agilent 7820A	Product: CO and CH_4 Production rate CH_4: 220 $\mu mol\ g^{-1}\ h^{-1}$ CO: 1500 $\mu mol\ g^{-1}\ h^{-1}$	[119]
$TiZrNbHfTaO_{11}$	400 W Hg lamp	GC-8A	Product: CO Production rate: 4.64 ± 0.30 $\mu mol\ g^{-1}$	[120]

combustion. N_2O has many negative impacts on the environment and living things, such as acid rain, photochemical smog, ozone layer depletion, respiratory illness, and lung function abnormalities [121, 122]. For a long time, N_2O was decomposed by a thermal catalytic method. In 1970, Photocatalysis under UV irradiation was developed to decompose N_2O into N_2 and O_2 used zinc oxides and ferric oxide [123]. While

Table 7 Photocatalystic systems for the reduction of N_2O

Photocatalyst	Light source	Analyzer	N_2O conversion (%)	References
TiO_2/g-C_3N_4	8 W Hg lamp	GC/BID	70.6	[124]
g-C_3N_4/ZnO	–	–	11	[125]
ZnS-MMT	8 W Hg lamp	GC/TCD	79	[126]
$BiVO_4$	Xenon lamp	GC/TCD	26.7	[127]
WO_3/g-C_3N_4	8 W Hg lamp	GC/BID	14	[128]
g-C_3N_4/$BiVO_4$	8 W Hg lamp	Gas Chromatograph	13	[127]

thermal catalytic decomposition of N_2O (without light) is suitable primarily for the abatement of N_2O generated by industry and combustion, photocatalytic decomposition of N_2O may be able to reduce the concentration in the atmosphere. The electrons generated by the UV irradiation of semiconductor photocatalysts can reduce N_2O. In Table 7, there are several researches about photocatalysts N_2O reduction.

4.1.3 NO_x

As major pollutant gases, nitrogen oxides (NO_x) have attracted much attention due to the substantial damage they cause to the environment and human well-being [129, 130]. NO_x emissions contribute to environmental problems such as acid rain, PM2.5 (particulate matter below 2.5 m), and photochemical smog [131]. Many technologies for NO_x purification have been developed, such as three-way catalysts, absorption, and selective catalytic reduction. Such strategies target ppm-level emission sources but are unfeasible for ppb-level NO_x in air temperatures [132, 133]. Photocatalysts have been used to solve environmental concerns due to their environmentally benign and ambient reaction conditions [134, 135]. Photocatalysis, powered by plentiful solar energy, has significant promise for removing NO_x at low concentrations from the environment.

The photocatalytic activity of the materials toward NO oxidation was investigated using a methodology similar to ISO 22197-1 and a standardized test method developed for characterizing air-purification performance. As NO is the most abundant NO_x component in ambient air (about 90%), numerous semiconductors, including g-C_3N_4 (CN), have been investigated as NO removal photocatalysts [136–138], LDHS [139], metal oxides [140–142], and their composites [143–146]. CN is a viable option due to its narrower energy band gap among these materials, stunning response to visible light, non-toxicity, and excellent stability [147, 148]. Creating a heterojunction combined with TiO_2 is an excellent method for improving photocarrier separation [149]. Under visible light, the composite photocatalyst outperformed pure CN and TiO_2 in photocatalytic activity. Hu et al. [150] demonstrated that Ti_3C_2 MXene, which has been somewhat oxidized, is an excellent candidate for fabricating photocatalysts for high-performance air purification. The composite materials have a high NO removal effectiveness and produce low NO_2 if used as NO_x purification photocatalysts. The synthesized photocatalyst exhibits a NO removal efficacy of up to 66.3% in visible light, which is significantly greater than g-C_3N_4 or TiO_2–Ti_3C_2.

4.1.4 N_2o

One of the "greenhouse gases" is nitrous oxide. It is a chemical compound discovered as one of the gases that damage the ozone layer. Human activity is estimated to emit 4.7–7 million tons of nitrous oxide annually, accounting for around two-fifths of the total emissions and natural sources (bogs, processes of putrefaction) [151]. For instance, N_2O has occurred in the commercial ammonia oxidation process byproducts using a catalyst made of platinum at temperatures more than 800 °C. Risk caused by a steady buildup in nitrous oxide concentrations in the atmosphere (approximately 0.2–0.3 annual percentage) necessitates a search for ways to minimize emissions [152]. As a result, controlling N_2O emissions has become an increasingly essential environmental subject. N_2O catalytic breakdown into N_2 and O_2 appears to be the most straightforward method of controlling N_2O emissions.

Various catalysts had been recognized to be involved in the breakdown of N_2O. Oi et al. [153] investigated Rh catalytic activity assisted by multiple metal oxides and determined that Rh/ZnO was the most active catalyst. Besides, Li and Armor [154] found that the Rh/ZSM-5 catalyst was particularly active in the catalysis of N_2O degradation when temperatures were low (e.g., 300 °C). Numerous research studies [155–157] on the catalytic breakdown of N_2O have been published using Al_2O_3-supported catalysts made of valuable metals (Ir, Ru, Rh). Meanwhile, other gases that coexist, such as CH_4, H_2O, and O_2, greatly influence the catalytic activity for N_2O breakdown [157–159].

4.2 Ground Level Smoke, Ozone, and Smogs

4.2.1 Photocatalytic Smoke Purifier

Smoke is a complex mixture of semi-volatile and non-volatile molecules as well as particulate matters (PMs)(known as tar), composed of numerous gaseous volatile substances [160]. Furthermore, it contains a variety of harmful compounds such as ammonia, 3-ethenylpyridine, acetone, ethylketones (MEK), polycyclic aromatic hydrocarbon (PAH), and acetaldehyde in addition to the typical compounds (nicotine, benzene, toluene, ethylbenzene, and xylenes) [161]. In addition, smoke contains 90% gaseous compounds, including CO, CO_2, nitrogen dioxide, VOCs, aldehyde, amines, and 10% PMs). Therefore, smoke causes carcinogens, respiratory toxicants, and cardiovascular toxicants. There are 200 different classes of harmful compounds, with approximately 60 being carcinogens made up of 4000 various compounds. According to the study by Wang et al., smoke is composed of three primary carbonyl substances: acetaldehyde, acetone, and formaldehyde. In addition, the concentration

Table 8 Major chemical composition of smoke [163]

Compound	Composition/cigarette (mg)	Compound	Concentration (mg)/cigarette (mg)
Carbon monoxide (CO)	14–23	Acetic acid	300–1700
Carbon dioxide (CO_2)	45–65	Formic acid	200–600
Nitrogen oxides (NOx)	100–680	Acetaldehyde	400–1400
Nicotine	100–3000	Formaldehyde	20–100
Nornicotine	5–150	Acetone	100–650
Total non-volatile HC	300–400	Methanol	80–100
Ammonia	10–130	Furan	20–40
Phenol	80–160	Pyridine	20–200

of benzene and toluene is also reasonably high in VOCs [162]. Europe institute also reported similar results for health and consumer protection. Table 8 shows the composition of each compound in smoke [163].

The toxic, volatile, and easily diffuse characteristics of smoke are critical factors to air pollution and pose a severe environmental and human health issue [164]. Furthermore, the inhalation of CO may result in an acute health effect caused by its strong affinity for hemoglobin, which impairs oxygen delivery. As a result, impaired vision and decreased brain function may occur at moderate dosage. Furthermore, CO poisoning can be fatal at greater concentrations. For example, exposure to a 0.1% CO atmosphere may lead to death due to the conversion of hemoglobin in blood to its useless CO derivative [165]. One cubic centimeter of the fresh cigarette mainstream smoke consists of around 4×10^9 spherical particles composed of different sized liquid and solid particles in source and composition with the arithmetic mean diameter of approximately 0.2 μm [166]. The significant risk of smoke in indoor air pollution includes lung cancer [167], cardiovascular and respiratory diseases [168], human skin disease [169], breast cancer mortality [170], and ischemic stroke [171]. The adverse effect of the risk will increase in relation to the exposure, especially the PM 10 and PM 2.5, which can deposit deeply inside the respiratory tract due to tiny size, which can penetrate deeper to the respiratory tract and cause harmful impacts. Moreover, it has the ability to adsorb toxic organic molecules and pass through the blood circulation, which could circulate through the whole body, including brain [172]. Moreover, exposure over a short period can irritate the eyes, nasal, and respiratory systems. The consequences to the respiratory system due to the exposures could include increased sneezing and coughing and could impact susceptible and sensitive individuals [162].

To enhance air quality and set standards, a variety of technologies have been deployed. Air quality limit values (AQLVs) are used in Europe to demarcate planning claims. The national ambient air quality standards (NAAQS) are used to set national air quality standards in the United States. The European directive establishes worldwide geographical air quality guidelines throughout Europe. The standards are established based on the danger of the exposed geographical area being used as monitoring/assessment zones for the recording of emissions and air pollutants. In the United States, limits are defined depending on the severity of the air quality problem in the area, and all sources of pollutants and their precursors are recorded. Although the standards are placed on different mechanisms, tremendous progress has been made in reducing overall emissions and health and environmental consequences [173].

The concentration of smoke pollutants and the residence time of VOCs, PMs, and other contaminants can be reduced by enhancing the ventilation systems. The American Association of Surgeons issued the health consequences due to the exposure to smoke in which they reported that purifying the air indoors or improving the room ventilation cannot eliminate the exposure to smoke. Centers for disease control and prevention in 2016 report also stated that spreading smoke could not be prevented even in a negative pressure smoking room in which the room is sealed and vented. Moreover, utilizing an air cleaning system can help reduce removing particles, although they are not managed to remove the gas. Furthermore, recent air-conditioning and ventilation systems may worsen as they can only spread and distribute the smoke throughout the building.

Generally speaking, technologies for smoke emission reduction can be broadly divided into two types: chemical conversion degradation, such as oxidation or decomposition, and adsorbing the VOCs, PMs, and other pollutants by activated carbon. Activated carbon removes VOCs by physical adsorption. However, desorption needs to be done as a post-treatment to recover the activated carbon, and at some points, the replacement cost would be needed afterward. Furthermore, adsorption with activated carbon also has limitations to the hydrophilic gas, such as acetaldehyde for hydrophobic activated carbon. Examples of chemical conversion methods include oxidation using ultraviolet (UV) radiation and plasma radiation with the employment of the photocatalyst. The photocatalysis plasma radiation method is usually employed in household air purifiers. Potent oxidizing agents such as O_3 oxidize VOCs to CO and CO_2. Consequently, O_3 is harmful to human health indoors. As a result, before the air is vented, O_3 must be removed from it. Manganese (Mn)-based catalysts are believed to be one of the most effective due to their excellent activity for O_3 breakdown at ambient temperature. For example, Kim et al. and Zhao et al. in 2020 [174, 175] examined the usage of O_3 to create oxidants at room temperature by using a MnO_x catalyst (25 °C). Kim used chemical vapor condensation (CVC) to make MnO_x/TiO_2 and compared it to Degussa P25 TiO_2 to totally oxidize air containing 45 ppm of formaldehyde (HCHO) to concentrations as low as 25 ppm and 20 ppm for CO_2 and CO, respectively, by adding 225 ppm of O_3. The result showed that 100% removal efficiency was achieved when the ratio of O_3 to HCHO was >5 by MnOx/CVC-TiO_2. It is reported that the removal efficiency of HCHO was significantly increased compared to the bare TiO_2 and MnO_2. This phenomenon could be ascribed to the difference in the concentration of O_3 on the catalyst surface, which resulted in different concentrations of the reactive oxygen species. Furthermore, they also investigated the effect of relative humidity (RH) and the ratio of O_3 to HCHO. The results showed that 100% removal efficiency could be achieved at 50% RH.

Purifying smoke in the indoor room through photocatalytic oxidation is a relatively new approach. UV light activates the photocatalyst, which creates hydroxyl radicals (oxidants) from adsorbed water to convert VOCs contained in the smoke to CO_2 and H_2O. However, the oxidation reaction rates are relatively slow for application in air purification systems [174].

Due to its low cost and relatively remarkable efficiency, metal oxide-based catalysts are widely employed for smoke degradation under photothermocatalytic conditions. Among these catalysts, an effective method is synthesizing the photothermocatalytic nanocomposites by combining metal oxides with the semiconductor TiO_2. For instance, Kim et al. fabricated MnO_x/TiO_2 nanocomposites to decompose aldehydes, VOCs, and ozone in an actual smoking room under a UV lamp. Figure 3 shows the evaluation over MnO_x/TiO_2 catalyst for the decomposition of acetaldehyde. The catalyst showed notable activity for acetaldehyde and VOCs removal, in which the removal capabilities were up to 95% within 30 min and 96%, respectively, in the prototype smoke purifier. For a real-scale smoke purifier, the study was conducted in an actual smoking room in which the acetaldehyde and VOCs removal was 70% (75.3–23 $\mu\gamma$ m^{-3}) and 80% (from 2200 to 493 $\mu g\ m^{-3}$), respectively. Furthermore, 100% nicotine removal efficiency from an initial of 106 to 0 $\mu g\ m^{-3}$. Generally, the removal rate will correspond to the material evaporation temperature. The higher the evaporation temperature, the better the removal. For example, in the case of activated carbon, the BTEX, which has a molecular weight of 78–106, has ~70% removal efficiency.

On the other hand, the removal efficiency of nicotine with a higher molecular weight is 100%. Moreover, the removal efficiency of volatile components such as acetaldehyde and formaldehyde is ~30%. However, from the above result, the

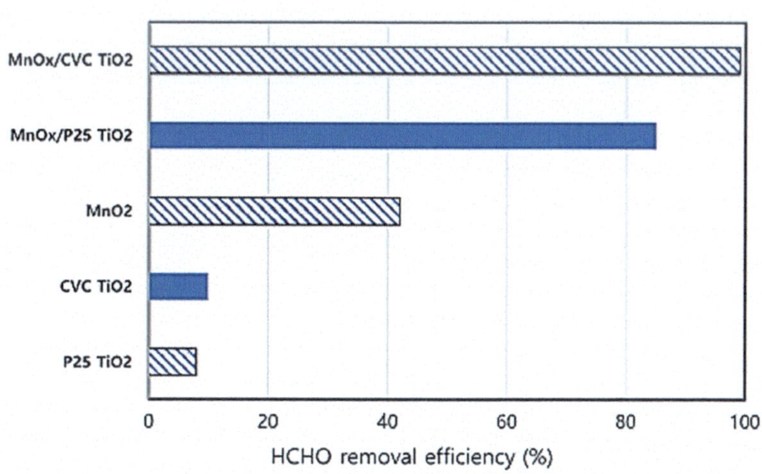

Fig. 3 The removal efficiency of CVC TIO_2, P25 TiO_2, MnO_x, MnO_x/CVC TiO_2, and MnO_x/P25 TiO_2.. Reproduced with permission. [176] Copyright 2022, Elsevier (License Number: 5332950029322)

MnO_x/TiO_2 shows high removal efficiency regardless of the molecular weight of the components. This outstanding performance is ascribed to the synergistic effect of the metal oxide MnO_x with the photocatalysis TiO_2 to remove the pollutants by decomposition, not by surface adsorption [174]. Notably, this technology showed excellent performance compared to the study conducted by Ochiai et al. employing the plasma-photocatalyst composite processing device in which only 30% of acetaldehyde removal could be achieved [177].

The ability of a semiconductor to undergo a photocatalytic reaction is determined by the relevant redox potential of the acceptor chemical species, which must be more favorable than the potential of the conduction band electrons. Besides the band edge positions, the suitable band gap energies also play an essential role in the performance of the photocatalyst. Rosli et al. tuned the band gap of the bare TiO_2, which is 3.06–2.9 eV, by employing recycled graphite. When the energy band gap is reduced, the TiO_2/graphite may decompose smoke 2 min faster under visible radiation [178].

4.2.2 Photocatalytic Smog and Ozone

The photochemical smog and ozone are a mixture of hazardous secondary pollutants resulting from the reaction of sunlight with primary pollutants such as NO_x and VOCs, which are trapped near the ground by temperature inversion. It has an unpleasant smell which causes adverse effects on human health. Among the pollutants, NO_2 and NO_x can be regarded as the primary pollutants produced by the dissociation in the combustion reaction. Ozone and peroxy acyl nitrate (PAN) that accumulates at ground level is the secondary pollutant that plays a vital role in air pollution [179]. Apart from the ground-level ozone, ozone also presents in the ozone layer (stratosphere) to protect the ground by filtering out the harmful UV rays from the sun, protecting lives [180]. In brief, ozone is formed by the photochemical photolysis of NO, which presents in the troposphere to generate nitrate (NO_3) radicals.

$$NO + O_3 \rightarrow NO_2 + O_2$$
$$NO_2 + O_3 \rightarrow NO_3 + O_2$$

The formation of nitrate radical can rapidly oxidize to NO_2, which leads to O_3 formation:

$$NO_2 + hv \rightarrow NO + O$$
$$O + O_2 + \text{third molecule} \rightarrow O_3$$

The third molecule could be nitrogen, oxygen, or other particles with a significant contribution to photochemical smog to captivate the heat from the reaction. Without this third molecule, O_3 would not be generated by only combining O and O_2. This cycle results in no accumulation of ozone concentration. Despite its role in protecting the earth from UV exposure, it can be considered as a severe risk to human health and the environment due to its strong oxidizing capability. Some studies reported that high exposure to ozone could lead to headaches, chest pain, and irritation of the eyes, nose, and throat [181]. Besides, even a low ozone concentration of 0.1 ppm has been demonstrated to induce skin irritation, coughing, and lacrimation, according to research [182]. Therefore, the emission of ozone is regulated by many organizations. Current national ambient air quality standards (NAAQS) were last updated in 2020. Based on the updated database, the recommended ozone level is 70 ppb for an 8-h daily maximum. However, WHO has a lower standard for 8-h daily max ozone, which is 50 ppb. High exposure to ozone will cause headaches, chest pain, and irritations of the eyes, nose, and throat. Therefore, it is necessary to remove ozone. Ozone can be decomposed to O_2 with the help of photocatalysis by the following equation [183].

$$O_3 + e^- \rightarrow O_3^-$$
$$OH + O_3 \rightarrow O_4^- + H^+$$
$$O_3^- + O_3 \rightarrow O_4^- + O_2$$

Various ozone purification methods have been explored, including adsorption, thermal decomposition, and catalytic removal. Catalytic removal of ozone includes catalytic decomposition and photocatalytic decomposition methods. Photocatalysis is considered a promising approach for ozone purifiers due to its mild reaction condition, high effectiveness, and low energy consumption. The catalytic reaction condition plays an essential role in the catalytic performance of the catalyst during the ozone degradation process. Furthermore, TiO_2 is the most common material and has been widely used for photocatalysis due to its good chemical and photonic stability [184, 185]. For instance, Lu et al. explored the impact of reaction circumstances (such as flow velocity, humidity, starting concentration, and UV light wavelength) on ozone degradation. The results show that the rate-limiting step will influence the degradation rate. When the flow velocity was set at 6.78 cm/s, the maximum removal efficiency of 64% could be achieved. However, further increasing the velocity to above 6.78 cm/s will have an insignificant performance for ozone removal. This phenomenon can be caused by the fixed number of electrons and holes generated by TiO_2. Therefore, when the flow velocity is above 6.78 cm/s, a low electron and hole can participate in the reaction. Furthermore, they also investigated that the amount of initial ozone concentration can affect the ozone degradation rate. This phenomenon was mainly due to the increased ozone molecule being in contact with the surface of the photocatalyst, which led to a rapid reaction between ozone and the photoinduced electron and holes [185]. However, it is widely known that the bare TiO_2 performance has low performance. Therefore, researchers have mainly focused on enhancing the performance of TiO_2. For example, Wang et al. introduced chlorine radicals to the bare TiO_2, and the performance could improve 2.5 times that of bare TiO_2. The enhanced performance is mainly due to partially replacing the HO-Ti group with the Cl-Ti group, resulting in a more vital absorption ability. In addition, the chlorine radical reaction rate with ozone is much faster than that of hydroxyl. Besides its superior performance, chlorinated TiO_2 also mimics the reaction of TiO_2 with the ozone in the atmosphere. The chlorofluorocarbons, mainly used as refrigerants, leaked into the atmosphere, producing chlorine-free radicals under the radiation of UV light. These generated radicals will react with ozone to generate oxygen. Besides, TiO_2 also generated free-radical under the radiation of UV light. Therefore, introducing chlorine-free radicals represents the real atmospheric reaction on the catalyst surface [184].

The main contribution of the smog formation is the reaction between nitrogen oxides and sunlight with the presence of ozone. However, smog cannot be formed from just VOCs and sunlight; NO_x and an oxidizer (ozone) are required to form smog. The emission of the harmful NO_x to the atmosphere contributes to acid rain and photochemical smog, which cause a severe problem for the environment. Thus, this implies that it is necessary to significantly remove the NO_x from the air because smog is created solely from the reaction of NO_x and sunlight. Apart from smog formation, NO_x may lead to the formation of acid rain, which begins with the oxidation of NO_x to form HNO_3, which is caused harmful effects and vegetation.

To date, numerous attempts have been attempted to reduce NO_x levels by converting it to N_2 or oxidizing it to HNO_3. Photo-decomposition, photo-oxidation, and photo-selective catalytic reduction are three methods that heterogeneous photocatalysis can use to convert NO_x. Due to the need to purify huge volumes of outdoor, photocatalytic oxidation is the most suitable approach among these three possibilities [183]. Therefore, the removal of NO_x by oxidation is regarded as suitable and hopeful technology to remove NO_x and has received tremendous attention due to the use of solar energy and ambient working condition [186]. The purpose of this process is to convert NO into HNO_3. However, before converting to HNO_3, the NO first need to be oxidized into higher oxidation state nitrogen oxides with the assistance of a photocatalyst. Upon irradiation, the photocatalysts generate e^-/h^+ pair, which can oxidize O_2, H_2O to form radicals (e.g., superoxide radical or hydroxyl radical). As a result, NO_x reacted with the radicals to finally produces innocuous nitrite/nitrate ions (NO_2^-/NO_3^-). The following equation can describe the process [122].

$$
\begin{aligned}
&NO + OH \rightarrow HNO_2 \\
&HNO_2 + OH_{ads} \rightarrow NO_2 + H_2O \\
&NO_2 + OH_{ads} \rightarrow HNO_3 \\
&NO + O^-_{2\,ads} \rightarrow NO^-_3 \\
&2NO + O^-_{2\,ads} + 3e^- \rightarrow 2NO_2 \\
&3NO_2 + 2OH^- \rightarrow 2NO^-_3 + NO + H_2O
\end{aligned}
$$

Recently, Huang et al. [187]. fabricated g-C_3N_4/TiO_2 composite via direct spraying as the photocatalytic air-purifying pavement and demonstrated that the g-C_3N_4/TiO_2 has a superior adhesion force, self-cleaning properties, light-induced super hydrophilicity which was appropriate for NO_x degradation. These outstanding properties could be ascribed to the role of TiO_2 as the chemical glue which can avoid the additional binders which can limit the performance of g-C_3N_4. The NO and NO_x removal rates were 70.7% and 35.7%, respectively. Furthermore, compared to the pristine metal oxide catalyst, the doped or mixed metal oxide catalysts are also studied by numerous researchers [187]. Regarding metal doping, the photocatalyst defect formation and chemical state can be controlled, significantly enhancing photocatalyst activity [188]. Incorporating the unbalanced valence ions can induce the formation of oxygen vacancies. Furthermore, the presence of oxygen vacancies can suppress the recombination of the photogenerated carriers due to its ability to trap the photogenerated electron [189]. Oxygen vacancies are also reported to adjust the band structure, which can increase the light-harvesting capability. More importantly, oxygen vacancies can boost photocatalytic performance by serving as the active sites by generating more superoxide radicals. Therefore, it is worth noting that oxygen vacancies can boost photocatalytic performance. For example, Nie et al. prepared a series of La-doped $ZnWO_4$ with different La concentrations. They investigated that introducing La into the $ZnWO_4$ exhibited a removal efficiency of 46%. In contrast, the pristine $ZnWO_4$ has a removal efficiency of 36.5. This enhanced performance can be ascribed to that La^{3+} doping can induce the formation of oxygen vacancies which can boost light utilization and suppress the recombination rate [190]. Besides, Li et al. prepared a series of Bi^{3+}-doped $ZnWO_4$ for photocatalytic NO removal. The catalytic performance decreased in a sequence of Bi^0/ZBWO > Bi^0/ZWO > ZBWO > ZWO. The reason can be ascribed to the synergistic effect of the SPR effect of Bi^0 loaded onto the ZBWO and the bandgap narrowing and formation of oxygen vacancies by introducing Bi^{3+}. The activities are beneficial for the electronic excitation and transformation, resulting in enhanced separation of photoinduced charge carrier and prolonging the life of the photoinduced electron. Moreover, Rao et al. induced the formation of oxygen vacancies by doping the BiOI with Zn^{2+} to remove NO under visible light irradiation. In addition, introducing Zn^{2+} can reduce the thickness of the nanosheet, which results in a highly exposed (001) facet. It is widely known that (001) can improve photocatalytic performance [191]. The 3% Zn-BIOI showed superior catalytic performance compared to the undoped BIOI photocatalyst, in which 53.6% NO removal can be achieved. Additionally, this catalyst was catalytically stable for NO removal after five consecutive cycles (Fig. 4). It is reported that introducing Zn^{2+} could increase the density of photo-active sites, resulting in better NO adsorption.

Bacteria and Viruses

Some information about bacteria and viruses is essential for designing air purifiers. Bacteria are microscopic single-celled organisms that lack a cell nucleus membrane (prokaryotic) [193]. A microscope is required to see it. Bacteria can survive in various environmental conditions [194], including extreme conditions. Bacteria, after all, can live in places where humans stayed, even in a location with high radioactive signals. Viruses, on the other hand, are not made of cells and require a patient's body to function and survive. Viruses are significantly smaller than bacteria. All but the most potent viruses are smaller than the smallest bacteria.

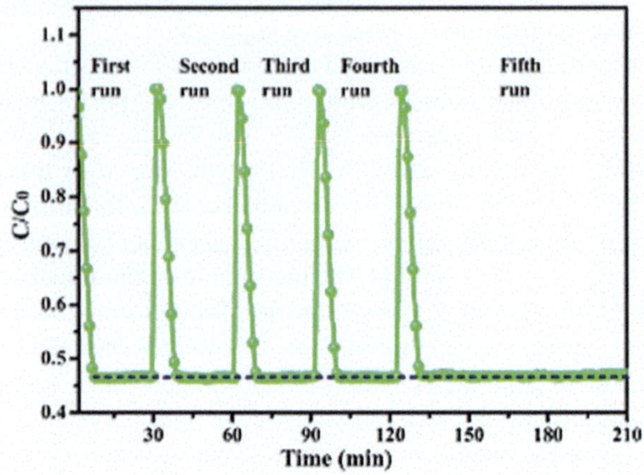

Fig. 4 Stability performance of 3% Zn-BIOI. [192]

Bacteria

Bacteria is derived from the Latin word bacterium and refers to a group of organisms that lack a cell nucleus membrane [195]. This organism is microscopic and belongs to the domain of prokaryotes. Because of this, these organisms were complicated to detect, especially before the invention of the microscope.

Bacteria are unicellular multi-celled microorganisms that lack chlorophyll pigments. Since there are no nucleus or membrane-bound organelles, the cell structure is more straightforward than other organisms. Bacteria maintain a fixed shape due to a rigid cell wall, even though they vary in contour, size, and layout [196, 197] (see Fig. 5).

These pathogens often appear in three main shapes when viewed under a light microscope: the rod (bacillus), the sphere (coccus), and the spiral type (vibrio). The bacterial structure is composed of two components: arrangement and shape. In terms of the arrangement, it can be paired (diplo), grape-like clusters (staphylo), or chains (strepto). Rods (bacilli), spheres (cocci), and cascades are the most common shapes (spirillum).

Bacteria are the most common germs found both indoors and outdoors. The design of a building and its circulation systems can influence the growth of germs in a room. In a particular investigation, the same building design and natural ventilation throughout 12 office spaces. It is examined that the bacterial community features, pathogen diversity, and source of bacteria in HEPA (high-efficiency particulate air) filters in air purifiers used in indoor environments. The findings aid us in taking the necessary actions to lessen the hazards of using air purifiers to remove PM. [194] In the HEPA filter of the air purifier, there is a variety of bacterial diversity and sources of bacteria, one of which is the bacterial taxa in the filter sample and dust. Taxa abundances at the class and genus levels were compared between filter and dust samples. In the following presentation, the abundance of most bacteria classes in the filter sample is demonstrated: Alphaproteobacteria (51.8%); Actinobacteria (17.2%).

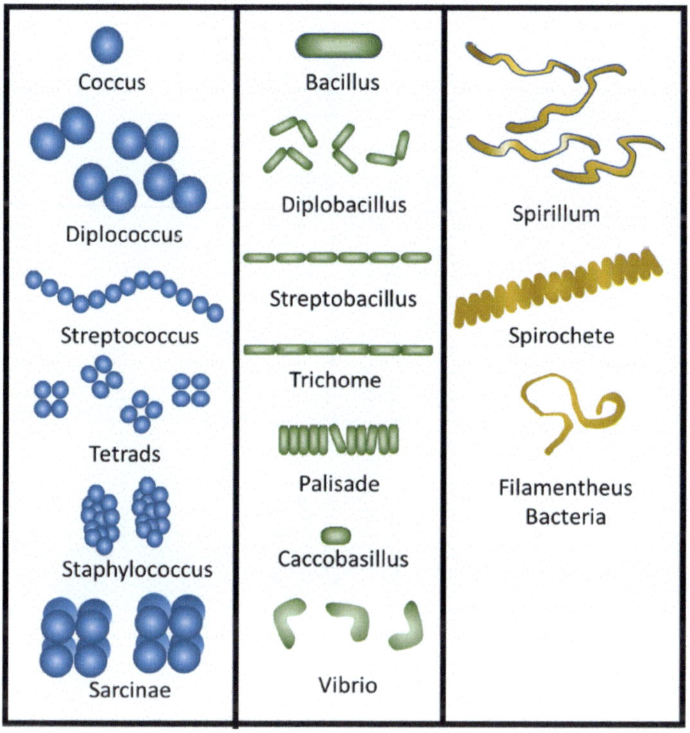

Fig. 5 Different shapes of bacteria [198–200]

Those in the dust samples were bacteroidia (25.6%), clostridia (13.9%), bacilli (15.9%), gammaproteobacteria (11.7%), and alphaproteobacteria (11.3%).

In general, some microbes that are filtered in large numbers can be classified into three most prominent groups, namely,

Spirochaetas

Spirochete (order Spirochaetales), also spelled Spirochaeta, is almost all of a group of spiral-shaped bacterium that causes diseases such as syphilis, tilts, Lyme disease, and relapsing fever. Spirochaeta, mainly sold, Borrelia, and *Leptospira* are some spirochete genera. Spirochetes are gram-negative, motile, spiral bacteria that range in length from 3 to 500 m (1 m = 0.001 mm). Spirochetes are marked by the appearance of endocellular flagella (axial fibrils or axial filaments), which range in number from 2 to more than 100 per organism, depending on the species. Each axial fibril connects at one end and winds around the cell body, which is surrounded by an envelope. Spirochetes are typically found in liquid environments (e.g., mud and water, blood and lymph) (see Table 9).

These bacteria cannot spread through the air and are trapped in the air purifier filtering system, which is aided by water vapor. *Leptospira* bacteria, a type of aerobic

Table 9 Characteristics of *Spirochaetes* [201, 202]

	Dimension (µm/flagella)	G + C content (mol%)	Oxygen relationship	Carbon + Energy source	Habitats
Spirochaetes	0.2–0.75 X 5–250; 2–40 Periplasmic Flagella (Almost = 2)	51–65	Facultatively anaerobic or anaerobic	Carbohydrates	Aquatics and free living

Spirochaeta bacteria that requires oxygen to live, are motile (could indeed move), gram-negative, and has a wrinkled shape. The bacteria *Leptospira* can also be found in the environment, water, rainwater, or plants that have been infected by rodent urine, blood, or tissue. When there is a flood, this happens quite frequently, and the Leptospirosis disease can spread quickly. In rare cases when it is transmitted through breast milk or from a mother to an unborn child, leptospirosis does not spread from person to person.

1. *Fusobacteria*

Fusobacteria is very difficult to identify in the clinical laboratory. Although the nomenclature and classification of fusobacteria have been clarified. There are few stable and specific tests for routine laboratory differentiation of the species. Since Fusobacteria have always been present in mixed culture with other obligatory anaerobes and facultative species, the use of selective media has been required to isolate the others. The ileal epithelium in small mammals is the widely accepted environment of Fusobacteria, and it has exciting life forms. This fibrous endospore-forming bacterium causes an endospore to a well septum or two major cells when it reproduces.

These bacteria, including *Fusobacterium* produced from *E. coli, Pseudomonas, Lactobacillus crispatus*, and *Lactobacillus gasseri*, can aid colonization by other species that produce and remove the by-products of their metabolism. Besides being suspected in the oral cavity, *Fusobacteria* bacteria can harm a woman's body's ability to produce breast milk. *Streptococci* are the most abundant bacteria in breast milk and can also attach to mucous membranes. *F. nucleatum* has infected the amniotic fluid and the fetus. Since fetal death is currently reported, it is most likely that *F. nucleatum* causes fetal death by local infection in the uterus.. The following is an experiment using mice to introduce Fusobacterium bacteria from the Han study in 2004. Only a few bacteria were identified from the liver and spleen of non-pregnant mice six hours after injection, but none were detected after 18, 24, 48, or 72 h and In numerous ways, this mouse model substantially resembled human intrauterine infection by *F. nucleatum*. According to the findings, *F. nucleatum* isolated from either amniotic fluid infection or an oropulmonary source caused fetal death. This is the first proof that hematogenous injection of an orally associated species can lead to adverse pregnancy outcomes. [203].

Fusobacterium nucleatum

Fusobacterium nucleatum is an oral bacteria and also has 700 oral bacteria with anaerobic properties, making it difficult to work without oxygen. It is gram-negative (but has a gram + genome), has a long pointed rod, is commensal, and does not sporulate. Colonization of colorectal cancer by *Fusobacterium nucleatum* (CRC).

Figure 6 indicates *Fusobacterium nucleatum* tumorigenic systems in colorectal cancer. The respective tumorigenic mechanisms are proposed in studies: Colorectal cancer and *F. nucleatum* (CRC). (1) Increased tumor cell proliferation: *F. nucleatum* promotes tumor formation by increasing the expression of tumorigenesis such as c-Myc, Cyclin D1, and microRNA-21. (2) Confiscation of Natural Killer (NK)

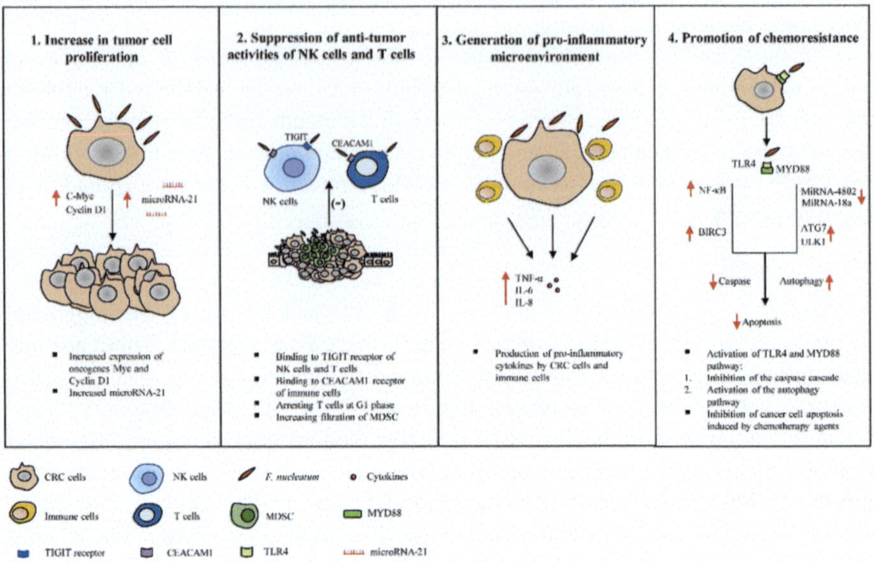

Fig. 6 Tumorigenic mechanisms of *Fusobacterium nucleatum* in colorectal cancer [195]

and T-cell anti-tumor activity: Activity of NK cells and T. cells nucleatum via contact with the T cell immunoglobulin receptor and the ITIM domain (TIGIT), which is expressed on NK cells and T cells, binds and fires up carcinoembryonic antigen-related cell adhesion particle (CEACAM) 1, restrains T cells in the G1 phase, and increases myeloid-derived in (3) The development of a pro-inflammatory microenvironment: *Fusobacterium nucleatum* causes cancer and immune cells to produce pro-inflammatory cytokines. (4) Chemoresistance promotion: F. nucleatum promotes chemoresistance to 5-fluorouracil (5-FU) by regulating the apoptotic protein baculoviral inhibitors repeat 3 (BIRC3) in CRC cells via the Toll-like receptor 4 (TLR4)/Nuclear factor-kappa B (NF-kB) pathway, which inhibits cancer cell apoptosis but also decreases chemosensitivity to 5-FU. By modulating autophagy, *F. nucleatum* promotes chemoresistance. TLR4 and MYD88 innate immune signaling pathways are activated in CRC cells infected with *F. nucleatum*, resulting in the loss of microRNAs miR-18a and miR-4802, as well as upregulation of autophagy elements ULK1 and ATG7.

4.2.3 Antibacterial Filtration

Environmental factors will considerably impact the quality and durability of an air filter from air purification equipment. As bacteria and fungi are involved, the role of filter material performance is critical. The antibacterial substance inhibits the growth of bacteria and fungi, improving the air filter's performance [204]. Escherichia (E) coli is a bacteria found in both animals and humans. Although most strains of this enterobacterial species are harmless commensals, they can induce diarrhea [205]. Diarrhea is just one of the many disorders caused by poor water quality. *Escherichia coli* is a microbe that has a mutually beneficial connection with other bacteria and only rarely causes disease. On the other hand, *E. coli* are highly evolved animals that have evolved into modern diseases. *E. coli* strains are divided into pathotypes of zoonotic intestinal pathogenic *E. coli* (IPEC) or extraintestinal pathogenic *E. coli* (EPEC) based on the kind of virulence determinants present and the host's clinical symptoms (ExPEC) [205]. *Staphylococcus aureus* is a Gram-positive bacterium that causes pneumonia and bacteremia. According to a recent study, patients with *S. aureus* infections had a 20.2% higher one-year mortality rate than matched non-infected inpatients [206]. *S. aureus* aerosols developed in the lab have particle sizes ranging from 0.542 to 1.197 m and aerodynamic particle sizes ranging from 0.723 to 0.777 m [207]. After incubation on the conventional filters, viral particles retained their ability to infect mammalian cells, and no infection was found after incubation on the treated filters. Compared to the unpasteurized filter, there was no considerable rise in cell infections when the filters were aged [208].

Based on Table 10, antibacterial behavior was evaluated against *E. coli* and S. aureus, and the results of the disc diffusion assay revealed effective bactericidal performance against bacterial strains (inhibition zones of 19 and 18 mm for S. aureus and *E. coli*, respectively). Ag nanoparticles and graphene oxide fillers increased the material's bactericidal, adsorptive, and mechanical properties. The prepared antimicrobial air filter could be used as a commercial anti-pollution mask in hospitals or industries to clean up airborne pollutants [209, 210].

4.2.4 Virus

The term "virus" is derived from virion's Latin word "poison." The virus itself is perpetually developing and growing. It is pretty hard to die or vanish. Viruses are constantly changing due to mutations, and all these mutations can sometimes result in a new virus strain. A new strain of the virus arrived and then faded, while others remained. There will be more versions in the future. New viruses may emerge, causing humans and other living things to get ill.

Microorganisms include viruses. Because 1 micron equals 0.001 mm, They are called microorganisms as the entities are typically a few microns in size or even smaller. Based on the preceding explanation of viruses, it is plausible to conclude that viruses. Viruses can be defined as parasitic microorganisms that require a home to survive. These germs must locate a host, which can include human cells.

It cannot self-replicate until it latches onto its host's body. Some of these creatures are capable of killing their host cells to proliferate. The virus will not be strong enough to survive long if it does not locate a host. In the United States, the CDC and other public health institutions record all strains of the virus that causes COVID-19. They are classified as variants to be monitored, variants of interest, variants to be concerned about, and variants of great consequence. Several variants spread more quickly and easily than others, perhaps leading to an increase in COVID-19 cases. Even if a

Table 10 Effect of bacteria on filter material during incubation time and efficiency

No.	Bacteria	Materials	Incubation	Filtration efficiency/%	Refs
1	*Staphylococcus aureus*	SiG	24 – 48 min	97–86	[204]
	Escherichia coli	chi-SiG	24 – 48 min	99.7 95.4	
		AgNPs- [chi-SiG]	24 – 48 min	100 98.8	
		AgNPs- [chi-SiG] contained 1.027% water	24 – 48 min	100 100	
2	*Escherichia coli*	PDMS/PMMA-chitosan	60 min	96.5%	[205]
	Staphylococcus aureus		60 min	95.2%	
3	*Escherichia coli*	Ag-MOFs	24 h	86.9%	[206]
		ZIF-8	24 h	65.2%	
4	*Escherichia coli* (ATCC 8739) and *Staphylococcus aureus* (CMCC 26,003)	PA6 ENM	48 h	97.36%	[207]
		PA6@Ag ENM	24 h	99.07%	
5	*Escherichia coli*	Ag/PAN	6 h	98.6%	[208]
6	*Escherichia coli*	AgNPs/PVDF	1 week-2 weeks	99.4%	[209]
	Staphylococcus aureus			99.6%	
7	*Escherichia coli*	HNTs-ZnO/PCL	24 h	95.9%	[210]
	Staphylococcus aureus			97.9%	
8	*Escherichia coli and Staphylococcus aureus*	PLLA/Ag-2MI	30 min	99.99%	[211]
	Escherichia coli and Staphylococcus aureus	*PLLA/Ag-2MI*		99.99%	
9	*Escherichia coli, Pseudomonas aeruginosa, Gram-positive Staphylococcus aureus*	Ag NPs	24 h	95%	[212]
	Staphylococcus aureus			99.9%	
10	*Escherichia coli*	PCL/zein/AgNP	24 h	98.1%	[213]
	Staphylococcus aureus			97.4%	

(continued)

Table 10 (continued)

No.	Bacteria	Materials	Incubation	Filtration efficiency/%	Refs
11	*Staphylococcus aureus*	SN/PVB/BH5	24 h	96.4%	[214]
12	*Escherichia coli and Staphylococcus aureus*	natural sea salt (NSS)	24 h	95%	[215]
13	*Escherichia coli*	rGZ2	18 h	99.9%	[49]
	Staphylococcus aureus			100%	

mutation causes less severe infection, a rise in the overall number of cases could lead to more hospitalizations, stretching healthcare resources, and possibly more deaths.

In air purifiers, interpreting the continually changing droplet size concerning SARS-CoV-2 is essential. A droplet was supposed to be a long air arc (greater than 5 microns). Since the beginning of the COVID epidemic, airborne (or aerosol) transmission viruses have been suggested as a route of infection. Aerosol particles are small compared to drops, but the size is not the sole distinction: Aerosols can travel for hours with wind currents, whereas droplets quickly fall to the ground. As an outcome, aerosol viruses are far more contagious and difficult to avoid than viruses linked to respiratory droplets. Airborne transmission has the potential to infect a huge number of hosts far away from the virus source. Here are several examples of viruses that can spread through the air, particularly in enclosed areas (see Tables 11 and 12).

Aerosol test procedures were developed, including the best time to generate aerosols, nebulization, and collection fluid. With low log reductions and no bacterial aerosol leakage into the environment, the HEPA filter air purification system has been shown effective against bacterial aerosols. HEPA air filters can filter viruses and other infectious particles [215]. In a closed space with insufficient ventilation, the pandemic hazard is unavoidable, according to the findings acquired thus far. The production of antiviral and antimicrobial air filters necessitates the use of a variety of ingredients. In this case, antiviral ions such as Ag, ZnO, TiO_2, CuO, and Cu, and metal/metal oxide nanoparticles with antiviral and antibacterial capabilities, are employed. Carbon nanomaterials such as carbon nanotubes, graphene, and their derivatives have also been used to develop effective air filters [209]. Furthermore, effective air filters can be made with natural substances such as biopolymers and herbal extracts having antiviral characteristics (see Table 13).

Table 11 Characterization of airborne viruses and incubation times

Type of virus			Shape	Size	RNA genome size	Incubation time	References
Influenza virus	Influenza A	H1N1	Spherical or filamentous	14 and 600 nm	13.5 kb	48 h	[223, 224]
		H3N2	Surface glycoproteins and spherical shape	15.3–5 μm	13.5 kb	≤4 days	[225 –228]
Rhino virus (RV)			Spheroidal	28–30 nm	7.3 Kb	2–4 days	[229 –231]
SARS-CoV			Spherical	80–160 mm	26–32 Kb	2–10 days	[232]
SARS-CoV-2			Spherical	60–140 nm	30 kb	14 days	[232, 233]
Human adenovirus (AdV)			Icosahedral	65–80 nm	26–46 kb	2–14 days	[234, 235]
Human respiratory syncytial virus (RSV)			Spherical, asymmetric, and filamentous	130 nm	15.2 kb	1 h through 36 h	[236]
Enteroviruses (EVs)			Spherical	28–30 nm	7.5–8.0 kb	5–14 days	[193, 229]
Noroviruses (NoVs)			Icosahedral	38–40 nm in diameter	7.4–8.3 kb	3–7 days	[237, 238]

Table 12 The effect of air filters on viruses that are transmitted through the air

Filter	Temperature operation	Virus	Shape	Effectiveness	References
HEPA	Outdoor temperature	H1N1; influenza A virus (IAV)	Disk with a diameter of 6 mm	99.97%	[239]
MERV 10	Outdoor temperature	SARS-CoV-2 or COVID-19	1–3 um size	50%	[239]
MERV 13	Outdoor temperature	SARS-CoV-2 or COVID-19	–	85%	[239]
Filtering facepiece respirators (FFRs) or N95 respirators	<100 °C (75% ± 5% and 22 °C ± 2 °C.)	1. influenza virus 2. Staphylococcus aureus 3. MS2 bacteriophage	0.3 μm	95%	[197]
Portable air cleaners (PAC) with HEPA	21–23 °C	SARS CoV-2	–	80% device effectiveness	[240]

Table 13 The duration of the covid virus in the open air [241]

No.	Virus	Day	Medium	Temperature (°C)	Reference
1	SARS-COV-2	3 days	Liquid medium or on dry filter paper	27	[242]
		7 days	Dry surfaces at room temperature	22	–
		1 day		37	–

5 Current Possible Materials and Performance for Photocatalytic Air Purifier

5.1 TiO_2-Based Photocatalysis

TiO_2 has been one of the most commonly used materials in photocatalysis applications due to its chemically stable nature, high oxidative ability, abundance, inexpensiveness, and non-toxic [243–245]. These reasons, with the antibacterial properties of the material, are making TiO_2 rose in popularity in air purification applications, assisting further in purifying not only gaseous air pollutants but also viruses and bacteria [246]. It also has ideal positions of VB and CB that are suitable for both oxidation and reduction reactions, with 3.2 and 3.0 eV bandgap for anatase and rutile phases, respectively, which promotes electron movement under UV irradiation ($\lambda < 380$ nm).

Upon light radiation, positive holes and negative electrons can easily recombine, resulting in no reaction in this instance. As a result, one of the most challenging tasks for a successful process is to prevent charge-carrier recombination to keep the photocatalyst active. The mixture of anatase and rutile phases may further aid charge separation by confining holes and electrons across the various crystal phases [247]. In addition, since undoped TiO_2 can only be activated by the spectrum of UV light, doping with other metal and non-metal ions which are introduced into the crystal lattice of TiO_2 without altering its basic crystallographic, can be a way to improve the activity of TiO_2 under a wider range of spectrum to include visible light [243–245]. The broadening of the light spectrum for TiO_2 activation from coupling it with other semiconductor materials might expand its application range to include indoor conditions. This part will compare and see how TiO_2 can photocatalytically break down different air pollutants.

5.1.1 Volatile Organic Compounds (VOCs) Degradation

A BTX photocatalytic air purification was also conducted by Zhang [248] using platinized (1 wt.%) TiO_2. The catalyst was synthesized with a conventional impregnation method [248]. The catalysts are tested by placing them in a quartz tube, forming a catalyst bed [248]. The highest photocatalytic degradation efficiency of the BTX in their mixture was found in m-xylene (95.9%), toluene (68.2%), and benzene (46.5%)

[248]. They found out that even a small content of precious metal such as platinum can improve the efficiency of TiO_2 in the photocatalytic degradation BTX [248].

Benzene is one of the four toxic gases in the group BTEX (Benzene, Toluene, Ethylbenzene, and Xylene). Selischev in 2021 researched the photocatalytic degradation of benzene using palladium-loaded, surface-modified TiO_2 catalyst [249]. Through conventional impregnation and a mixture of thermal- and photo-deposition, the catalyst Pd/TiO_2 is synthesized and deposited on a 9.1 cm^2 glass plate for a further test [249]. They found that a small percentage of 1 wt.% of palladium content can help improve the photocatalytic air purification activity of TiO_2, especially towards benzene [249].

Toluene is another toxic compound in the VOC group BTEX. Photocatalytic degradation of toluene has been done by Ho in 2022, exhibiting a staggering performance of 97.5% efficiency [250]. They synthesized the iridium-doped TiO_2 by a hydrothermal process and obtained a spherical morphology with a uniform size of 10–15 nm [250]. It also has a large surface area, with 156–170 m^2/g of their sample [250]. They tested this material by filling the photocatalyst in a packed-bed reactor made of a quartz tube [250]. This research discovered that the iridium mole fraction and air humidity had a significant impact on the photocatalystic toluene decomposition [250]. Another study that focuses on toluene degradation is published by Wang [251]. They synthesized silver-loaded TiO_2 nanorod composite using solvothermal and photo-reduction methods [251]. After testing it by electrospinning the fibrous material, they reached a 90% degradation rate of both toluene and PM particulates, with the utilization of optical energy in the 900 nm range [251].

A mixture of toluene/xylene and formaldehyde was the choice of VOC pollutant by Wu [252]. The catalyst, palladium-doped $TiO_{2,}$ was manufactured using a modified photo-deposition method [252]. The catalyst was deposited at the bottom of a glass bottle and a mixture of formaldehyde and toluene/xylene was dropped. The loading of Pd cocatalyst removes the surface OH groups on TiO_2. It speeds up the immediate activation of O_2 into ·OH radicals, which promotes the oxidation of toluene and xylene methyl group, favoring the formation of aromatic aldehydes and their breakdown to CO_2, resulting in a much better overall performance [252].

Formaldehyde is also another typical example of VOC. In a research done by Chen [253], they managed to synthesize $Cu–TiO_2$ through the impregnation method, resulting in nano-clusters of copper oxide grafted onto the TiO_2 [253]. The catalyst was deposited into a round dish made of stainless steel with a diameter of 80 mm for the testing of formaldehyde degradation [253]. They found that CuO_x clusters on TiO_2 have strong redox reversibility between Cu (II) and Cu (I), making the hole and electron separation much more accessible. Similar research was conducted by Wang [254], in which nano-copper oxide modified TiO_2 was synthesized through a one-step hydrothermal method [254]. The catalyst was dispersed onto a glass carrier and 3% (around 180 ppm on the initial volatilization) of formaldehyde dropped into the container [254]. A 90.3% degradation efficiency was achieved using CuO_x/TiO_{2-x}-6% with dioxymethylene as an intermediate [254].

Gaseous methyl ethyl ketone (MEK) is also often used as a VOC pollutant in photocatalytic air purification. In a study done by Shayegan [255], they engineered a two-phased (anatase and brookite) surface fluorinated iron-doped TiO_2 for removal of MEK [255]. The material was synthesized using a solvothermal method, displaying a crystal shape of quasi oval and cubic [255]. The catalyst was placed in a bed-like structure with a 7 cm × 9 cm cross-section area in an aluminum reactor. The results show that fluorinated-FeTi-0.4% and non-fluorinated FeTi-0.4% offer the highest and second-highest removal efficiency of MEK, respectively [255]. Another study by Mamaghani [256], also shows an excellent photocatalytic MEK reduction by 7 morphologically-different TiO_2 catalysts: solid, hollow, mesoporous spheres, nanotubes, nanosheets, 3d-hierarchically porous, and 3d sea-urchin-like [256]. The catalysts were coated on a nickel filter with a cross-section area of 10 cm × 10 cm and injected the VOC perpendicular to the filter and show that the TiO_2 nanosheets show the best catalytic performance with 71.3% removal efficiency of MEK [256]. They theorized that this happened because of the abundance of exposed [001] facets of their nanosheets, as well as their surface fluorination [256].

Furthermore, A study on chloroform done by Abedi [257] showed that Mo:TiO_2/GAC can exhibit a photocatalytic degradation with their best performance that was under 30% humidity [257]. Granular Activated Carbon (GAC) is granular activated carbon made by chemical grafting a nano-porous polymer network for iron impregnation with nanoscale zero valence and water contamination removal [258]. Mo:TiO_2/GAC was prepared with a fast sol–gel technique and tested in a honeycomb Teflon plate reactor [257]. They found that even though the result of photocatalytic degradation under visible light is not as high as UV light, it is still possible to be done by this catalyst [257].

5.1.2 Greenhouse Gas Reduction

CO_2 is one of the greenhouse gases that is currently posing a climate crisis due to our overdependency on combusting fossil fuels. There are a couple of reaction pathways that can reduce CO^{-2} to form formic acid (HCOOH), methanol (CH_3OH), methane (CH_4), and carbon monoxide (CO). The work done by Ku et al. [259] showed that NiO/TiO_2 synthesized through the conventional impregnation method could reduce CO_2 to methane [259]. The optimized 10 wt.% NiO/TiO_2 suggested that the high surface area (58.8 m^2/g) and p–n junction significantly improved the separation of produced charges on the photocatalyst surface, leading to the increased photocatalytic activity [259]. Meanwhile, Zhou et al. [260] synthesized ruthenium-doped TiO_2 through the hydrothermal method to follow the same methane formation pathway [260]. The catalysts are dispersed at the bottom of a glass reactor to further test their performance [260]. With a selectivity of 90.93%, the 1% Ru-TiO_{2-x} photocatalyst obtained the maximum CO_2 reduction selectivity [260].

Another pathway that can be gone through for a CO_2 reduction reaction is one that ended up with methanol production. Cheng et al. [261] have investigated on Cu_2O/TiO_2 heterojunctions as a photocatalyst for CO_2 reduction through this pathway [261]. The material was synthesized through a facile soft chemical method, resulting in a mesoporous structure [261]. They used a closed cylindrical quartz reactor to test the efficiency of the catalyst in reducing CO_2 into a lower-carbon molecule [261]. They obtained a high yield of methanol at around 9–13 μmol/g/h in UV light and 12–70 μmol/g/h in visible light, proving that their material is effective in the photocatalytic breakdown of CO_2 [261]. Another study published by Xi et al. [262] shows that a staggering methanol selectivity of 98.85% was observed [262]. The catalyst is synthesized through a hydrothermal method, followed by a deposition–precipitation method [262]. The other possible pathway is the production of acetaldehyde from CO_2 reduction. In a study by Qian et al. [263], a niobium-doped TiO_2 catalyst was synthesized through anodization and then uniformly dispersed in a quartz reactor. A selectivity of acetaldehyde and methanol by 98.1% and 1.9%, respectively, is obtained [263]. Substitutional doping of Nb^{5+} to replace Ti^{4+} resulted in the development of acid centers with positive charges and the creation/maintenance of a Ti^{3+} defect, which improved CO_2 adsorption and activation on the Nb-doped TiO_2 surface [263].

5.1.3 Viruses and Bacteria Removal

As mentioned before, TiO_2 has an antibacterial property, making it suitable for the inactivation of viruses and bacteria in an attempt to enhance the air quality. In a work done by Hernández-Gordillo and Arriaga [264], TiO_2 monoliths were synthesized and infused with CdS and CuO through a sol–gel method [264]. The maximal bacteria inactivation of monoliths m-TiO_2, m-TiO_2/CdS, and m-TiO_2/CuO was 64.8%, 99.9%, and 76.5%, respectively [264]. While a study by Moon et al. [246] shows that their synthesized Cu-doped TiO_2 can deactivate human norovirus (HuNoV) [246]. They prepared the catalyst using a sol–gel method and tested the catalytic activity by placing the catalyst on a 9 cm × 2 cm dish [246]. The virus is then dispersed by a nebulizer that is connected to the testing chamber [246]. These findings suggested that viral particles of HuNoVs might be successfully destroyed utilizing a Cu/TiO_2 nano-woven fiber irradiated with UVA-LED light; however, prolonged exposure is required [246].

5.2 *Bi-Based Photocatalysis*

Bismuth (Bi), a metallic element with an atomic configuration of $6s^2 6p^3$, has been attracting much attention to develop this kind of material for photocatalyst. The lone-pair distortion of the Bi 6 s orbital in bismuth-based complex oxides may cause the O 2p and Bi 6 s orbitals to overlap in the valence band, which reduces the band gap and enhances photoinduced charge mobility while also improving visible light responsiveness [265]. As shown in Fig. 7, various new bismuth-based semiconductors with various crystal structures have been explored for photocatalysis. They can be classified as bismuth metal, bismuth oxides, bismuth oxyhalides, bismuth sulfides, and multi-component oxides.

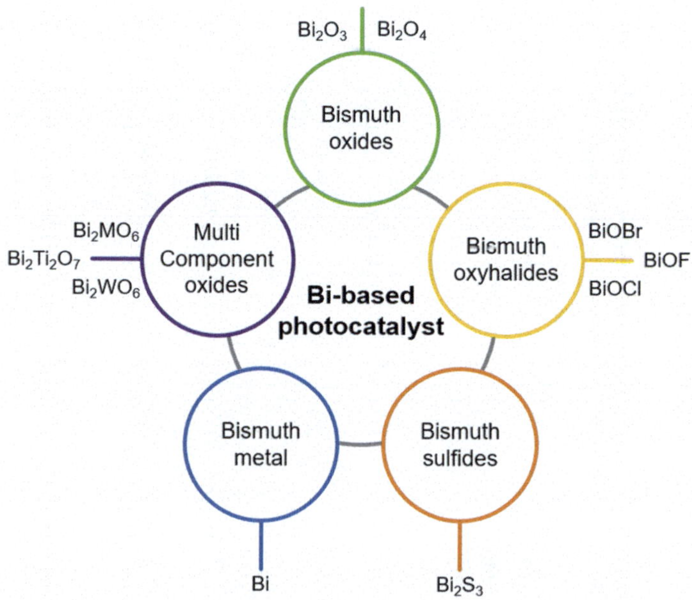

Fig. 7 A variety of classified and chemical formulations for Bi-based photocatalysts

Most of the Bi-based photocatalysts have an excellent response under visible light irradiation owing to the bandgap of less than 3.0 eV, except for BiOF and BiOCl, as shown in Fig. 8 [265]. The band gap will have an impact on light absorption capacity and carrier generation [266]. Particularly, Bi_2S_3, BiOI, and $LiBiO_3$ have an ability to absorb visible light with longer wavelengths, likely due to the bandgap less than <2.0 eV. The band structure of Bi-based semiconductors is known to be mainly constituted by Bi, O, and other elements. Bi 6p states are used to construct the conduction band minimum (CBM) in most cases [267].

Many bismuth-based photocatalysts have demonstrated photocatalytic effectiveness in decomposing air pollutants when exposed to visible light [269]. Table 14 summarizes the development of bismuth-based photocatalyst for air purification in recent years. The exploration of catalysts is to improve performance by increasing degradation extents and decreasing degradation times while preserving stability and uniformity.

NO_x is the most frequently degraded air pollutant with bismuth, followed by VOCs. Bi-based photocatalyst can degrade NO_x at a low level and convert it into nitrate to promote plant growth and enhance soil carbon storage. The alkaline solution can remove a substantial amount of NO_2 due to the poor solubility of NO and the high solubility of NO_2 [268]. Oxygen vacancy can form the middle gap level in the band structure to make the bismuth active at the visible light [272, 280]. Three-dimensional (3D) Bi metal@defective BiOBr hierarchical microspheres were produced by Dong et al. to activate O_2 with photo-induced electrons and create superoxide radicals. On defective BiOBr, the adsorption energy (E_{ads}) of O_2 molecules is -0.38 eV, which is

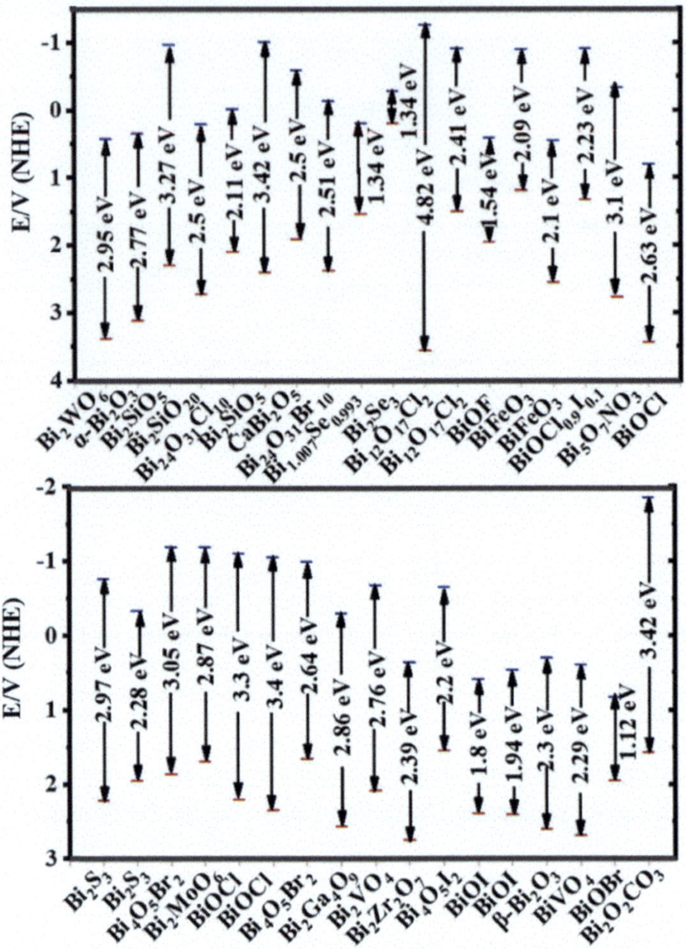

Fig. 8 The band structure positions of Bi-based photocatalyst. Copyright 2022. Reproduced with permission. [268] Copyright 2022, Elsevier (License Number: 5333360716952)

0.08 eV lower than on BiOBr without OVs. The amount of Oxygen Vacancy (OV) in a material affects its optical and photocatalytic capabilities. OV seeks to localize many electrons, increasing the activation of inert gas molecules. The formation of additional oxygen vacancies is accompanied by the deposition of Bi metal on BiOBr (Fig. 9a.) [270]. Chen et al. fabricated Bi metal nanoparticle and oxygen vacancies decorated defective $Bi_2O_2CO_3$ photocatalysts. Aside from significantly improving photocatalytic NO removal stability, Bi metal and oxygen vacancies can also promote electron transfer between intermediates and catalyst surfaces (Fig. 9b) [273]. In addition, the synergistic effects of Bi metal and oxygen vacancies were found on Bi in defective BiOI nanosheets and Bi metal co-modified $Bi_{12}TiO_{20}$ nanofibers photocatalyst. Because electrons from the valence band tend to congregate at vacancy states,

Table 14 Photocatalytic activity of bismuth-based photocatalysts for air purification

Photocatalyst	Preparation methods	Characterization	Air pollutant	Removal efficiency	References
Bi/BiOBr	Solvothermal	XRD, XPS, SEM, TEM, UV–vis DRS, N2 adsorption–desorption, PL, ESR, and EPR	NOx	63.53% for 20 min	[270]
$Bi_3O_4Cl_{0.5}Br_{0.5}$	Solid-state reaction	XRD, XPS, SEM, TEM, HRTEM, UV–vis DRS, and PL	NOx	60.00% for 10 min	[271]
Bi/BiOSi	Chemical reduction	XRD, XPS, FTIR, SEM, TEM, UV–vis DRS, PL, ESR, and EPR	NOx	50.20% for 10 min	[272]
$Bi/Bi_2O_2CO_3$	One-step hydrothermal	XRD, XPS, SEM, TEM, UV–vis DRS, PL, ESR, and EPR	NOx	40.80% for 5 min	[273]
α-$Bi_2O_3/CuBi_2O_4$	Solvothermal and calcination	XRD, XPS, SEM, TEM, TGA, UV–vis DRS, N2 adsorption–desorption, PL, and ESR	NOx	30.00% for 10 min	[274]
Bi/CdS	Hydrothermal	XRD, XPS, FESEM, EDS, TEM, UV–vis DRS, N2 adsorption–desorption, PL, and ESR	NOx	58.10% for 10 min	[275]
Bi/BiOI	Partial reduction	XRD, XPS, SEM, TEM, UV–vis DRS, N_2 adsorption–desorption, PL, ESR, and EPR	NOx	40.80% for 10 min	[276]

(continued)

Table 14 (continued)

Photocatalyst	Preparation methods	Characterization	Air pollutant	Removal efficiency	References
CQDs/Bi_2WO_6	In situ hydrothermal	XRD, XPS, FTIR, SEM, TEM, UV–vis DRS, N2 adsorption–desorption, and PL	Toluene and formaldehyde	96.90% (toluene) and 97.1% (formaldehyde) for 120 min	[277]
$LaVO_4$/BiOBr	In situ hydrothermal	XRD, XPS, SEM, TEM, UV–vis DRS, PL, and ESR	Acetone and toluene	95.4% (acetone) for 180 min and 87.1% (toluene) for 240 min	[25]
Bi2O2CO3	Hydrothermal	XRD, XPS, FTIR, FESEM, TEM, UV–vis DRS, UPS, and N2 adsorption–desorption	Toluene	99.00% for 210 min	[278]
Bi_2MoO_6/diatomite	Solvothermal	XRD, XPS, FTIR, SEM, EDS, TEM, UV–vis DRS, N2 adsorption–desorption, and PL	Formaldehyde	94.00% for 180 min	[279]
Pt/Bi-Bi_2WO_6	Chemical reduction	XRD, XPS, FTIR, FESEM, TEM, and EPR	Toluene	90.00% for 90 min	[280]

the increased charge density would lead the adsorbed oxygen to convert into superoxide radicals more rapidly, and then into hydroxyl radicals. The surface plasmon resonance (SPR) effect of elemental Bi leads to greater visible light absorption efficiency and promotes charge carrier separation, both of which are important aspects in photocatalysis (Fig. 9c, d) [276, 281].

In recent decades, VOCs have been eliminated using direct oxidation technologies such as direct combustion, catalytic oxidation, and photocatalytic oxidation [282, 283]. CQDs/Bi_2WO_6 was fabricated by Liu et al. for VOCs (toluene and formaldehyde) removal under visible light. Bi_2WO_6 is stimulated to create e^-/h^+ couples, where h^+ is a powerful oxidant that can mineralize VOCs and CQDs will increase the conversion properties (Fig. 10a) [277]. Yang et al. prepared diatomite and Bi self-doping Bi_2MoO_6, which can remove VOCs by direct oxidation and indirect oxidation of HCHO methods. Direct oxidation is the electrons in the valence band

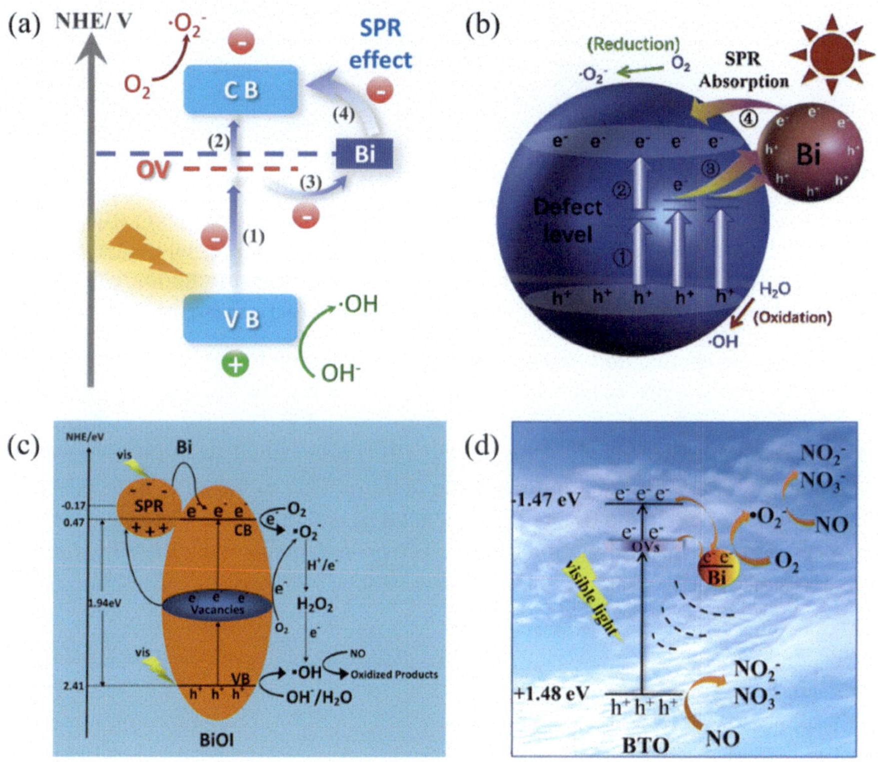

Fig. 9 Proposed Bi-based photocatalyst mechanism for NO_x purification by **a** Bi/BiOBr. Reproduced with permission. [270] Copyright 2022, Elsevier (License Number: 5327540795798); **b** Bi/$Bi_2O_2CO_3$, **c** Bi/BiOI. Reproduced with permission. [276] 2022, Elsevier (License Number: 5327541362694), and **d** Bi/$Bi_{12}TiO_{20}$. Reproduced with permission. [281] 2022, Elsevier (License Number: 5327550373139)

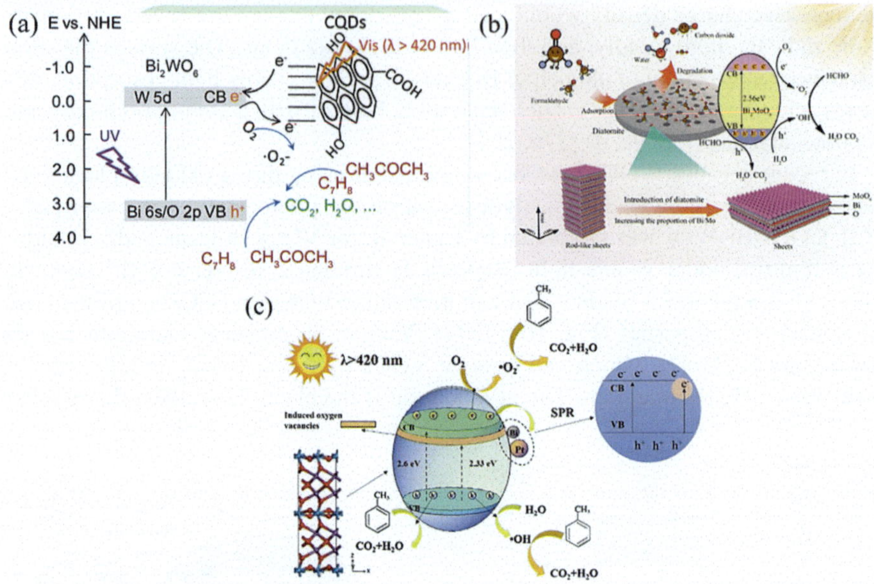

Fig. 10 Proposed Bi-based photocatalyst mechanism for VOCs degradation by **a** CQDs/Bi_2WO_6. *Reproduced with permission.* [284] *Copyright 2022, Elsevier (License Number:* 5327580057364*),* **b** Bi_2MoO_6/diatomite. *Reproduced with permission.* [279] *Copyright 2022, Elsevier (License Number:* 5327561138504*),* **c** Pt/Bi_2WO_6. *Reproduced with permission.* [280] *Copyright 2022, Elsevier (License Number:* 5327570018928*)*

of Bi_2MoO_6 are excited and transported to the conduction band, leaving photogenerated h^+ in the valence band with significant oxidative ability. Indirect oxidation means that the produced e^- and h^+ react with O_2 and H_2O in the air to form $\cdot O_2$ and ·OH radicals, respectively (Fig. 10b) [279]. Pt-deposited Bi_2WO_6 also shows VOC degradation with high photocatalytic activity. Due to the SPR-induced local electromagnetic field in the Bi and Pt, the loaded Bi and Pt could serve as an electron trap, decreasing the recombination rate of electron–hole pairs in the Bi_2WO_6 surface (Fig. 10b) [280].

5.3 WO_3-Based Photocatalysis

Despite the numerous researches that mainly focused on TiO_2 as the photocatalyst for air pollution, researchers never give up on exploring novel photocatalysts. WO_3 has been reported to have a similar ability to the TiO_2 photocatalyst with a bandgap lies between 2.6 and 2.8 eV. In addition, the VB potential lies between 2.7 and 3.44 eV. Therefore it has a similar ability to TiO_2 photocatalyst [285]. Therefore, it is widely known that WO_3 is a promising material for visible light adsorption in the photocatalytic application field as an n-type semiconductor. WO_3 has been explored in many photocatalytic applications, such as decontaminating polluted water [286], detecting hazardous gases [287], and energy conversion [287]. Unlike the photocatalytic degradation of wastewater in the liquid phase, a photocatalytic air purifier that employs gas pollutant restricts the even distribution on the catalyst surface. Therefore, a strong adsorption ability of photocatalyst is required. WO_3 is considered an ideal candidate for gas pollutants degradation due to its strong adsorption toward gas-phase pollutants, remarkable photochemical stability, and durability. The crystalline structure and defects have a strong influence on its performance. WO_3 has been reported to have five crystal structures strongly dependent on the temperature. Monoclinic II occurs at the temperature of <-43 °C. However, Wang et al. stabilized the monoclinic II at room temperature by incorporating Cr doping [288]. The triclinic WO_3 occurs at -43 to 17 °C, monoclinic I occur between 17 and 330 °C, orthorhombic WO_3 occurs at 330–740 °C, and tetragonal WO_3 occurs at >740 °C [289]. Many factors can influence the performance of photocatalysts, such as the degree of crystallinity and surface lattice defect. Different synthesis methods, precursor reagents, structure-directing agents, and synthesis conditions resulted in various performances. Thus, several parameters need to be optimized to get an optimized catalyst. Several morphologies of WO_3, such as sphere [290], plate [291], rod [292], and tube [293] can be synthesized using different salt precursors, capping agents through several synthesis methods. Several directing agents have been used to produce desired morphology, such as Triton X-100 [294], sodium chloride [295], sodium sulfate [295], and cetyltrimethyl ammonium bromide (CTAB) [296]. Furthermore, the reaction condition and the synthesis methods are the parameters that are commonly used to control the final morphology of the product. For example, a study by Nagy et al. demonstrated that the pH of the reaction affected the bandgap of WO_3 due to the difference in crystal phases and morphology. According to their results, preparation at pH = 0.1 to 1.05 will generate monoclinic WO_3, whereas preparation at pH above 1.05 will generate hexagonal WO_3 [297].

However, pristine WO_3 is limited due to its relatively low energy density and smaller surface area. Besides, the pristine WO_3 has low reduction potential, suppressing the separation of photogenerated electrons and holes pair. Therefore, the photogenerated electron in the CB of WO_3 restricted the adsorbed oxygen molecules to form the superoxide anion free radicals. Thus, to address this problem, several strategies have been made to overcome these drawbacks, such as metal doping, formation of composites with carbon materials, and formation of heterojunction with a metal oxide or other semiconducting material. For example, Wang et al. used WO_3 as the supporting material to isolate single atom Au by electrochemical method for toluene degradation. The results showed that anchoring Au single atom in WO_3/TiO_2 exhibited 1.8 mA cm^{-2}, which was 17 times higher than pristine WO_3. It is demonstrated that anchoring Au atoms could assist the formation of superoxide radicals which can increase the production of superoxide radicals to attack the methyl group of toluene, resulting in benzyl alcohol, benzaldehyde, and benzoic acid. Thus, it improves the separation and the photogenerated electron and holes pathways. In addition, the heterojunction of WO_3/TiO_2 promoted better electron transfer from the CB of WO_3 to the VB of TiO_2. Thus, the photogenerated electron will aggregate in the CB of WO_3, and the holes will aggregate in the VB of TiO_2, suppressing the recombination rate [298].

Besides, a similar observation was reported by Luévano-Hipólito [299] and co-workers in which WO_3/TiO_2 heterojunction showed enhanced photocatalytic activity for the NO removal due to better electron transport provided by the WO_3/TiO_2 heterojunction leads to a longer lifetime of the formed electron and holes [299]. In 2019, Lu et al. [300] reported the synthesized of nonstoichiometric tungsten oxides ($WO_{3\text{-}x}$) with abundant amount of oxygen vacancies for nonmetallic plasmonic photocatalyst to promote ethanol dehydrogenation under UV–visible light radiation. However, it was found that irradiation under UV and visible light performed better than those irradiated under visible light only. Furthermore, irradiation under UV light is also 12 times higher than irradiation under visible light. This enhanced performance is due to the unique electron band structure and the Z-scheme heterojunction that the nonstoichiometric tungsten oxide under UV visible light irradiation facilitates the light-harvesting ability of full-spectrum solar light irradiation. Furthermore, the oxygen vacancies also provide active sites for improving ethanol adsorption and degradation. The defect band may serve as the pathway to stimulate and transfer hot electrons and hot holes by the oxygen vacancy of $WO_{3\text{-}x}$, which can form the plasmonic thermal effect under the whole spectrum irradiation [300].

In comparison with pristine WO_3, the photocatalytic activity of different crystal phases of WO_3 has been significantly improved to some extent. In 2017, for acetaldehyde oxidation, Fukumara et al. [301] produced ϒ-WO_3 and ε-WO_3. At ϒ -WO_3, oxygen reduction was seen, while acetaldehyde oxidation was shown at ε-WO_3. This is owing to the ferroelectricity characteristics of ε-WO_3, which boost its photocatalytic activity by preventing the produced holes and electrons from recombination. Furthermore, it is demonstrated that acetaldehyde prefers to interact with ε-WO_3 via dipole–dipole interactions. In addition, CeO_2, which has the same refractive index as WO_3, has been utilized as an inorganic binder to immobilize WO_3 in thin transparent layers on the glass. CeO_2 also has a scavenging feature that improves charge separation. Due to the synergistic effect of WO_3/CeO_2, photocatalytic activity is increased [301].

Furthermore, it has been investigated by many researchers that different morphology of photocatalyst influenced its performances. For example, Kong et al. [302] synthesized a series of nanoplate WO_3 using oxalic acid to control the morphology of the nanoplate film. The catalytic performance of this oxalic acid-assisted WO_3 was much higher than the pristine WO_3. The photocatalytic degradation of acetaldehyde to CO_2 and H_2O was carried out under the radiation of 100 mW/cm^2. The generation of CO_2 with the assistance of oxalic acid is 2.3 times higher than pristine WO_3. In addition, the oxalic acid assistant-prepared WO_3 has a narrower band gap and higher visible light absorption. Also, the addition of oxalic acid influenced the crystallinity and the thickness of WO_3. Increased film thickness resulted in better use of sunlight. Meanwhile, the exposed crystallinity of the WO_3 may suppress the recombination rate of the photogenerated electron and hole pairs [302]. Sayama et al. [303] prepared a series of WO_3 samples with different morphology. The WO_3 prepared from amorphous peroxo-tungstic acid WO_3 (PA) showed the highest catalytic activity over WO_3 prepared by calcination, homemade WO_3 powder, and commercial WO_3. It is found that the WO_3 prepared from amorphous peroxo-tungstic acid performed six times better than the commercial WO_3 for hexane degradation. The better performance is due to the flat and smooth morphology possessed by the WO_3 (PA) leads to more efficient visible light absorption and suppresses the light reflection.

On the contrary, the other WO_3 samples have large aggregates, and a very rough surface leads to considerable light reflection, resulting in lower photocatalytic activity [303]. Shi et al. [304] synthesized a series of the plate-like $WO_3/CuBi_2O_4$ and highlighted the S-scheme charge transfer in the heterojunction for photocatalytic reduction of CO_2. The n-p junction was reported to be located at the interface between WO_3 and $CuBi_2O_4$. They further claimed that the charge transfer follows an S-scheme mechanism, in which holes remain in the VB of WO_3 and the electron in the CB of WO_3 is transferred to the VB of $CuBi_2O_4$, resulting in effective charge transfer separation. Comprehensive valence band-X-Ray Spectroscopy (VB-XPS) characterization backed up this suggested mechanism. The two peaks of W 4f in light shifted to the higher binding energy in the deconvoluted XPS spectra. In this heterojunction, however, the Cu 2p in light reversed shifted, indicating photogenerated electron transport from WO_3 to $CuBi_2O_4$. An optimal yield of 1115.8 μmol/m^2 was achieved with $WO_3/CuBi_2O_4$, whereas the pristine WO_3 and pristine $CuBi_2O_4$ obtained no yield [304].

Lately, research about carbon-based material has sparked due to its high reduction ability, chemical stability, and visible-light absorption. Katsumata et al. [305] have synthesized composites g-C_3N_4/WO_3 to improve the charge separation in photocatalytic gaseous acetaldehyde degradation. The results show that g-C_3N_4/WO_3 composites exhibited 45% gaseous acetaldehyde removal, which was better than pristine WO_3 (23%) and pristine g-C_3N_4 (16%). The reason could be ascribed to the synergistic effect between this composite. The presence of g-C_3N_4 enhanced the adsorption ability, and higher WO_3 content leads to higher photodegradation rate, which showed that WO_3 plays an important role in the photodegradation of acetaldehyde [305]. Moreover, Jin et al. also made the composite of carbon-based material with WO_3 for photodegradation of acetaldehyde. His research mainly focused on synthesizing g-C_3N_4 and WO_3 composite and mixed using a planetary mill to improve the photocatalytic activity for acetaldehyde degradation. The more positive valence band (VB) potential of WO_3 makes it a suitable material for oxidation, while g-C_3N_4 has a more negative CB potential which is ideal for the reduction reaction. Thus, a reduction reaction is preferred. Therefore, this composite follows the Z-scheme charge separation mechanism, which improves the complete oxidation of acetaldehyde [306]. Wang et al. synthesized oxygen vacancies rich WO_3 and 3D doped carbon matrix composite for the degradation of VOCs, in which the reaction rates for formaldehyde degradation rate constant of WO_3/NC were calculated to 0.039 min^{-1} over pristine WO_3, which is only 0.011 min^{-1}. The enhanced performance of WO_3/NC due to the larger surface area enhances the light absorption of WO_3/NC and VOC molecules and suppresses the recombination of photogenerated electrons and holes due to the presence of oxygen vacancies. It exhibited outstanding catalytic durability even after eight successive cycles, and the degradation remains at 92%, which was appropriate for practical application [307]. Kim et al. [308] utilized the earth-abundant carbon in the form of nanodiamonds as a co-catalyst loaded on WO_3. It exhibited comparable performance to noble-metal Pt, which is 17 times higher compared to the pristine WO_3 for the degradation of VOC under visible light. They reported that the structure of the synthesized nanodiamonds, which contain the diamond core structure (sp^3) and the graphitic carbon layer (sp^2) on their surface regions, allows the better electron transfer to the reactant. Thus, it enabled enhanced charge separation, electron transfer, and lower overpotential of O_2 reduction leads to better efficiency [308].

In addition, numerous methods for improving the photocatalytic performance of WO_3 have been presented. Doping is a well-known method for modifying semiconductor adsorption edges and improving photoactivity under low-energy irradiation, which causes a redshift to the visible light region [309]. The critical effect of the impurities doping is to alter the position of energy level as well as the electronic structure of semiconductor. In WO_3, it has been reported that transition metal doping, noble metal doping, rare earth metals, and nonmetal could narrow the band gap and suppress the electron–hole recombination rate [310]. For example, Sheng et al. [311] reported the influence of transition metal Fe doping to WO_3 resulted in 98.21% degradation rate of formaldehyde after 6 h of reaction. These characterization results show that the Fe^{3+} acts as a scavenger and traps the electrons and hole of WO_3 to prevent the recombination of excited charge carriers. Furthermore, the Fe^{3+} reacts with hydroxyl ions and oxygen to form hydroxyl radicals and superoxide radicals. These radicals will attack the formaldehyde to form the final product, such as CO_2 and H_2O [311].

Apart from photocatalytic oxidation, several attempts have been made to enhance efficiency. For example, Li et al. [312] utilized WO_3 for photothermooxidation in the removal of volatile organic compounds in the air. They synthesized the WO_3 with various amounts of oxygen vacancies by controlling the time of hydrogen reduction treatment. As a result, the photothermocatalytic activity in acetaldehyde removal was 5.2 times higher over the photocatalytic and thermocatalytic activity, which originate from the synergistic effect of heat and light. Thus, photothermooxidation considered as a promising technology as it integrates both thermocatalysts and photocatalysts. Therefore, it can enhance the synergistic effect at relatively low temperatures [312].

5.4 Ag-Based Photocatalysis

Silver is classified as a noble and transition metal, belonging to the 5th period and group 11 of the periodic table. It has an atomic number of 47. As a cocatalyst, silver is excellent for environmental and heterogeneous catalysis applications. Due to its low coordination chemistry, Silver provides stability in aqueous media with its colloidal size. Furthermore, as the oxidation number of silver is +1 or 2+, it can work with d^{10} or d^9 orbitals [313].

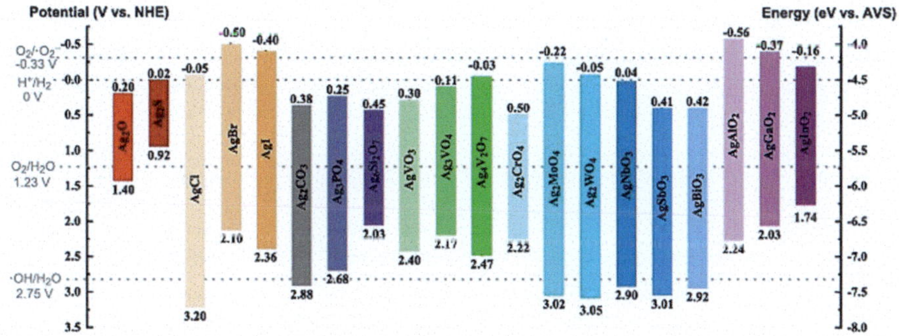

Fig. 11 Conduction band and valence band of Ag-based semiconductor photocatalysts. *Reproduced with permission.* photocatalysts [316] *Copyright 2022, Elsevier (License Number: 5327571188183)*

As visible-light-driven photocatalysts, silver-based compounds and composites have attracted considerable research interest for their synthesis, characterization, and performance. Silver has the ability to trap photogenerated electrons from semiconductors and allows holes to form hydroxyl radicals that degrade organic species [314]. In addition, photocatalytic activity silver-based can be enhanced by generating a local electric field [315]. Apart from AgCl, Ag_2WO_4, and Ag_2MoO_4, Ag-based photocatalysts showed an energy band gap of less than 3.0 eV, indicating that Ag-based Photocatalysts exhibited excellent visible light utilization. Apart from the bandgap, the photocatalytic activity is also affected by the CB and VB potentials. The band positions of Ag-based semiconductors both in the normal hydrogen electrode (NHE) are illustrated in Fig. 11.

There are three main reactive oxygen species in degradation pollutants: the superoxide radical (O_2^-), hydroxyl radical (·OH), and holes (h^+) with corresponding redox potential was as $O_2/{\cdot}O_2^-$ (−0.33 V vs. NHE), $H_2O/{\cdot}OH$ (2.75 V vs. NHE). Except for AgI and $AgXO_2$ (X = Al, Ga, In), most Ag-based semiconductors could not directly produce O_2^- via the reaction between photogenerated electrons and oxygen. The photogenerated electrons of silver-based photocatalyst were more negative than the evolution potential of H_2O_2 (O_2/H_2O_2: 0.68 V vs. NHE). Therefore, a direct one-step two-electron reduction route would produce H_2O_2, which may then be converted to ·OH to degrade pollutants [316].

Silver-based photocatalysts have demonstrated their ability to reform air pollutants as semiconductor photocatalysts. Table 15 summarizes several recent research about silver-based photocatalysts for removing air pollutants.

Table 15 Silver-based photocatalyst material for gas types

Photocatalyst	Preparation method	Experiment parameter	Main Product and Removal efficiency	References
$Ag/g\text{-}C_3N_4$	Simple photo deposition	Analyzer: Sabio 6042 Light source: xenon lamp	Products: NO (80% after 30 min)	[317]
$Ag/Ag_2O@TiO_2$	Chemical precipitation reduction method	VOC: toluene Analyzer: GC-FID Light source: UV lamp (50 W, 254 nm)	Products: Toluene (99.3% after 45 min)	[318]
$Ag_3PO_4/Ag/GdCrO_3$	Hydrothermal	VOC: toluene Analyzer: GC9790 Light source: 300 W xenon lamp	Product: Toluene (near 100% after 70 min)	[319]
$Ag_3PO_4/Ag/$ SrTiO3	Hydrothermal	Light source: xenon lamp	Product: Toluene >90% for 6 h	[320]
Ag/AgCl @ $BiOCl/Bi_{12}O_{17}C_{12}$	Deposition–precipitation	Analyzer: - Light source: 300 W Xe arc lamp	NOx (49.5% for 15 min)	[321]
$Ag_3PO_4/Cr\text{-}SrTiO_3$	Hydrothermal	Analyzer: GC-2014 Light source: 300 W Xe arc lamp	Isopropyl alcohol (97.00% for 5 min)	[322]
$Ag/g\text{-}C_3N_4$	Ultrasonic calcination	Analyzer: gas chromatography Light source: -	Ethyl acetate (81.3%)	[323]

5.5 *CN-Based Photocatalysis*

Graphitic carbon nitride ($g\text{-}C_3N_4$), an excellent photocatalyst, has elicited much interest because of its many remarkable properties. The $g\text{-}C_3N_4$ exhibits a bandgap of 2.7 eV, suitable position of the conduction band (CB) and valance band (VB) of −1.3 eV and +1.4 eV, respectively. Therefore, $g\text{-}C_3N_4$ is highly intriguing to absorb visible light, split water into H_2 and O_2, and reduce CO_2 into energy-rich compounds, and some organic and inorganic pollutants can be degraded into harmless compounds [324]. The $g\text{-}C_3N_4$ is composed of tri-s-triazine rings and nitrogen bridges. C and N atoms form covalent bonds, the tri-s-triazine rings polymers combine with hydrogen bonds, forming a graphite-like phase p-conjugated plane [325]. The tri-s-triazine ring structure and the high degree of condensation provide high stability against thermal

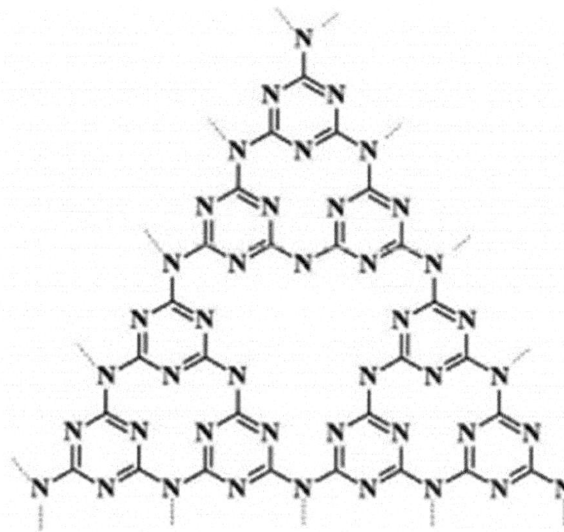

Fig. 12 Structure of g-C_3N_4 (tri-s-triazine ring) [326, 327]

(up to 600 °C in air) and chemical attack (e.g., acid, base, and organic solvents) [326]. (see Fig. 12)

Graphitic carbon nitride (g-C_3N_4) has garnered attention due to its stable allotrope. The g-C_3N_4 also simply derived by thermal polymerization of nitrogen-rich precursors like melamine, dicyandiamide, cyanamide, urea, thiourea [328]. Due to its wonderful characteristics such as high stability, nontoxicity, middle energy gap, inexpensive precursors, and easy preparation method, graphitic carbon nitride (g-C_3N_4) has been widely used in many photocatalytic applications. However, the pristine g-C_3N_4 has many shortcomings due to its lack of visible light response, few active sites, and rapid recombination of photoinduced carriers [329]. Therefore, a remarkable number of attempts have been made to overcome these limitations by improving g-C_3N_4 absorption efficiency in the visible area, its specific surface area, and the rate of charge migration.

The heterostructure strategy of g-C_3N_4 is one of the efficient ways to improve absorption efficiency in the visible area and specific surface area of a photocatalyst. The g-C_3N_4/Bi_2WO_6/NCQs Z-scheme heterojunction was synthesized through in situ calcination and physical deposition. According to the UV–vis spectroscopy analysis results, compared with pristine g-C_3N_4, g-C_3N_4/Bi_2WO_6/NCQs show enhanced light absorption capabilities. g-C_3N_4/Bi_2WO_6/NCQs have a specific surface area higher than that of pristine g-C_3N_4, which proves that g-C_3N_4/Bi_2WO_6/NCQs can provide more active sites to promote photocatalytic activity [330].

Doping with metal or non-metal also can improve visible light response and surface area of photocatalyst. C and O co-doped g-C_3N_4 (denoted as CNB) photocatalyst was fabricated through simple and facile method. The UV–vis absorption spectra compare with pristine CN show that The CNB sample exhibited much better light-harvesting ability, especially in the visible light region. The BET surface area of

CNB is 107.19 m^2/g, which is much higher than CN (41.47 m^2/g) [331]. As a potential and transformational photocatalyst, Graphitic Carbon Nitride (g-C_3N_4) might be used for reforming air pollutants. Different semiconductors have been attempted for photocatalytic CO_2 reduction using g-C_3N_4. N-doped C dot/CoAl-layered double hydroxide/g-C_3N_4 (NCD/LDH/CN) hybrid heterojunction photocatalyst was fabricated by a facile one-pot hydrothermal. Under simulated solar light, the photocatalyst showed high efficiency at reducing CO_2 with water into CH_4. In the NCD/LDH/CN hybrid photocatalyst, the optimum production rate was 25.69 mol g 1 h 1, and the quantum yield was 0.62%, with 99% selectivity for CH_4 [332].

The g-C_3N_4 doped with carbon quantum dots (CQDs) could also improves CO_2 adsorption, reaction kinetics, and CO_2 photoreduction pathways to produce CH_4. In comparison with pristine g-C_3N_4, CQDs/g-C_3N_4 can generate six times more CO and CH_4. CQDs/g-C_3N_4 can enhance the surface adsorption of nonpolar CO_2 molecules and alter the CO_2 photoreduction pathways, resulting in the production of a considerable amount of CH_4 (20.78 $mmol{\cdot}h^{-1}{\cdot}g^{-1}$) and CO (23.38 $mmol{\cdot}h^{-1}{\cdot}g^{-1}$) in the presence of H_2O. Under similar conditions, pristine g-C_3N_4 could only produce a fractional amount of CO (4.14 $mmol{\cdot}h^{-1}{\cdot}g^{-1}$) and H_2 (4.57 $mmol{\cdot}h^{-1}{\cdot}g^{-1}$) without generating CH_4 [333].

In addition to CO_2 reduction, g-C_3N_4 photocatalyst also shows an ability to remove NO gas. Pure g-C_3N_4 as a photocatalyst for removing NO at high temperatures (above 400 °C) has been shown to exhibit 15.06% of NO conversion [334]. Using urea (CN-U) and thiourea (CN-T) as precursors in the same conditions [335], prepared g-C_3N_4 with different structures using heat treatment. In comparison with CN-T, CN-U has a porous structure, high specific surface area, and a large bandgap. The result showed NO removal ratio for CN-U (40.4%) is higher than that of CN-T (34.9%) due to the different microstructures.

Several attempts have been made by researchers to increase the ability of g-C_3N_4 photocatalyst for NO removal. Photoreactivity of g-C_3N_4 such as the creation of nitrogen vacancies, doping g-C_3N_4 with metals or nonmetals, modification of g-C_3N_4 with carbon materials, and coupling g-C_3N_4 with other semiconductors to form a heterojunction. Flower-like g-C_3N_4 assembly from porous nanosheets with nitrogen vacancies was fabricated for NO removal. g-C_3N_4 has been synthesized by calcination melamine-cyanuric acid (MCA) supramolecular aggregates at 500 °C for different times (2–14 h). As a light source, a visible LED lamp ($\lambda \geq 400$ nm) was used to test the photoreactivity of g-C_3N_4 prepared via NO oxidation in a continuous-flow reactor. The flower-like g-C_3N_4 displays increased photocatalytic activity against NO oxidation under visible light illumination compared to bulk g-C_3N_4. Flower-like g-C_3N_4 was exhibited NO removal rate (59.7%), much higher than bulk g-C_3N_4 (45.8%) [336].

Wu in 2019 [337], improved the photocatalytic activity of g-C_3N_4 by repeated calcination of g-C_3N_4. The g-C_3N_4 nanosheets (g-C_3N_4-NSs) were fabricated after repeated calcination of bulk g-C_3N_4. After repeated calcination for seven times, the visible photoreactivity of g-C_3N_4/NSs toward NO oxidation increased from 20.6% to 35.8% in a continuous reactor. Doping g-C_3N_4 with other materials also increase the ability to remove NO. The Au-g-C_3N_4 nanocomposite was fabricated by in situ wet-chemical reduction method. Photocatalytic removal of NO with the Au-g-C_3N_4 composite was carried out under visible-light irradiation ($\lambda \geq 420$ nm). The NO conversion ratio increases from 24.0% on pure g-C_3N_4 to 41.0% on Au-g-C_3N_4 [338]. The g-C_3N_4 also shows a great potential for photocatalytic oxidation desulfurization (PODS) applications and attracting considerable attention from researchers. However, PCN still has limitations to achieve efficient PODS activity. Therefore, the research of g-C_3N_4 photocatalysts for desulfurization has been relatively rare.

Cerium fluoride (CeF_3) semiconductor was decorated the graphite carbonitride (g-C_3N_4) nanosheets to form heterojunction nanocomposites by hydrothermal method. Photocatalytic degradation of dibenzothiophene (DBT) was used to evaluate the effect of the CeF_3/g-C_3N_4 nanocomposite. Nanocomposite CeF_3/g-C_3N_4 demonstrated that the desulfurization rate reached 84.2% for DBT after 3 h [339].

CuO-ZnO@graphitic carbon nitride (Cu/Zn/g) triplex heterojunction nanocomposite is fabricated with ultrasound-assisted hydrothermal. According to the results with ultrasound-assisted desulfurization experiments, the best sulfur removal was obtained using Cu/Zn/g at a photocatalyst dose of 0.2 g/L, initial dibenzothiophene (DBT) concentration of 250 ppm, and H_2O_2 loading of 250 ppm at 298 K within 60 min treatment in the presence of ultrasonic waves (80 W/m^2) and light irradiation (150 W). An optimal desulfurization rate of 99.1% was achieved with Cu/Zn/g and g-C_3N_4 concentration of 10%, compared with the pristine CuO-ZnO catalyst (76.2%) [340] trying to improve by harvesting visible light, layered/defective structures, functional groups, appropriate/adjustable band positions, and Lewis base sites, which exist, and many studies are trying to improve by harvesting visible light, layered/defective structures, functional groups, appropriate/adjustable band positions, and Lewis base sites. Due to poor visible light usage, sluggish charge kinetics, and severe aggregation, photocatalyst PCNs are still ineffective in eliminating VOCs. Electronic architectures and heterostructure constructions in PCNs are rapidly evolving, allowing PCN photocatalysts to be used for VOC elimination in new ways. Some recent investigations on g-C_3N_4-based photocatalysts for VOC elimination are listed in Table 16.

5.6 Other Metal Oxide (MOx)

ZnO

Numerous studies have synthesized ZnO with and without dopants because zinc oxide is a low-cost photocatalytic material with excellent performance. However, it still has limitations due to high electron–hole recombination. Co doping zinc oxide with silver, platinum, titanium, and gold inhibited electron and hole recombination has been reported [346–349]. As mentioned by Abdelsamad et al., an Ag-doped zinc oxide sol–gel film performed well in organic degradation due to lower bandgap energy and increased surface area [350]. Furthermore, as mentioned by Lam et al., excessive silver content could reduce photocatalytic efficiency. The excessive doping

Table 16 g-C_3N_4-based photocatalyst for VOC removal

Photocatalyst	Experiment parameter	Main product	References
Porous g-C_3N_4	Reactor: 1.5 L Cat.a): 0.5 g Conc.b): 20 L VOC: HCHO Analyzer: Spectrophotometer Light Source: LED light	Products: CO_2, H_2O (HCHO): 56.32%	[341]
H3PW12O40/PCN	Reactor: 300 mL Cat.: 135 mg Conc.: 500 L VOC: m-xylene Analyzer: Agilent 1200 HPLC Light Source: 300 W Xe lamp	Products: BZH, CO_2 (m-xylene): 90.1%	[342]
Ni-gC_3N_4	VOC: toluene and nitrobenzene Analyzer: TOC analyzer	Product: CO_2 (toluene): 85.2% (nitrobenzene): 92.8%	[343]
Bi_2MoO_6/g-C_3N_4	Reactor: 25 mL Cat.: 40 mg Conc.: 0.1 ppm VOC: toluene Analyzer: GCMS-QP2010 Light Source: 300 W Xe lamp	Products: BZH (toluene): ≈90%	[344]
Pt-Cox/CN	Cat.: 0.1 g Conc.: 210 ppm VOC: toluene Analyzer: GC 9160 Light Source: 300 W Xe lamp	Product: CO_2 (toluene): 83%	[345]

concentrations cause the production of Ag precipitates, which coat the surface of particles and impede reaction sites [351].

In a study, ZnO was synthesized with the combustion method, doped with 2.5, 5, and 10% mol Ag. The doping concentrations will affect the material chemical properties, microstructure, and bandgap energy of their photocatalytic efficacy. Zinc oxide powders containing 2.5 mol% silver indicated the highest photocatalytic performance in the study. The improved efficiency was attributed to proper secondary phase concentration, particle sizes, and bandgap energy. Less than 5 mol% Ag secondary phase was produced with 2.5 mol% Ag doping. The presence of a small amount of secondary phase may operate as a charge carrier recombination inhibitor. The 2.5 mol% Ag-doped powder also has particle refinement and tiny cluster sizes, contributing to increased reactive sites and thus improved photocatalytic effects. The narrow bandgap in the ZnO powder containing 2.5 mol% Ag could accommodate charge carrier formation, resulting in enhanced photocatalytic activities, specifically dye degradation and antibacterial performance. The particles agglomerated into clusters ranging in size from 99 to 255 nm. The best groups were discovered in ZnO doped with 2.5 mol% Ag. Optical bandgap values varied from 3.40 to 3.22 eV, with a smaller bandgap seen in ZnO doped with 2.5 and 5 mol% Ag. Methylene blue photocatalytic degradation was most significant in ZnO with 2.5 mol% Ag. The antimicrobial activity of the powders followed the same pattern as the photocatalytic degradation. Fine cluster sizes, a low amount of Ag secondary phase, and a relatively narrow bandgap may explain the ZnO with 2.5 mol% Ag powder with outstanding photocatalytic and antibacterial activity [352].

SnO_2

On the other hand, photocatalysts based on SnO_2 are developed. Environmental science and energy use photocatalytic materials with intense photocatalytic activity and prospective future applications. These years, SnO_2 has attracted the attention of researchers because of its immense oxidation potential and inertness of chemicals, non-toxicity, resistance to corrosion, economic productivity, environmental preservation, and long-term stability. On the contrary, conventional SnO_2 photocatalytic effectiveness is far from expectations, which can be related to SnO_2 physical properties, which are: (i) wide bandgap of around 3.6 eV, as a result of solar insufficiency exposure and UV light absorption; (ii) severe recombination of photogenerated electrons and holes [353].

The structural architecture of photocatalysts that are based on SnO_2 can be adjusted in numerous ways, including doping, creation of a solid solution, and stoichiometric modification to enhance its photocatalytic performance. Pure SnO_2 cannot be stimulated via visible light because of its large bandgap, which is 3.6 eV. Impurity doping is a systematic and successful way to narrow SnO_2 bandgap and improve its light absorption. As a result of the interaction between impure ions, the impurity energy level will frequently appear in the bandgap under high doping values. This band of impurity energy reacts with the lattice of the host valence band (VB) or conduction band (CB) to produce an energy band tail above VB (or below CB), which will reduce the inherent bandgap. The bandgap structure could be enriched further by using the co-doping method with two or more specific impurities and reducing the isolated impurity states induced by single ion doping with carrier recombination centers [354].

On the other hand, introducing Sn atoms into oxygen vacancies of various valence states will form multiple defect centers, resulting in continual bandgap modulation [355, 356], and thus attain visible light absorption performance while avoiding impurity doping structural defects [357]. In contrast to doping, a solid solution is a uniform, single and solid crystal that is continuously generated from the mutual disintegration of at least two components under stable conditions. Furthermore, the bandgap between the reliable solution and the original semiconductors could be modified to improve photocatalytic performance.

The situation now for indoor air pollution is unfavorable, and gas-phase organics, particularly trichloroethylene (TCE), formaldehyde (HCHO), and aromatic compounds, have emerged as significant environmental concerns affecting human well-being. Oxidation of organic gases into CO_2 and H_2O by SnO_2-based photocatalysts has naturally attracted people's curiosity. For example, Pt@SnO_2 core–shell NPs degrade formaldehyde at 93.2% in 180 min [358]. In addition, Pt interaction with the metal oxide framework (MOF) improves formazan H_2O desorption, encouraging aldehyde oxidation [359]. Accordingly, SnO_2 was added to $ZnSn(OH)_6$, demonstrating very effective photocatalytic activity for removing aromatic chemicals from the gas environment via benzene degradation and mineralization [360]. During the 200-h test period, gaseous trichloroethylene (TCE) in the dry syngas circulation stream can be completely degraded to up to 400 ppm concentration by using ZnO/SnO_2 with no catalyst deactivation [361].

6 Conclusion

The HEPA Filter is an excellent filter for dealing with mold, dust, germs, viruses, and other gases in indoor and outdoor systems. Moreover, it is prevalent among allergy and asthma sufferers. These filters are made up of multiple layers of filter material and must filter out at least 99.97% of particles larger than 3 microns, as required by law. TiO_2 has been employed as a filter precursor in several investigations. In most photocatalyst applications, TiO_2 is the same chemical molecule used as a pigment. However, it serves an entirely different purpose in terms of air cleaning. Many hospital-grade filtration systems, like those in the TRIO Field Controls unit, use TiO_2 filters as many filters. The key to air purifier technology developments is adding numerous groups such as the transition metals, alkali metals, and post-transition metals to improve the filter function. Owing to technological advancements, this filter is ideal for germs, smoke, and smells. Air filtration is not always as straightforward as it appears. Air purifiers come in various shapes and sizes with various working methods. By learning how different filtration systems work, we may figure out which one is appropriate for humans.

References

1. Lu X, Chen Y, Yan W, Wang K, Zhou Y, Gao C (2022) Amphiphobic polytetrafluoroethylene membrane with a ring-on-string-like micro/nano structure for air purification. J Memb Sci 652(18):120476. https://doi.org/10.1016/j.memsci.2022.120476
2. Hu F, Guo Y (2021) Health impacts of air pollution in China. Front Environ Sci Eng 15(4):1–18
3. Oluwatoyin M, Sakugawa H (2020) Science of the total environment short communication a simple, inexpensive method for gas-phase singlet oxygen generation from sensitizer-impregnated Fi lters: potential application to bacteria/virus inactivation and pollutant degradation. Sci Total Environ 746:141186. https://doi.org/10.1016/j.scitotenv.2020.141186
4. He F, Muliane U, Weon S, Choi W (2020) Applied catalysis B : environmental substrate-specific mineralization and deactivation behaviors of TiO_2 as an air-cleaning photocatalyst. Appl Catal B Environ 275:119145. https://doi.org/10.1016/j.apcatb.2020.119145

5. Xu Y, Chen W, Jin R, Shen J, Smallbone K, Yan C, Hu L (2018) Experimental investigation of photocatalytic effects of concrete in air purification adopting entire concrete waste reuse model. J Hazard Mater 353:421–430. https://doi.org/10.1016/j.jhazmat.2018.04.030
6. Papailias I, Todorova N, Giannakopoulou T, Karapati S, Boukos N, Dimotikali D, Trapalis C (2018) Applied surface science enhanced NO_2 abatement by alkaline-earth modified g-C_3N_4 nanocomposites for efficient air purification. Appl Surf Sci 430(2):225–233. https://doi.org/10.1016/j.apsusc.2017.08.084
7. Papailias I, Todorova N, Giannakopoulou T, Karapati S, Boukos N, Dimotikali D, Trapalis C (2018) Enhanced NO_2 abatement by alkaline earth modified g-C_3N_4 nanocomposites for efficient air purification. Appl Surf Sci 430(2):225–233. https://doi.org/10.1016/j.apsusc.2017.08.084
8. Mavrikos A, Papoulis D, Todorova N, Papailias I, Trapalis C, Panagiotaras D (2021) Journal of photochemistry & photobiology, a: chemistry synthesis of Zn/Cu metal ion modified natural palygorskite clay–TiO_2 nanocomposites for the photocatalytic outdoor and indoor air purification. J Photochem Photobiol A Chem 2022(423):113568. https://doi.org/10.1016/j.jphotochem.2021.113568
9. Liu J, Cui S, Chen G, Zhang Y, Wang X, Wang Q, Gao P, Hang J (2021) Science of the total environment the in Fl uence of solar natural heating and NOx–O_3 photochemistry on Fl Ow and reactive pollutant exposure in 2D street canyons. Sci Total Environ 759(x):143527. https://doi.org/10.1016/j.scitotenv.2020.143527
10. Roegiers J, Denys S (2018) CFD-modelling of activated carbon fibers for indoor air purification. Chem Eng J 2019(365):80–87. https://doi.org/10.1016/j.cej.2019.02.007
11. Wu SY, Xu L, Xiao L (2020) Air purification and thermal performance of photocatalytic-trombe wall based on multiple physical fields coupling. Renew Energy 148:338–348. https://doi.org/10.1016/j.renene.2019.10.039
12. Han R, Andrews R, Rourke CO, Hodgen S, Mills A (2021) Photocatalytic air purification: effect of HNO_3 accumulation on NOx and VOC removal. Catal Today 380(x):105–113. https://doi.org/10.1016/j.cattod.2021.04.017
13. Ligotski R, Sager U, Schneiderwind U, Asbach C, Schmidt F (2018) Prediction of VOC adsorption performance for estimation of service life of activated carbon based filter media for indoor air purification. Build Environ 2019(149):146–156. https://doi.org/10.1016/j.buildenv.2018.12.001
14. Swamy GSNVKSN (2021) Heliyon development of an indoor air puri Fi cation system to improve ventilation and air quality. Heliyon 7(October):e08153. https://doi.org/10.1016/j.heliyon.2021.e08153
15. Mamaghani AH, Haghighat F, Lee CS (2020) Role of titanium dioxide (TiO_2) structural design/morphology in photocatalytic air purification. Appl Catal B Environ 269(February). https://doi.org/10.1016/j.apcatb.2020.118735
16. Le Pivert M, Kerivel O, Zerelli B, Leprince-Wang Y (2021) ZnO nanostructures based innovative photocatalytic road for air purification. J Clean Prod 318(March):128447. https://doi.org/10.1016/j.jclepro.2021.128447
17. Uyar C, Tokur ME, Perc D (2022) Effectiveness of air purifiers in intensive care units: an intervention study. J Hosp Infect 120:14–22. https://doi.org/10.1016/j.jhin.2021.10.011
18. Liu DT, Phillips KM, Speth MM, Besser G, Mueller CA, Sedaghat AR (2021) Portable HEPA purifiers to eliminate airborne SARS-CoV-2: a systematic review. Otolaryngol Head Neck Surg 166(4):615–622. https://doi.org/10.1177/01945998211022636
19. Chakladar A, Mbbs CGJ, Siu J, Mbbs MOH, Mbbs MK (2021) American journal of infection control microbial contamination of powered air purifying respirators (PAPR) used by healthcare staff during the COVID-19 pandemic: an in situ microbiological study. AJIC Am J Infect Control 49(6):707–712. https://doi.org/10.1016/j.ajic.2021.02.006
20. Kumar A, Singh J (2022) Applied surface science advances photoelectrochemical oxidation assisted air purifiers; perspective as potential tools to control indoor SARS-CoV-2 exposure. Appl Surf Sci Adv 9(March):100236. https://doi.org/10.1016/j.apsadv.2022.100236

21. Kaushik A (2021) Expert opinion on drug delivery manipulative magnetic nanomedicine: the future of COVID-19 pandemic/endemic therapy therapy. Expert Opin Drug Deliv 18(5):531–534. https://doi.org/10.1080/17425247.2021.1860938
22. Jayakantha DNPR, Bandara HMN, Gunawardana NM, Rajapakse RPVJ, Thilakarathne DS, Comini E, Gunawardhana N, Karunarathne SMML (2022) HardwareX design and construction of a low cost air purifier for killing harmful airborne microorganisms using a combination of a strong multi-directional electric-field and an ultra violet light. HardwareX 11:e00279. https://doi.org/10.1016/j.ohx.2022.e00279
23. Tiwari S, Juneja S, Ghosal A, Bandara N, Khan R, Wallen SL, Ramakrishna S, Kaushik A (2022) Antibacterial and antiviral high-performance nanosystems to mitigate new SARS-CoV-2 variants of concern. Curr Opin Biomed Eng 21:100363. https://doi.org/10.1016/j.cobme.2021.100363
24. Kang S, Hwang J (2022) Hierarchical ZnO nano-spines grown on a carbon fiber seed layer for efficient VOC removal and airborne virus and bacteria inactivation. J Hazard Mater 424(PA):127262. https://doi.org/10.1016/j.jhazmat.2021.127262
25. Zou X, Yuan C, Dong Y, Ge H, Ke J, Cui Y (2019) Lanthanum orthovanadate/bismuth oxybromide heterojunction for enhanced photocatalytic air purification and mechanism exploration. Chem Eng J 2020(379):122380. https://doi.org/10.1016/j.cej.2019.122380
26. Gai H, Liu X, Feng B, Gai C, Huang T, Xiao M (2020) An alternative scheme of biological removal of ammonia nitrogen from wastewater–Highly dispersed Ru cluster @ mesoporous TiO_2 for the catalytic wet air oxidation of low-concentration ammonia. Chem Eng J 2021(407):127082. https://doi.org/10.1016/j.cej.2020.127082
27. Yu B, He W, Li N, Zhou F, Shen Z, Chen H (2017) Experiments and kinetics of solar PCO for indoor air puri fi cation in PCO/TW system. Build Environ 115:130–146. https://doi.org/10.1016/j.buildenv.2017.01.026
28. Tiwari D, Lee S, Kim D (2022) Photocatalytic degradation of amoxicillin and tetracycline by template synthesized nano-structured Ce^{3+}@TiO_2 thin film catalyst. Environ Res 210(January):112914. https://doi.org/10.1016/j.envres.2022.112914
29. Hu Z, Xu T, Liu P, Oeser M (2020) Microstructures and optical performances of nitrogen-vanadium co-doped TiO_2 with enhanced purification efficiency to vehicle exhaust. Environ Res 2021(193):110560. https://doi.org/10.1016/j.envres.2020.110560
30. Zhong C, Xiong X (2021) Preparation of a composite coating film via vapor induced phase separation for air purification and real-time bacteria photocatalytic inactivation. Prog Org Coatings 161(July):106486. https://doi.org/10.1016/j.porgcoat.2021.106486
31. Gao X, Zheng K, Zhang Q, Cao X, Wu S, Su J (2021) Applied surface science self-assembly TiO_2-RGO/LDHs nanocomposite: photocatalysis of VOCs degradation in simulation air. Appl Surf Sci 2022(586):152882. https://doi.org/10.1016/j.apsusc.2022.152882
32. Tiwari D, Lee S, Kim D (2021) Photocatalytic degradation of amoxicillin and tetracycline by template synthesized nano-structured Ce^{3+}@TiO_2 thin film catalyst. Environ Res 2022(210):112914. https://doi.org/10.1016/j.envres.2022.112914
33. Xu T, Zheng H, Zhang P (218) Performance of an innovative VUV-PCO puri Fi Er with nanoporous TiO_2 Fi Lm for simultaneous elimination of VOCs and by-product ozone in indoor air 142(April):379–387. https://doi.org/10.1016/j.buildenv.2018.06.047
34. Kim M, Ph D, Jung H, Ph D, Park E, Ph D, Jurng J, Ph D (2020) Performance of an air purifier using a MnO_x/TiO_2 catalyst-coated filter for the decomposition of aldehydes, VOCs and ozone: an experimental study in an actual smoking room. Build Environ 186(June):107247. https://doi.org/10.1016/j.buildenv.2020.107247
35. Prakash J, Cho J, Kumar Y (2022) Micro and nano engineering photocatalytic TiO_2 nanomaterials as potential antimicrobial and antiviral agents: scope against blocking the SARS-COV-2 spread. Micro Nano Eng. 14:100100. https://doi.org/10.1016/j.mne.2021.100100
36. Liu S, Zou Q, Ma Y, Chi D, Chen R, Fang H, Hu W, Zhang K, Chen L (2022) Inorganica chimica acta metal-organic frameworks derived TiO_2/carbon nitride heterojunction photocatalyst with efficient catalytic performance under visible light. Inorg Chim Acta 536(January):120918. https://doi.org/10.1016/j.ica.2022.120918

37. Sanzone G, Zimbone M, Cacciato G, Ru F, Carles R, Privitera V (2018) Superlattices and microstructures 123(July):394–402. https://doi.org/10.1016/j.spmi.2018.09.028
38. Carabineiro AC, Lv K, Zhang L, Li Y, Li Q, Fan J (2021) Recent advances on bismuth-based photocatalysts: strategies and mechanisms 419 (March). https://doi.org/10.1016/j.cej.2021.129484
39. Ma H, He Y, Chen P, Wang H, Sun Y, Li J, Dong F, Xie G, Sheng J (2021) Ultrathin two-dimensional bi-based photocatalysts: synthetic strategies, surface defects, and reaction mechanisms 417(March). https://doi.org/10.1016/j.cej.2021.129305
40. Xu M, Yang J, Sun C, Liu L, Cui Y, Liang B (2019) Performance enhancement strategies of Bi-based photocatalysts: a review on recent progress. Chem Eng J 2020(389):124402. https://doi.org/10.1016/j.cej.2020.124402
41. Chen P, Dong F (2020) Bi-based photocatalysts for light-driven environmental and energy applications: structural tuning, reaction mechanisms, and challenges, no June, 1–31. https://doi.org/10.1002/eom2.12047
42. Zuarez-chamba M, Rajendran S, Herrera-robledo, M, Priya AK, Navas C (2022) Bi-based photocatalysts for bacterial inactivation in water: inactivation mechanisms, challenges, and strategies to improve the photocatalytic activity 209(January). https://doi.org/10.1016/j.envres.2022.112834
43. Ashfaq M, Talreja N, Chauhan D, Rodríguez CA, Mera AC, Ramalinga M (2022) Journal of industrial and engineering chemistry synthesis of reduced graphene oxide incorporated bimetallic (Cu/Bi) nanorods based photocatalyst materials for the degradation of gallic acid and bacteria. J Ind Eng Chem. https://doi.org/10.1016/j.jiec.2022.03.023
44. Boon C, Yong L, Wahab A (2016) A review of ZnO nanoparticles as solar photocatalysts: synthesis, mechanisms and applications. Renew Sustain Energy Rev 2018(81):536–551. https://doi.org/10.1016/j.rser.2017.08.020
45. Castillo C, Cabello G, Chornik B, Huentupil Y, Buono-core GE (2020) Characterization of photochemically grown Pd loaded WO_3 thin Fi Lms and its evaluation as ammonia gas sensor. J Alloys Compd 825:154166. https://doi.org/10.1016/j.jallcom.2020.154166
46. Tarar A, Peng C (2022) Jo Ur n Pr Oo F. J Environ Chem Eng 107796. https://doi.org/10.1016/j.jece.2022.107796
47. Yang J, Yu F, Chen A, Zhao S, Zhou Y, Zhang S, Sun T, Hu G (2022) Ur Na l P F. Adv Powder Mater 100045. https://doi.org/10.1016/j.apmate.2022.100045
48. Cheek E, Guercio V, Shrubsole C, Dimitroulopoulou S (2021) Science of the total environment portable air puri Fi cation: review of impacts on indoor air quality and health. Sci Total Environ 766:142585. https://doi.org/10.1016/j.scitotenv.2020.142585
49. Luo Y, Zhai F, Zhang Y, Chen Z, Ding M, Qin D, Yang J, Feng G, Li L (2021) A superfine glass fibre air filter with rapid response to photocatalytic antibacterial properties under visible light by loading RGO/ZnO
50. Organization, W. H. Air Pollution, April 20, 2022
51. Kampa M, Castanas E (2008) Human health effects of air pollution. Environ Pollut 151(2):362–367. https://doi.org/10.1016/j.envpol.2007.06.012
52. Organization WH (2021) WHO global air quality guidelines: particulate matter (PM2.5 and PM10), ozone, nitrogen dioxide, sulfur dioxide and carbon monoxide; World Health Organization
53. Organization WH (April 12, 2022) Household air pollution and health
54. Organization WH (April 12, 2022) Ambient (Outdoor) air pollution
55. Akimoto H (2003) Global air quality and pollution. Science 302(5651):1716–1719. https://doi.org/10.1126/science.1092666
56. Administration EP (April 20, 2020) Environmental protection administration-Taiwan air quality monitoring network
57. Annex VII government regulation no. 22 on environmental protection and management; central government of indonesia, 2021
58. Taiwan Air Quality Index (AQI) and Taiwan Air Pollution | IQAir, April 20, 2022
59. Indonesia Air Quality Index (AQI) and Air Pollution Information | IQAir, April 20, 2022

60. Wang M, Liu F, Zheng M (2021) Air quality improvement from COVID-19 lockdown: evidence from China. Air Qual Atmos Heal 14(4):591–604. https://doi.org/10.1007/s11869-020-00963-y
61. Naqvi HR, Datta M, Mutreja G, Siddiqui MA, Naqvi DF, Naqvi AR (2021) Improved air quality and associated mortalities in india under COVID-19 lockdown. Environ Pollut 268:115691. https://doi.org/10.1016/j.envpol.2020.115691
62. Pan S, Jung J, Li Z, Hou X, Roy A, Choi Y, Gao HO (2020) Air Quality implications of COVID-19 in california. Sustainability 12(17):7067. https://doi.org/10.3390/su12177067
63. Ravindra K, Singh T, Vardhan S, Shrivastava A, Singh S, Kumar P, Mor S (2022) COVID-19 pandemic: what can we learn for better air quality and human health? J Infect Public Health 15(2):187–198. https://doi.org/10.1016/j.jiph.2021.12.001
64. Ching J, Kajino M (2020) Rethinking air quality and climate change after COVID-19. Int J Environ Res Public Health 17(14):5167. https://doi.org/10.3390/ijerph17145167
65. Agarwal N, Meena CS, Raj BP, Saini L, Kumar A, Gopalakrishnan N, Kumar A, Balam NB, Alam T, Kapoor NR, Aggarwal V (2021) Indoor air quality improvement in COVID-19 pandemic: review. Sustain Cities Soc 70:102942. https://doi.org/10.1016/j.scs.2021.102942
66. Koppmann R (2007) Volatile organic compounds in the atmosphere, 1st ed. Blackwell Pub, Oxford, Ames, Iowa
67. Wang N, Ernle L, Bekö G, Wargocki P, Williams J (2022) Emission rates of volatile organic compounds from humans. Environ Sci Technol 56(8):4838–4848. https://doi.org/10.1021/acs.est.1c08764
68. BTEX: Risks and Control Measures. Phoslab Environmental Laboratories, June 8, 2016
69. CDC | Facts About Benzene, June 12, 2019
70. Benzene | 71-43-2. ChemicalBook, June 12, 2022
71. Graziano G (2006) Benzene solubility in water: a reassessment. Chem Phys Lett 429(1):114–118. https://doi.org/10.1016/j.cplett.2006.08.006
72. PubChem. Benzene, June 12, 2022
73. PubChem. Toluene, June 12, 2022
74. Toluene | 108-88-3. ChemicalBook, June 12, 2022
75. Filley CM, Halliday W, Kleinschmidt-DeMasters BK (2004) The effects of toluene on the central nervous system. J Neuropathol Exp Neurol 63(1):1–12. https://doi.org/10.1093/jnen/63.1.1
76. Toluene | Medical Management Guidelines | Toxic Substance Portal | ATSDR, June 12, 2022
77. PubChem. Ethylbenzene, June 12, 2022
78. Ethylbenzene | 100-41-4. ChemicalBook, June 12, 2022
79. Ethylbenzene | Toxicological Profile | ATSDR, June 12, 2022
80. Xylene | Chemical Compound | Britannica, June 12, 2022
81. Xylene | 1330-20-7. ChemicalBook, June 12, 2022
82. Xylenes | Medical Management Guidelines | Toxic Substance Portal | ATSDR, June 12, 2022
83. Safarvand D, Naser I, Samipourgiri M, Arjmand M (2020) Efficient photoelectrocatalytic degradation of BTEX using TiO_2/CuO/Cu_2O nanorod-array film as the photoanode and MWCNT/GO/graphite felt as the photocathode. Electrocatalysis 11(2):188–202. https://doi.org/10.1007/s12678-019-00576-9
84. Dixon WT, Norman ROC (1964) Electron spin resonance studies of oxidation. Part IV. Some benzenoid compounds. J Chem Soc 4850–4856
85. Denisov ET, Metelitsa DI (1968) Oxidation of benzene. Russ Chem Rev 37:1547–1566
86. Cheng ZW, Peng-fei S, Jiang YF, Yu JM, Chen JM (2013) Ozone-assisted UV254nm photodegradation of gaseous ethylbenzene and chlorobenzene: effects of process parameters, degradation pathways, and kinetic analysis. Chem Eng J 228:1003–1010. https://doi.org/10.1016/j.cej.2013.05.076
87. Kenley RA, Davenport JE, Hendry DG (1981) Gas-phase hydroxyl radical reactions. Products and pathways for the reaction of OH with aromatic hydrocarbons. J Phys Chem 85(19):2740–2746

88. Rakibuddina M, Ananthakrishnan R (2016) Effective photocatalytic dechlorination of 2,4-dichlorophenol by a novel graphene encapsulated ZnO/Co_3O_4 core-shell hybrid under visible light. Photochem Photobiol Sci 15:86–98
89. PubChem. Chloroform, June 12, 2022
90. Chloroform | 67-66-3. ChemicalBook, June 12, 2022
91. Chloroform | Chemical Compound | Britannica, June 12, 2022
92. Chloroform | NIOSH | CDC, June 12, 2020
93. Milieu (2004) Assesment of the effectiveness of european air quality policies and measures-case study 2 comparison of the EU and US air quality standards & planning requirements 6:1–3
94. Cohen LR, Peña LA, Seidl AJ, Olsen JM, Wekselbaum J, Hoggard PE (2009) The photocatalytic decomposition of chloroform by tetrachloroaurate(III). Monatshefte für Chemie-Chem Mon 140(10):1159–1165. https://doi.org/10.1007/s00706-009-0175-x
95. PubChem. Formaldehyde, June 12, 2022
96. Formaldehyde | 50-00-0. ChemicalBook, June 12, 2022
97. US EPA, O. Facts about formaldehyde, June 12, 2013
98. Formaldehyde in Your Home: What You Need to Know | Formaldehyde and Your Health | ATSDR, June 12, 2020
99. Formaldehyde-Cancer-Causing Substances-NCI, June 12, 2015
100. Wang MY, Lu YW, Wu F, Zhang XJ, Yang CX (2015) Photocatalytic decomposition of formaldehyde by combination of ozone and AC network with UV365nm, UV254nm and UV254+185nm. Procedia Eng 121:521–527. https://doi.org/10.1016/j.proeng.2015.08.1023
101. Methyl Ethyl Ketone | NIOSH | CDC, June 8, 2020
102. Agency USEP (June 8, 1989) Updated health effects assessment for methyl ethyl ketone | US EPA
103. Agency USEP (June 8, 2003) Toxicological review of methyl ethyl ketone | US EPA
104. Scientific T (June 8, 2020) Methyl ethyl ketone (MEK) material safety data sheet-ThermoFisher scientific
105. Shayegan Z, Lee C-S, Haghighat F (2019) Effect of surface fluorination of P25-TiO_2 coated on nickel substrate for photocatalytic oxidation of methyl ethyl ketone in indoor environments. J Environ Chem Eng 7(5):103390. https://doi.org/10.1016/j.jece.2019.103390
106. Kongboon R, Gheewala SH, Sampattagul S (2022) Greenhouse gas emissions inventory data acquisition and analytics for low carbon cities. J Clean Prod 343(February):130711. https://doi.org/10.1016/j.jclepro.2022.130711
107. Sharma K, Park YK, Nadda AK, Banerjee P, Singh P, Raizada P, Banat F, Bharath G, Jeong SM, Lam SS (2022) Emerging chemo-biocatalytic routes for valorization of major greenhouse gases (GHG) into industrial products: a comprehensive review. J Ind Eng Chem 109:1–20. https://doi.org/10.1016/j.jiec.2022.02.001
108. Vadikkeettil Y, Subramaniam Y, Murugan R, Ananthapadmanabhan PV, Mostaghimi J, Pershin L, Batiot-Dupeyrat C, Kobayashi Y (2022) Plasma assisted decomposition and reforming of greenhouse gases: a review of current status and emerging trends. Renew Sustain Energy Rev 161(February):112343. https://doi.org/10.1016/j.rser.2022.112343
109. Huang X, Hu Y, Zhou L, Lei J, Wang L, Zhang J (2021) Fabrication of CuS-modified inverse opal g-C_3N_4 photocatalyst with enhanced performance of photocatalytic reduction of CO_2. J CO_2 Util 54(July):101779. https://doi.org/10.1016/j.jcou.2021.101779
110. deRichter R, Caillol S (2011) Fighting global warming: the potential of photocatalysis against CO_2, CH_4, N_2O, CFCs, tropospheric O_3, BC and other major contributors to climate change. J Photochem Photobiol C Photochem Rev 12(1):1–19. https://doi.org/10.1016/j.jphotochemrev.2011.05.002
111. Shen H, Peppel T, Strunk J, Sun Z (2020) Photocatalytic reduction of CO_2 by metal-free-based materials: recent advances and future perspective. Sol RRL 4(8). https://doi.org/10.1002/solr.201900546
112. Ávila-López MA, Luévano-Hipólito E, Torres-Martínez LM (2022) Optimizing the CO_2 reduction to produce CH_3OH using flexible $NiMoO_4$ coatings as a photocatalyst. J Alloys Compd 918:165549. https://doi.org/10.1016/j.jallcom.2022.165549

113. Prajapati PK, Malik A, Nandal N, Pandita S, Singh R, Bhandari S, Saran S, Jain SL (2022) Morphology controlled fe and ni-doped CeO_2 nanorods as an excellent heterojunction photocatalyst for CO_2 reduction. Appl Surf Sci 588(February):152912. https://doi.org/10.1016/j.apsusc.2022.152912
114. Dai Y-M, Li C-Y, Ting W-H, Jehng J-M (2022) Plasmon $Ag/AgVO_3/TiO_2$-nanowires S-scheme heterojunction photocatalyst for CO_2 reduction. J Environ Chem Eng 10(May):108045. https://doi.org/10.1016/j.jece.2022.108045
115. Wang Y, Xu J, Wan J, Wang J, Wang L (2022) A tube-like dual Z-scheme indium oxide@indium phosphide/cuprous oxide photocatalyst based on metal-organic framework for efficient CO_2 reduction with water. J Colloid Interface Sci 616:532–538. https://doi.org/10.1016/j.jcis.2022.02.101
116. Chen P, Du T, Jia H, Zhou L, Yue Q, Wang H, Wang Y (2022) A novel Bi_2WO_6/Si heterostructure photocatalyst with fermi level shift in valence band realizes efficient reduction of CO_2 under visible light. Appl Surf Sci 585(February):152665. https://doi.org/10.1016/j.apsusc.2022.152665
117. Zhao J, Xiong Z, Wang J, Qiu Y, Liu P, Zhao Y, Zhang J (2022) $SnTa_2O_6$: a novel CO_2 reduction photocatalyst with nearly 100% CO selectivity. Chem Eng J 446(P3):137242. https://doi.org/10.1016/j.cej.2022.137242
118. Tang Z, Zhu F, Zhou J, Chen W, Wang K, Liu M, Wang N, Li N (2022) Monolithic NF@ZnO/Au@ZIF-8 photocatalyst with strong photo-thermal-magnetic coupling and selective-breathing effects for boosted conversion of CO_2 to CH_4. Appl Catal B Environ 309(2):121267. https://doi.org/10.1016/j.apcatb.2022.121267
119. Yu M, Lv X, Mahmoud Idris A, Li S, Lin J, Lin H, Wang J, Li Z (2022) Upconversion nanoparticles coupled with hierarchical $ZnIn_2S_4$ nanorods as a near-infrared responsive photocatalyst for photocatalytic CO_2 reduction. J Colloid Interface Sci 612:782–791. https://doi.org/10.1016/j.jcis.2021.12.197
120. Obalová L, Reli M, Lang J, Matějka V, Kukutschová J, Lacný Z, Kočí K (2013) Photocatalytic decomposition of nitrous oxide using TiO_2 and Ag-TiO_2 nanocomposite thin films. Catal Today 209:170–175. https://doi.org/10.1016/j.cattod.2012.11.012
121. Zhao K, Sun X, Wang C, Song X, Wang F, Li K, Ning P (2021) Supported catalysts for simultaneous removal of SO_2, NO_x, and Hg_0 from industrial exhaust gases: a review. Chin Chem Lett 32(10):2963–2974. https://doi.org/10.1016/j.cclet.2021.03.023
122. Lasek J, Yu YH, Wu JCS (2013) Removal of NO_x by photocatalytic processes. J Photochem Photobiol C Photochem Rev 14(1):29–52. https://doi.org/10.1016/j.jphotochemrev.2012.08.002
123. Cunningham J, Kelly JJ, Penny AL (1970) Reactions involving electron transfer at semiconductor surfaces. I. Dissociation of nitrous oxide over n-type semiconductors at 20°. J Phys Chem 74(9):1992–2000. https://doi.org/10.1021/j100704a029
124. Troppová I, Šihor M, Reli M, Ritz M, Praus P, Kočí K (2018) Unconventionally prepared TiO_2/g-C_3N_4 photocatalysts for photocatalytic decomposition of nitrous oxide. Appl Surf Sci 430:335–347. https://doi.org/10.1016/j.apsusc.2017.06.299
125. Kočí K, Reli M, Troppová I, Šihor M, Bajcarová T, Ritz M, Pavlovský J, Praus P (2019) Photocatalytic decomposition of N_2O by using nanostructured graphitic carbon nitride/zinc oxide photocatalysts immobilized on foam. Catalysts 9(9):1–13. https://doi.org/10.3390/catal9090735
126. Obalová L, Šihor M, Praus P, Reli M, Kočí K (2014) Photocatalytic and photochemical decomposition of N_2O on ZnS-MMT catalyst. Catal Today 230:61–66. https://doi.org/10.1016/j.cattod.2013.09.047
127. Wang L, Liu J, Song W, Wang H, Li Y, Liu J, Zhao Z, Tan J, Duan Z, Deng J (2019) Experimental and DFT insights of $BiVO_4$ as an effective photocatalytic catalyst for N_2O decomposition. Chem Eng J 366(January):504–513. https://doi.org/10.1016/j.cej.2019.02.038
128. Reli M, Troppová I, Šihor M, Pavlovský J, Praus P, Kočí K (2018) Photocatalytic decomposition of N_2O over g-C_3N_4/$BiVO_4$ composite. Appl Surf Sci 2019(469):181–191. https://doi.org/10.1016/j.apsusc.2018.10.255

129. Boningari T, Smirniotis PG (2016) Impact of nitrogen oxides on the environment and human health: Mn-based materials for the NOx abatement. Curr Opin Chem Eng.https://doi.org/10.1016/j.coche.2016.09.004
130. Delfino RJ, Staimer N, Gillen D, Tjoa T, Sioutas C, Fung K, George SC, Kleinman MT (2006) Personal and ambient air pollution is associated with increased exhaled nitric oxide in children with asthma. Environ Health Perspect 114(11). https://doi.org/10.1289/ehp.9141
131. Wang X, Su J, Chen H, Li GD, Shi Z, Zou H, Zou X (2017) Ultrathin In_2O_3 nanosheets with uniform mesopores for highly sensitive nitric oxide detection. ACS Appl Mater Interfaces 9(19). https://doi.org/10.1021/acsami.7b04395
132. Heo I, You YW, Lee JH, Schmieg SJ, Yoon DY, Kim CH (2020) Urealess NO_x reduction by carbon monoxide in simulated lean-burn exhausts. Environ Sci Technol 54(13). https://doi.org/10.1021/acs.est.9b07935
133. Seneque M, Can F, Duprez D, Courtois X (2016) NO_x selective catalytic reduction (NO_x-SCR) by urea: evidence of the reactivity of HNCO, including a specific reaction pathway for NO_x reduction involving NO^+ NO_2. ACS Catal 6(7). https://doi.org/10.1021/acscatal.6b00785
134. Akhundi A, Habibi-Yangjeh A, Abitorabi M, Rahim Pouran S (2019) Review on photocatalytic conversion of carbon dioxide to value-added compounds and renewable fuels by graphitic carbon nitride-based photocatalysts. Catal Rev Sci Enghttps://doi.org/10.1080/01614940.2019.1654224
135. Habibi-Yangjeh A, Asadzadeh-Khaneghah S, Feizpoor S, Rouhi A (2020) Review on heterogeneous photocatalytic disinfection of waterborne, airborne, and foodborne viruses: can we win against pathogenic viruses? J Colloid Interface Sci. https://doi.org/10.1016/j.jcis.2020.07.047
136. Dong G, Zhao L, Wu X, Zhu M, Wang F (2019) Photocatalysis removing of NO based on modified carbon nitride: the effect of celestite mineral particles. Appl Catal B Environ 245. https://doi.org/10.1016/j.apcatb.2019.01.013
137. Li J, Dong X, Sun Y, Jiang G, Chu Y, Lee SC, Dong F (2018) Tailoring the rate-determining step in photocatalysis via localized excess electrons for efficient and safe air cleaning. Appl Catal B Environ 239. https://doi.org/10.1016/j.apcatb.2018.08.019
138. Li X, Dong G, Guo F, Zhu P, Huang Y, Wang C (2020) Enhancement of photocatalytic NO removal activity of g-C_3N_4 by modification with illite particles. Environ Sci Nano 7(7). https://doi.org/10.1039/d0en00415d
139. Rodriguez-Rivas F, Pastor A, Barriga C, Cruz-Yusta M, Sánchez L, Pavlovic I (2018) Zn-Al layered double hydroxides as efficient photocatalysts for NO_x abatement. Chem Eng J 346. https://doi.org/10.1016/j.cej.2018.04.022
140. Duan Y, Zhang M, Wang L, Wang F, Yang L, Li X, Wang C (2017) Plasmonic Ag-TiO_{2-x} nanocomposites for the photocatalytic removal of NO under visible light with high selectivity: the role of oxygen vacancies. Appl Catal B Environ 204. https://doi.org/10.1016/j.apcatb.2016.11.023
141. Pirhashemi M, Habibi-Yangjeh A, Rahim Pouran S (2018) Review on the criteria anticipated for the fabrication of highly efficient ZnO-based visible-light-driven photocatalysts. J Ind Eng Chem 62:1–25. https://doi.org/10.1016/j.jiec.2018.01.012
142. Shekofteh-Gohari M, Habibi-Yangjeh A, Abitorabi M, Rouhi A (2018) Magnetically separable nanocomposites based on ZnO and their applications in photocatalytic processes: a review. Crit Rev Environ Sci Technol 48(10–12). https://doi.org/10.1080/10643389.2018.1487227
143. Hu J, Chen D, Li N, Xu Q, Li H, He J, Lu J (2017) In situ fabrication of $Bi_2O_2CO_3/MoS_2$ on carbon nanofibers for efficient photocatalytic removal of NO under visible-light irradiation. Appl Catal B Environ 217. https://doi.org/10.1016/j.apcatb.2017.05.088
144. Huang Y, Zhu D, Zhang Q, Zhang Y, Cao JJ, Shen Z, Ho W, Lee SC (2018) Synthesis of a $Bi_2O_2CO_3/ZnFe_2O_4$ heterojunction with enhanced photocatalytic activity for visible light irradiation-induced NO removal. Appl Catal B Environ 234. https://doi.org/10.1016/j.apcatb.2018.04.039

145. Li Y, Lv K, Ho W, Zhao Z, Huang Y (2017) Enhanced visible-light photo-oxidation of nitric oxide using bismuth-coupled graphitic carbon nitride composite heterostructures. Cuihua Xuebao/Chin J Catal 38(2). https://doi.org/10.1016/S1872-2067(16)62573-1
146. Liu D, Chen D, Li N, Xu Q, Li H, He J, Lu J (2020) Surface engineering of g-C_3N_4 by Stacked BiOBr sheets rich in oxygen vacancies for boosting photocatalytic performance. Angew Chemie Int Ed 59(11). https://doi.org/10.1002/anie.201914949
147. Akhundi A, Badiei A, Ziarani GM, Habibi-Yangjeh A, Muñoz-Batista MJ, Luque R (2020) Graphitic carbon nitride-based photocatalysts: toward efficient organic transformation for value-added chemicals production. Mol Catalhttps://doi.org/10.1016/j.mcat.2020.110902
148. Asadzadeh-Khaneghah S, Habibi-Yangjeh A (2020) g-C_3N_4/carbon dot-based nanocomposites serve as efficacious photocatalysts for environmental purification and energy generation: a review. J Clean Prodhttps://doi.org/10.1016/j.jclepro.2020.124319
149. Giannakopoulou T, Papailias I, Todorova N, Boukos N, Liu Y, Yu J, Trapalis C (2017) Tailoring the energy band gap and edges' potentials of g-C_3N_4/TiO_2 composite photocatalysts for NO_x removal. Chem Eng J 310. https://doi.org/10.1016/j.cej.2015.12.102
150. Hu X, Wang Y, Ling Z, Song H, Cai Y, Li Z, Zu D, Li C (2021) Ternary g-C_3N_4/TiO_2/Ti_3C2 MXene S-scheme heterojunction photocatalysts for NOx removal under visible light. Appl Surf Sci 556. https://doi.org/10.1016/j.apsusc.2021.149817
151. Kapteijn F, Rodriguez-Mirasol J, Moulijn JA (1996) Heterogeneous catalytic decomposition of nitrous oxide. Appl Catal B. https://doi.org/10.1016/0926-3373(96)90072-7
152. Krawczyk K, Mlotek M (2001) Combined plasma-catalytic processing of nitrous oxide. Appl Catal B Environ 30(3–4). https://doi.org/10.1016/S0926-3373(00)00243-5
153. Oi J, Obuchi A, Bamwenda GR, Ogata A, Yagita H, Kushiyama S, Mizuno K (1997) Decomposition of nitrous oxide over supported rhodium catalysts and dependency on feed gas composition. Appl Catal B Environ 12(4). https://doi.org/10.1016/S0926-3373(96)00079-3
154. Li Y, Armor JN (1992) Catalytic decomposition of nitrous oxide on metal exchanged zeolites. Appl Catal B Environ 1(3). https://doi.org/10.1016/0926-3373(92)80019-V
155. Yuzaki K, Yarimizu T, Aoyagi K, Ito SI, Kunimori K (1998) Catalytic decomposition of N_2O over supported Rh catalysts: effects of supports and Rh dispersion. Catal Today 45(1–4). https://doi.org/10.1016/S0920-5861(98)00259-4
156. Zeng HC, Pang XY (1997) Catalytic decomposition of nitrous oxide on alumina-supported ruthenium catalysts Ru/Al_2O_3. Appl Catal B Environ 13(2). https://doi.org/10.1016/S0926-3373(96)00096-3
157. Ohnishi C, Iwamoto S, Inoue M (2008) Direct decomposition of nitrous oxide in the presence of oxygen over iridium catalyst supported on alumina. Chem Eng Sci 63(20). https://doi.org/10.1016/j.ces.2007.08.011
158. Wang XF, Zeng HC (1998) Decomposition of water-containing nitrous oxide gas using Ru/Al_2O_3 catalysts. Appl Catal B Environ 17(1–2). https://doi.org/10.1016/S0926-3373(98)00004-6
159. Satsuma A, Maeshima H, Watanabe K, Suzuki K, Hattori T (2000) Effects of methane and oxygen on decomposition of nitrous oxide over metal oxide catalysts. Catal Today 63(2–4). https://doi.org/10.1016/S0920-5861(00)00478-8
160. Slimen H, Ochiai T, Nakata K, Murakami T, Houas A, Morito Y, Fujishima A (2012) Photocatalytic decomposition of cigarette smoke using a TiO_2-impregnated titanium mesh filter. Ind Eng Chem Res 51(1). https://doi.org/10.1021/ie2016762
161. Talhout R, Opperhuizen A, van Amsterdam JGC (2007) Role of acetaldehyde in tobacco smoke addiction. Eur Neuropsychopharmacol. https://doi.org/10.1016/j.euroneuro.2007.02.013
162. Wang B, Hang Ho SS, Fai Ho K, Huang Y, Sing Chan C, Yau Feng NS, Sai Ip SH (2012) An environmental chamber study of the characteristics of air pollutants released from environmental tobacco smoke. Aerosol Air Qual Res 12(6). https://doi.org/10.4209/aaqr.2011.11.0221
163. Kotzias D, Geiss O (2007) Tobacco, cigarettes and cigarette smoke

164. Guo X, Yao Y, Zhao H, Chi C, Zeng F, Qian F, Liu Z, Huo L, Lv Y (2021) Environmental impacts of functional fillers in polylactide (PLA)-based bottles using life cycle assessment methodology. Sci Total Environ 788:147852. https://doi.org/10.1016/j.scitotenv.2021.147852
165. Townsend CL, Maynard RL (2002) Effects on health of prolonged exposure to low concentrations of carbon monoxide. Occup Environ Med 59(10):708–711. https://doi.org/10.1136/oem.59.10.708
166. Borgerding M, Klus H (2005) Analysis of complex mixtures-cigarette smoke. Exp Toxicol Pathol 57(SUPPL. 1):43–73. https://doi.org/10.1016/j.etp.2005.05.010
167. Avino P, Scungio M, Stabile L, Cortellessa G, Buonanno G, Manigrasso M (2018) Second-hand aerosol from tobacco and electronic cigarettes: evaluation of the smoker emission rates and doses and lung cancer risk of passive smokers and vapers. Sci Total Environ 642:137–147. https://doi.org/10.1016/j.scitotenv.2018.06.059
168. Chi MC, Guo SE, Hwang SL, Chou CT, Lin CM, Lin YC (2017) Exposure to indoor particulate matter worsens the symptoms and acute exacerbations in chronic obstructive pulmonary disease patients of southwestern taiwan: a pilot study. Int J Environ Res Public Health 14(1):1–11. https://doi.org/10.3390/ijerph14010004
169. Ngoc LTN, Park D, Lee Y, Lee YC (2017) Systematic review and meta-analysis of human skin diseases due to particulate matter. Int J Environ Res Public Health 14(12). https://doi.org/10.3390/ijerph14121458
170. Tagliabue G, Borgini A, Tittarelli A, Van Donkelaar A, Martin RV, Bertoldi M, Fabiano S, Maghini A, Codazzi T, Scaburri A, Favia I, Cau A, Barigelletti G, Tessandori R, Contiero P (2016) Atmospheric fine particulate matter and breast cancer mortality: a population-based cohort study. BMJ Open 6(11):1–6. https://doi.org/10.1136/bmjopen-2016-012580
171. Zhang C, Meng Q, Zhang X, Wu S, Wang S, Chen R, Li X (2016) Role of astrocyte activation in fine particulate matter-enhancement of existing ischemic stroke in sprague-dawley male rats. J Toxicol Environ Health Part A Curr Issues 79(9–10):393–401. https://doi.org/10.1080/15287394.2016.1176615
172. Kim KH, Kabir E, Kabir S (2015) A review on the human health impact of airborne particulate matter. Environ Int 74:136–143. https://doi.org/10.1016/j.envint.2014.10.005
173. Manisalidis I, Stavropoulou E, Stavropoulos A, Bezirtzoglou E (2020) Environmental and health impacts of air pollution: a review. Front Public Heal 8(February):1–13. https://doi.org/10.3389/fpubh.2020.00014
174. Kim M, Jung H, Park E, Jurng J (2020) Performance of an air purifier using a MnO_x/TiO_2 catalyst-coated filter for the decomposition of aldehydes, VOCs and ozone: an experimental study in an actual smoking room. Build Environ 186(June):107247. https://doi.org/10.1016/j.buildenv.2020.107247
175. Wang Y, Zhao J, Chen M, Huang X, Xu Y (2021) Journal of colloid and interface science fast organic degradation on Ti- and Bi-based photocatalysts via co- deposited Pt and Ni_3 $(PO_4)_2$ nanoparticles. J Colloid Interface Sci 600:629–638. https://doi.org/10.1016/j.jcis.2021.05.091
176. Kim M, Park E, Jurng J (2018) Oxidation of gaseous formaldehyde with ozone over MnO_x/TiO_2 catalysts at room temperature (25 °C). Powder Technol 325:368–372. https://doi.org/10.1016/j.powtec.2017.10.031
177. Ochiai T, Hayashi Y, Ito M, Nakata K, Murakami T, Morito Y, Fujishima A (2012) An effective method for a separation of smoking area by using novel photocatalysis-plasma synergistic air-cleaner. Chem Eng J 209:313–317. https://doi.org/10.1016/j.cej.2012.07.139
178. Rosli NS, Abdullah CAC, Hazan R (2018) Synthesis, characterization and investigation of photocatalytic activity of nano-titania from natural ilmenite with graphite for cigarette smoke degradation. Results Phys. 11(September):72–78. https://doi.org/10.1016/j.rinp.2018.08.032
179. Bejan A (2016) Advanced engineering thermodynamics. https://doi.org/10.1002/9781119245964
180. Hanrahan G (2012) Air pollutants and associated chemical and photochemical processes. Key Concepts Environ Chem 215–242. https://doi.org/10.1016/b978-0-12-374993-2.10007-x
181. Zhang JJ, Wei Y, Fang Z (2019) Ozone pollution: a major health hazard worldwide. Front Immunol 10(Oct):1–10. https://doi.org/10.3389/fimmu.2019.02518

182. Cao R, Zhang P, Liu Y, Zheng X (2019) Ammonium-treated birnessite-type MnO_2 to increase oxygen vacancies and surface acidity for stably decomposing ozone in humid condition. Appl Surf Sci 495(August):143607. https://doi.org/10.1016/j.apsusc.2019.143607
183. Boyjoo Y, Sun H, Liu J, Pareek VK, Wang S (2017) A review on photocatalysis for air treatment: from catalyst development to reactor design. Chem Eng J 310:537–559. https://doi.org/10.1016/j.cej.2016.06.090
184. Wang L, Guan J, Han H, Yao M, Kang J, Peng M, Wang D, Xu J, Hao J (2021) Enhanced photocatalytic removal of ozone by a new chlorine-radical-mediated strategy. Appl Catal B Environ 2022(306):121130. https://doi.org/10.1016/j.apcatb.2022.121130
185. Lu Y, Zhao X, Wang M, Yang Z, Zhang XJ, Yang C (2014) Feasibility analysis on photocatalytic removal of gaseous ozone in aircraft cabins. Build Environ 81. https://doi.org/10.1016/j.buildenv.2014.05.024
186. Zouzelka R, Rathousky J (2017) Photocatalytic abatement of NO_x pollutants in the air using commercial functional coating with porous morphology. Appl Catal B Environ 217:466–476. https://doi.org/10.1016/j.apcatb.2017.06.009
187. Huang Y, Zhang J, Wang Z, Liu Y, Wang P, Cao JJ, Ho W (2020) g-C_3N_4/TiO_2 composite film in the fabrication of a photocatalytic air-purifying pavements. Sol RRL 4(8). https://doi.org/10.1002/solr.202000170
188. Yang S, Lei G, Xu H, Xu B, Li H, Lan Z, Wang Z, Gu H (2019) A DFT study of CO adsorption on the pristine, defective, in-doped and sb-doped graphene and the effect of applied electric field. Appl Surf Sci 480. https://doi.org/10.1016/j.apsusc.2019.02.244
189. Chen T, Hao Q, Yang W, Xie C, Chen D, Ma C, Yao W, Zhu Y (2018) A honeycomb multilevel structure Bi_2O_3 with highly efficient catalytic activity driven by bias voltage and oxygen defect. Appl Catal B Environ 237. https://doi.org/10.1016/j.apcatb.2018.05.044
190. Nie J, Hassan QU, Jia Y, Gao J, Peng J, Lu J, Zhang F, Zhu G, Wang Q (2020) La-doped $ZnWO_4$ nanorods with enhanced photocatalytic activity for NO removal: effects of la doping and oxygen vacancies. Inorg Chem Front 7(2). https://doi.org/10.1039/c9qi01152h
191. Han J, Zhu G, Hojamberdiev M, Peng J, Zhang X, Liu Y, Ge B, Liu P (2015) Rapid adsorption and photocatalytic activity for rhodamine B and Cr(VI) by ultrathin BiOI nanosheets with highly exposed {001} facets. New J Chem 39(3). https://doi.org/10.1039/c4nj01765j
192. Rao F, Huang Y, Zhu G, Hojamberdiev M, Zhang W, Li S, Gao J, Zhang F, Huang Y (2019) Uniform Zn^{2+}-doped BiOI microspheres assembled by ultrathin nanosheets with tunable oxygen vacancies for super-stable removal of NO. J Phys Chem C 123(26):16268–16280. https://doi.org/10.1021/acs.jpcc.9b03961
193. Wang S, Wang K, Du J (2020) The structure, function, and mechanisms of action of enterovirus non-structural protein 2C. 11(December). https://doi.org/10.3389/fmicb.2020.615965
194. Guo J, Xiong Y, Kang T, Xiang Z, Qin C (2020) Bacterial community analysis of floor dust and HEPA filters in air purifiers used in office rooms in. Sci Rep1–11. https://doi.org/10.1038/s41598-020-63543-1
195. Lee SA, Liu F, Riordan SM, Lee CS, Zhang L (2019Global investigations of fusobacterium nucleatum in human colorectal cancer. Front Oncol 9(July)1–11https://doi.org/10.3389/fonc.2019.00566
196. Surata IW, Nindhia TGT, Yolanda WE (2020) Grain size effect on tensile and flexural strength of particulate composites reinforced with acropora waste. Mater Today Proc 22:156–161. https://doi.org/10.1016/j.matpr.2019.08.031
197. Max A, Schumm M (2021) Filtering facepiece respirator (N95 Respirator) reprocessing a systematic review. Jama 325(13):1296–1317. https://doi.org/10.1001/jama.2021.2531
198. Young KD (2007) Bacterial morphology: why have different shapes? Curr Opin Microbiol 10(6):596–600. https://doi.org/10.1016/j.mib.2007.09.009
199. Mohanna MT (2016) Morphology, and classification of bacteria. Res Gate 225–238
200. Constantino MA, Jabbarzadeh M, Fu HC, Bansil R (2016) Helical and rod-shaped bacteria swim in helical trajectories with little additional propulsion from helical shape. Sci Adv 2(11). https://doi.org/10.1126/sciadv.1601661

201. Živić T (2021) Encyclopædia Britannica Online. Stud Lexicogr 14(27). https://doi.org/10.33604/sl.14.27.6
202. Karami A, Sarshar M, Ranjbar R, Zanjani RS (2014) The phylum spirochaetaceae. In: The prokaryotes: other majo lineages of bacteria and the archaea. vol. 9783642389542. https://doi.org/10.1007/978-3-642-389542_156
203. Han YW, Redline RW, Li M, Yin L, Hill GB, McCormick TS (2004) Fusobacterium nucleatum induces premature and term stillbirths in pregnant mice: implication of oral bacteria in preterm birth. Infect Immun 72(4):2272–2279. https://doi.org/10.1128/IAI.72.4.2272-2279.2004
204. Ebadzadsahrai G, Keppler EAH, Soby SD (2020) Inhibition of fungal growth and induction of a novel volatilome in response to chromobacterium vaccinii volatile organic compounds. Front Microbiol 11(May). https://doi.org/10.3389/fmicb.2020.01035
205. Ramos S, Silva V et al (2020) Escherichia coli as commensal and pathogenic bacteria among food-producing animals: health implications of extended spectrum β-lactamase (ESBL) production. Anim Open Access J MDPI 1. https://doi.org/10.3390/ani10122239
206. Li X, Huang T, Xu K, Li C, Li Y (2019) Molecular characteristics and virulence gene profiles of staphylococcus aureus isolates in hainan, China, 1–12
207. Madsen AM, Kurdi I, Feld L, Tendal K (2018) Airborne MRSA and total staphylococcus aureus as associated with particles of different sizes on pig farms. Ann Work Expo Health 62(8):966–977https://doi.org/10.1093/annweh/wxy065
208. Watson R, Oldfield M, Bryant JA, Riordan L, Hill HJ, Watts JA, Alexander MR, Cox MJ, Stamataki Z, Scurr DJ (2022) Efficacy of antimicrobial and anti-viral coated air filters to prevent the spread of airborne pathogens. Sci Rep (0123456789):1–12https://doi.org/10.1038/s41598-022-06579-9
209. Mallakpour S, Azadi E, Hussain CM (2022) Fabrication of air filters with advanced filtration performance for removal of viral aerosols and control the spread of COVID-19. Adv Colloid Interface Sci 303(January). https://doi.org/10.1016/j.cis.2022.102653
210. Widyastuti E, Hsu J (2022) Insight on photocatalytic and photoinduced antimicrobial properties of ZnO thin films deposited by HiPIMS through thermal oxidation
211. Iqbal M, Adlim M, Maulana I, Suhartono S, Hayati Z, Hana N, Abu H (2022) Green synthesis of chitosan-stabilized silver- colloidal nanoparticles immobilized on white-silica-gel beads and the antibacterial activities in A. Arab J Chem 15(2):103596. https://doi.org/10.1016/j.arabjc.2021.103596
212. Liu H, Chen Z, Chen G, Liu H, Huang J, Mao J, Chen Z, Chen G, Lai Y (2019) Transparent antibacterial nanofiber air filters with highly efficient moisture resistance for sustainable particulate matter capture transparent antibacterial nanofiber air filters with highly efficient moisture resistance for sustainable particulate matt. Iscience 19:214–223. https://doi.org/10.1016/j.isci.2019.07.020
213. Ma S, Zhang M, Nie J, Tan J, Yang B, Song S (2019) Design of double-component metal–organic framework air filters with PM 2.5 capture, gas adsorption and antibacterial capacities. Carbohydr Polym 203:415–422. https://doi.org/10.1016/j.carbpol.2018.09.039. (September 2018)
214. Ju Y, Han T, Yin J, Li Q, Chen Z, Wei Z, Zhang Y, Dong L (2021) Science of the total environment bumpy structured nano fi brous membrane as a highly efficient air filter with antibacterial and antiviral property. Sci Total Environ 777:145768. https://doi.org/10.1016/j.scitotenv.2021.145768
215. Claudia A, Bortolassi C, Nagarajan S, De Araújo B, Giovana V, Lopes M, Huon V, Soussan L, Cornu D, Miele P, Bechelany M (2019) Materials science & engineering cefficient nanoparticles removal and bactericidal action of electrospun nano fibers membranes for air filtration. Mater Sci Eng C 102(April):718–729https://doi.org/10.1016/j.msec.2019.04.094
216. Xiao Y, Wang Y, Zhu W, Yao J, Sun C, Militky J, Venkataraman M, Zhu G (2020) Development of tree-like nanofibrous air filter with durable antibacterial property. Sep Purif Technol 2021(259):118135. https://doi.org/10.1016/j.seppur.2020.118135
217. Tian G, Huang Z, Wang H, Cui C (2022) Applied clay science polycaprolactone nanofiber membrane modified with halloysite and ZnO for anti-bacterial and air filtration. Appl Clay Sci 223(March):106512. https://doi.org/10.1016/j.clay.2022.106512

218. Yang D, Zhu Y, Li J, Yue Z, Zhou J, Wang X (2022) International journal of biological macromolecules degradable, antibacterial and ultrathin filtrating electrospinning membranes of Ag-MOFs/poly (L-Lactide) for air pollution control and medical protection. Int J Biol Macromol 212(January):182–192. https://doi.org/10.1016/j.ijbiomac.2022.05.112
219. Zhang S, Ye J, Sun Y, Kang J, Liu J, Wang Y, Li Y (2020) Electrospun Fi brous mat based on silver (I) metal-organic frameworks- polylactic acid for bacterial killing and antibiotic-free wound dressing. Chem Eng J 390(January):124523. https://doi.org/10.1016/j.cej.2020.124523
220. Liu Y, Li S, Lan W, Hossen A, Qin W (2021) Electrospun antibacterial and antiviral poly (ε-caprolactone)/zein/Ag bead-on-string membranes and its application in air filtration. Mater Today Adv 12:100173. https://doi.org/10.1016/j.mtadv.2021.100173
221. Qin M, Liu D, Meng X, Dai Z, Zhu S, Wang N, Yan X (2021) Materials letters: X electrospun polyvinyl butyral/berberine membranes for antibacterial air filtration. Mater Lett X 10:100074. https://doi.org/10.1016/j.mlblux.2021.100074
222. Jeong SB, Heo KJ, Lee BU (2020) Antimicrobial air filters using natural sea salt particles for deactivating airborne bacterial particles
223. Tapia R, García V, Mena J, Bucarey S, Medina RA, Neira V (2018) Infection of novel reassortant H1N2 and H3N2 swine influenza a viruses in the guinea pig model. Vet Res 1–8. https://doi.org/10.1186/s13567-018-0572-4
224. Baudon E, Chu DKW, Tung DD, Nga PT, Vu H, Phuong M, Le N, Hang K, Thanh LT, Thuy NT, Khanh NC, Mai LQ (2018) Swine in Fl uenza viruses in northern vietnam in 2013–2014. https://doi.org/10.1038/s41426-018-0109-y
225. Zhou J, Wei J, Choy K, Fun S, Rowlands DK, Yu D, Wu C (2018) Defining the sizes of airborne particles that mediate influenza transmission in ferrets. 115 (10). https://doi.org/10.1073/pnas.1716771115
226. Jhung MA, Epperson S, Biggerstaff M, Allen D, Balish A, Barnes N, Beaudoin A, Berman L, Bidol S, Blanton L, Blythe D, Brammer L, Mello TD, Danila R, Davis W, De Fijter S, Diorio M, Durand LO, Emery S, Fowler B, Garten R, Grant Y, Greenbaum A, Gubareva L, Havers F, Haupt T, House J, Ibrahim S, Jiang V, Jain S, Jernigan D, Kazmierczak J, Klimov A, Lindstrom S, Longenberger A, Lucas P, Lyn R, Mcmorrow M, Moll M, Morin C, Ostroff S, Page SL, Park SY, Peters S, Quinn C, Reed C, Richards S, Scheftel J, Simwale O, Shu B, Soyemi K, Stauffer J, Steffens C, Su S, Torso L, Uyeki TM, Vetter S, Villanueva J, Wong KK, Shaw M, Bresee JS, Cox N, Finelli L (2013) Outbreak of variant in Fl uenza a (H3N2) virus in the United States 57 (Cdc):1703–1712. https://doi.org/10.1093/cid/cit649
227. Abdelrahman Z, Li M, Wang X (2020) Comparative review of SARS-CoV-2, SARS-CoV, MERS-CoV, and influenza a respiratory viruses. Front. Immunol. 11(September). https://doi.org/10.3389/fimmu.2020.552909
228. Al-dalawi L, Dunham SP, Rauch C, Le SB, Rauch C (2022) Lipid biophysics and/or soft matter- inspired approach for controlling enveloped virus infectivity. J R Soc Interface (20210943):19. https://doi.org/10.1098/rsif.2021.0943
229. Yates MV (2014) Enterovirus, 2nd ed. Elsevier. https://doi.org/10.1016/B978-0-12-415846-7.00025-1
230. Hershenson M, Id O Article, O. DR. marc hershenson (Orcid ID: 0000-0001-9436-5593) Article Type: Original Article. 0–1. https://doi.org/10.1111/irv.12602
231. Lessler J, Reich NG, Brookmeyer R, Perl TM, Nelson KE, Cummings DAT (2009) Incubation periods of acute respiratory viral infections: a systematic review. Lancet Infect Dis 9(5):291–300. https://doi.org/10.1016/S1473-3099(09)70069-6
232. Yang Y, Xiao Z, Ye K, He X, Sun B, Qin Z, Yu J, Yao J (2020) SARS-CoV-2: characteristics and current advances in research 1–17
233. Bar-on YM, Flamholz AVI, Phillips ROB, Milo RON (2020) SARS-CoV-2 (COVID-19) by the numbers 2:1–15
234. King AMQ, Adams MJ, Carstens EB, Lefkowitz EJ (2012) Virus taxonomy: ninth report of the international committee on taxonomy of viruses. Family-totiviridae. Virus Taxon. (June) 2014

235. Ghebremedhin B (2014) Human adenovirus: viral pathogen with increasing importance. Eur J Microbiol Immunol 4:26–33. https://doi.org/10.1556/EuJM1.4.2014.1.2
236. Ke Z, Dillard RS, Chirkova T, Leon F, Stobart CC, Hampton CM, Strauss JD, Rajan D, Rostad CA, Taylor JV, Yi H, Shah R, Jin M, Hartert TV, Jr RS Peebles, Graham BS, Moore ML, Anderson LJ, Wright ER (2018) The morphology and assembly of respiratory syncytial virus revealed by cryo-electron tomography. https://doi.org/10.3390/v10080446
237. Yang L (2018) Detection and characterization of a novel norovirus in bats, China. Virol.Sin 12250:10–13. https://doi.org/10.1007/s12250-018-0010-9
238. Prasad BVV, Shanker S, Muhaxhiri Z, Choi J (216) Structural biology of noroviruses. Elsevier Inc. https://doi.org/10.1016/B978-0-12-802241-2/00016-X
239. Faulkner CA, Castellini JE, Zuo W, Lorenzetti DM, Sohn MD (2022) Investigation of HVAC operation strategies for office buildings during COVID-19 pandemic. Build Environ 207(PB):108519. https://doi.org/10.1016/j.buildenv.2021.108519
240. Rodríguez M, Palop ML, Seseña S, Rodríguez A (2021) Science of the total environment are the portable air cleaners (PAC) really effective to terminate airborne SARS-CoV-2? 785:0–3. https://doi.org/10.1016/j.scitotenv.2021.147300
241. Riddell S, Goldie S, Hill A, Eagles D, Drew TW (2020) The effect of temperature on persistence of SARS-CoV-2 on common surfaces. Virol J 17(1). https://doi.org/10.1186/s12985-020-01418-7
242. Sun Z, Cai X, Gu C, Zhang R, Han W, Qian Y, Wang Y, Xu W, Wu Y, Cheng X, Yuan Z, Xie Y, Qu D (2020) Survival of SARS-COV-2 under liquid medium, dry filter paper and acidic conditions 4–7. https://doi.org/10.1038/s41421-020-00191-9
243. Shayegan Z, Lee C-S, Haghighat F (2018) TiO_2 photocatalyst for removal of volatile organic compounds in gas phase–a review. Chem Eng J 334:2408–2439. https://doi.org/10.1016/j.cej.2017.09.153
244. Sedaghati N, Habibi-Yangjeh A, Pirhashemi M, Vadivel S (2019) Boosted visible-light photocatalytic performance of TiO_2-x decorated by BiOI and AgBr nanoparticles. J Photochem Photobiol A Chem 384:112066. https://doi.org/10.1016/j.jphotochem.2019.112066
245. Binas V, Venieri D, Kotzias D, Kiriakidis G (2017) Modified TiO_2 based photocatalysts for improved air and health quality. J Mater 3(1):3–16. https://doi.org/10.1016/j.jmat.2016.11.002
246. Moon EW, Lee H-W, Rok JH, Ha J-H (2020) Photocatalytic inactivation of viral particles of human norovirus by Cu-doped TiO_2 non-woven fabric under UVA-LED wavelengths. Sci Total Environ 749:141574. https://doi.org/10.1016/j.scitotenv.2020.141574
247. Ohtani B, Prieto-Mahaney OO, Li D, Abe R (2010) What is degussa (Evonik) P25? Crystalline composition analysis, reconstruction from isolated pure particles and photocatalytic activity test. J Photochem Photobiol A Chem 216(2–3):179–182. https://doi.org/10.1016/j.jphotochem.2010.07.024
248. Zhang J, Vikrant K, Kim K-H, Dong F (2022) Photocatalytic destruction of volatile aromatic compounds by platinized titanium dioxide in relation to the relative effect of the number of methyl groups on the benzene ring. Sci Total Environ 822:153605. https://doi.org/10.1016/j.scitotenv.2022.153605
249. Selishchev D, Svintsitskiy D, Kovtunova L, Gerasimov E, Gladky A, Kozlov D (2021) Surface modification of TiO_2 with Pd nanoparticles for enhanced photocatalytic oxidation of benzene micropollutants. Coll Surf A Physicochem Eng Asp 612:125959. https://doi.org/10.1016/j.colsurfa.2020.125959
250. Ho VTT, Chau DH, Bui KQ, Nguyen NTT, Tran TKN, Bach LG, Truong SN (2022) A high-performing nanostructured Ir doped-TiO_2 for efficient photocatalytic degradation of gaseous toluene. Inorganics 10(3):29. https://doi.org/10.3390/inorganics10030029
251. Wang S-J, Zhang X-Y, Su D, Wang Y-F, Qian C-M, Zhou X-R, Li Y-Z, Zhang T (2020) Electrospinning Ag-TiO_2 nanorod-loaded air treatment filters and their applications in air purification. Molecules 25(15):3369. https://doi.org/10.3390/molecules25153369

252. Wu Q, Ye J, Qiao W, Li Y, Niemantsverdriet JW (Hans), Richards E, Pan F, Su R (2021) Inhibit the formation of toxic methylphenolic by-products in photo-decomposition of formaldehyde–toluene/xylene mixtures by Pd cocatalyst on TiO_2. Appl Catal B Environ 291:120118. https://doi.org/10.1016/j.apcatb.2021.120118
253. Chen M, Wang H, Chen X, Wang F, Qin X, Zhang C, He H (2020) High-performance of Cu-TiO_2 for photocatalytic oxidation of formaldehyde under visible light and the mechanism study. Chem Eng J 390:124481. https://doi.org/10.1016/j.cej.2020.124481
254. Wang Y, Gao P, Li B, Yin Z, Feng L, Liu Y, Du Z, Zhang L (2022) Enhanced photocatalytic performance of visible-light-driven CuO_x/TiO_{2-x} for degradation of gaseous formaldehyde: roles of oxygen vacancies and nano copper oxides. Chemosphere 291:133007. https://doi.org/10.1016/j.chemosphere.2021.133007
255. Shayegan Z, Haghighat F, Lee C-S (2021) Anatase/brookite biphasic surface fluorinated Fe–TiO_2 photocatalysts to enhance photocatalytic removal of VOCs under visible and UV light. J Clean Prod 287:125462. https://doi.org/10.1016/j.jclepro.2020.125462
256. Mamaghani AH, Haghighat F, Lee C-S (2020) Role of titanium dioxide (TiO_2) structural design/morphology in photocatalytic air purification. Appl Catal B Environ 269:118735. https://doi.org/10.1016/j.apcatb.2020.118735
257. Abedi K, Shahmoradi B, Mohammadi E, Wantala K, Maleki A, Zandsalimi Y, Salimi S, Kohzadi S (2022) Photocatalytic degradation of VOCs from air stream using Mo: TiO_2/GAC nanocomposites. Mater Res Express 9(2):25502. https://doi.org/10.1088/2053-1591/ac4aa3
258. Mines PD, Uthuppu B, Thirion D, Jakobsen MH, Yavuz CT, Andersen HR, Hwang Y (2018) Granular activated carbon with grafted nanoporous polymer enhances nanoscale zero-valent iron impregnation and water contaminant removal. Chem Eng J 339(February):22–31. https://doi.org/10.1016/j.cej.2018.01.102
259. Ku Y, Lee P-C, Luong GKT (2021) Photocatalytic reduction of gaseous carbon dioxide over NiO/TiO_2 under UV light illumination. J Taiwan Inst Chem Eng 125:291–296. https://doi.org/10.1016/j.jtice.2021.06.036
260. Zhou Y, Zhang Q, Shi X, Song Q, Zhou C, Jiang D (2022) Photocatalytic reduction of CO_2 into CH_4 over ru-doped TiO_2: synergy of ru and oxygen vacancies. J Colloid Interface Sci 608:2809–2819. https://doi.org/10.1016/j.jcis.2021.11.011
261. Cheng S-P, Wei L-W, Wang H-P (2022) Photocatalytic reduction of CO_2 to methanol by Cu_2O/TiO_2 heterojunctions. Sustainability 14(1):374. https://doi.org/10.3390/su14010374
262. Xi H, Xu Y, Zou W, Ji J, Cai Y, Wan H, Dong L (2022) Enhanced methanol selectivity of Cu_xO/TiO_2 photocatalytic CO_2 reduction: synergistic mechanism of surface hydroxyl and low-valence copper species. J CO_2 Util 55:101825. https://doi.org/10.1016/j.jcou.2021.101825
263. Qian X, Yang W, Gao S, Xiao J, Basu S, Yoshimura A, Shi Y, Meunier V, Li Q (2020) Highly selective, defect-induced photocatalytic CO_2 reduction to acetaldehyde by the Nb-doped TiO_2 nanotube array under simulated solar illumination. ACS Appl Mater Interfaces 12(50):55982–55993. https://doi.org/10.1021/acsami.0c17174
264. Hernández-Gordillo A, Arriaga S (2022) Mesoporous TiO_2 monoliths impregnated with CdS and CuO nanoparticles for airborne bacteria inactivation under visible light. Catal Lett 152(3):629–640. https://doi.org/10.1007/s10562-021-03659-9
265. Qin K, Zhao Q, Yu H, Xia X, Li J, He S, Wei L, An T (2021) A review of bismuth-based photocatalysts for antibiotic degradation: insight into the photocatalytic degradation performance, pathways and relevant mechanisms. Environ Res 199(May):111360. https://doi.org/10.1016/j.envres.2021.111360
266. Chen P, Liu H, Cui W, Lee SC, Wang L, Dong F (2020) Bi-based photocatalysts for light-driven environmental and energy applications: structural tuning, reaction mechanisms, and challenges. EcoMat 2(3):1–31. https://doi.org/10.1002/eom2.12047
267. Ye L, Deng Y, Wang L, Xie H, Su F (2019) Bismuth-based photocatalysts for solar photocatalytic carbon dioxide conversion. Chemsuschem 12(16):3671–3701. https://doi.org/10.1002/cssc.201901196

268. Cui Y, Li M, Zhu N, Cheng Y, Lam SS, Chen J, Gao Y, Zhao J (2022) Bi-based visible light-driven nano-photocatalyst: the design, synthesis, and its application in pollutant gover nance and energy development. Nano Today 43:101432. https://doi.org/10.1016/j.nantod.2022.101432
269. Truong PL, Kidanemariam A, Park J (2021) A critical innovation of photocatalytic degradation for toxic chemicals and pathogens in air. J Ind Eng Chem 100:19–39. https://doi.org/10.1016/j.jiec.2021.05.012
270. Dong X, Zhang W, Sun Y, Li J, Cen W, Cui Z, Huang H, Dong F (2018) Visible-light-induced charge transfer pathway and photocatalysis mechanism on Bi semimetal@defective BiOBr hierarchical microspheres. J Catal 357:41–50. https://doi.org/10.1016/j.jcat.2017.10.004
271. Bai Y, Yang P, Wang P, Fan Z, Xie H, Wong PK, Ye L (2018) Solid phase fabrication of bismuth-rich $Bi_3O_4Cl_xBr_{1-x}$ solid solution for enhanced photocatalytic NO removal under visible light. J Taiwan Inst Chem Eng 82:273–280. https://doi.org/10.1016/j.jtice.2017.10.021
272. Li X, Zhang W, Li J, Jiang G, Zhou Y, Lee SC, Dong F (2018) Transformation pathway and toxic intermediates inhibition of photocatalytic NO removal on designed Bi metal@defective $Bi_2O_2SiO_3$. Appl Catal B Environ 2019(241):187–195. https://doi.org/10.1016/j.apcatb.2018.09.032
273. Chen P, Liu H, Sun Y, Li J, Cui W, Wang L, Zhang W, Yuan X, Wang Z, Zhang Y, Dong F (2019) Bi metal prevents the deactivation of oxygen vacancies in $Bi_2O_2CO_3$ for stable and efficient photocatalytic NO abatement. Appl Catal B Environ 2020(264):118545. https://doi.org/10.1016/j.apcatb.2019.118545
274. Chen Q, Long H, Chen M, Rao Y, Li X, Huang Y (2020) In situ construction of biocompatible Z-scheme α-Bi_2O_3/$CuBi_2O_4$ heterojunction for NO removal under visible light. Appl Catal B Environ 272(January):119008. https://doi.org/10.1016/j.apcatb.2020.119008
275. Li R, Ou X, Zhang L, Qi Z, Wu X, Lu C, Fan J, Lv K (2021) Photocatalytic oxidation of NO on reduction type semiconductor photocatalysts: effect of metallic Bi on CdS nanorods. Chem Commun 57(78):10067–10070. https://doi.org/10.1039/d1cc03516a
276. Sun M, Zhang W, Sun Y, Zhang Y, Dong F (2019) Synergistic integration of metallic bi and defects on BiOI: enhanced photocatalytic NO removal and conversion pathway. Chin J Catal 40(6):826–836. https://doi.org/10.1016/S1872-2067(18)63195-X
277. Liu Y, Zhu C, Sun J, Ge Y, Song F, Xu Q, Liu Y, Zhu C, Song F, Zhu C, Sun J, Ge Y, Xu Q (2020) In situ assembly of CQDs/Bi_2WO_6 for highly efficient photocatalytic degradation of VOCs under visible light. New J Chem 44(8):3455–3462. https://doi.org/10.1039/c9nj04957f
278. Ding J, Wang H, Luo Y, Xu Y, Liu J, Lin Y (002) Oriented $Bi_2O_2CO_3$ nanosheets with enhanced photocatalytic performance for toluene removal in air. Catalysts 10(4). https://doi.org/10.3390/catal10040389
279. Yang R, Li C, Yuan F, Wu C, Sun Z, Ma R (2022) Synergistic effect of diatomite and Bi self-doping Bi_2MoO_6 on visible light photodegradation of formaldehyde. Microporous Mesoporous Mater 339(April):112003. https://doi.org/10.1016/j.micromeso.2022.112003
280. Zhang S, Pu W, Chen A, Xu Y, Wang Y, Yang C, Gong J (2019) Oxygen vacancies enhanced photocatalytic activity towards VOCs oxidation over Pt deposited Bi_2WO_6 under visible light. J Hazard Mater 2020(384):121478. https://doi.org/10.1016/j.jhazmat.2019.121478
281. Liu H, Mei H, Miao N, Pan L, Jin Z, Zhu G, Gao J, Wang J, Cheng L (2021) Synergistic photocatalytic NO removal of oxygen vacancies and metallic bismuth on $Bi_{12}TiO_{20}$ nanofibers under visible light irradiation. Chem Eng J 414:128748. https://doi.org/10.1016/j.cej.2021.128748. (November 2020)
282. Sepehri A, Sarrafzadeh MH (2017) Effect of nitrifiers community on fouling mitigation and nitrification efficiency in a membrane bioreactor. Chem Eng Process Process Intensif 2018(128):10–18. https://doi.org/10.1016/j.cep.2018.04.006
283. Piumetti M, Fino D, Russo N (2015) Mesoporous manganese oxides prepared by solution combustion synthesis as catalysts for the total oxidation of VOCs. Appl Catal B Environ 163:277–287. https://doi.org/10.1016/j.apcatb.2014.08.012

284. Qian X, Yue D, Tian Z, Reng M, Zhu Y, Kan M, Zhang T, Zhao Y (2016) Carbon quantum dots decorated Bi_2WO_6 nanocomposite with enhanced photocatalytic oxidation activity for VOCs. Appl Catal B Environ 193:16–21. https://doi.org/10.1016/j.apcatb.2016.04.009
285. Chen S, Hu Y, Meng S, Fu X (2014) Study on the separation mechanisms of photogenerated electrons and holes for composite photocatalysts g-C_3N_4–WO_3. Appl Catal B Environ 150–151. https://doi.org/10.1016/j.apcatb.2013.12.053
286. Simelane S, Ngila JC, Dlamini LN (2017) The fate, behaviour and effect of WO_3 nanoparticles on the functionality of an aerobic treatment unit. Environ Nanotechnol Monit Manage 8. https://doi.org/10.1016/j.enmm.2017.07.007
287. Staerz A, Somacescu S, Epifani M, Kida T, Weimar U, Barsan N (2020) WO_3-based gas sensors: identifying inherent qualities and understanding the sensing mechanism. ACS Sens 5(6). https://doi.org/10.1021/acssensors.0c00113
288. Wang L, Teleki A, Pratsinis SE, Gouma PI (2008) Ferroelectric WO_3 nanoparticles for acetone selective detection. Chem Mater 20(15). https://doi.org/10.1021/cm800761e
289. Peleyeju MG, Viljoen EL (2021) WO_3-based catalysts for photocatalytic and photoelectrocatalytic removal of organic pollutants from water–a review J Water Process Eng. https://doi.org/10.1016/j.jwpe.2021.101930
290. He C, Li X, Li Y, Li J, Xi G (2017) Large-scale synthesis of Au-WO_3 porous hollow spheres and their photocatalytic properties. Catal Sci Technol 7(17). https://doi.org/10.1039/c7cy01399j
291. Shi W, Wang J (1998) Controllable synthesized step-scheme heterojunction of $CuBi_2O_4$ decorated WO_3 plates for visible-light-driven CO_2 reduction
292. Ikram M, Sajid MM, Javed Y, Afzal AM, Shad NA, Sajid M, Akhtar K, Yousaf MI, Sharma SK, Aslam H, Hussain T, Hussain D, Razaq A (2021) Crystalline growth of tungsten trioxide (WO_3) nanorods and their development as an electrochemical sensor for selective detection of vitamin C. J Mater Sci Mater Electron 32(5). https://doi.org/10.1007/s10854-021-05351-5
293. Morais PV, Suman PH, Silva RA, Orlandi MO (2021) High gas sensor performance of WO_3 nanofibers prepared by electrospinning. J Alloys Compd 864. https://doi.org/10.1016/j.jallcom.2021.158745
294. Memar A, Phan CM, Tade MO (2014) Controlling particle size and photoelectrochemical properties of nanostructured WO_3 with surfactants. Appl Surf Sci 305. https://doi.org/10.1016/j.apsusc.2014.03.194
295. Rashid AA, Mara UT, Berhad N (2015) Hydrothermal synthesis of tungsten oxide (WO_3) nanostructures using sodium chloride as structure directing agent 2(6):1572–1576. https://doi.org/10.13140/RG.2.1.3242.7126
296. Ilager D, Seo H, Shetti NP, Kalanur SS (2020) CTAB modified Fe-WO_3 as an electrochemical detector of amitrole by catalytic oxidation. J Environ Chem Eng 8(6). https://doi.org/10.1016/j.jece.2020.104580
297. Nagy D, Nagy D, Szilágyi IM, Fan X (2016) Effect of the morphology and phases of WO_3 nanocrystals on their photocatalytic efficiency. RSC Adv 6(40):33743–33754. https://doi.org/10.1039/c5ra26582g
298. Wang X, Pan H, Sun M, Zhang Y (2021) Au single atom-anchored WO_3/TiO_2 nanotubes for the photocatalytic degradation of volatile organic compounds. J Mater Chem A 10(11):6078–6085. https://doi.org/10.1039/d1ta08143h
299. Luévano-Hipólito E, Martínez-De La Cruz A, López-Cuellar E, Yu QL, Brouwers HJH (2014) Synthesis, characterization and photocatalytic activity of WO_3/TiO_2 for NO removal under UV and visible light irradiation. Mater Chem Phys 148(1–2):208–213. https://doi.org/10.1016/j.matchemphys.2014.07.034
300. Lu C, Li J, Chen G, Li B, Lou Z (2019) Self-Z-Scheme plasmonic tungsten oxide nanowires for boosting ethanol dehydrogenation under UV-visible light irradiation. Nanoscale 11(27). https://doi.org/10.1039/c9nr03741a
301. Fukumura T, Sambandan E, Yamashita H (2017) Synthesis and VOC degradation ability of a CeO_2/WO_3 thin-layer visible-light photocatalyst. Mater Res Bull 94:493–499. https://doi.org/10.1016/j.materresbull.2017.07.003

302. Kong L, Guo X, Xu J, Mo Z, Li L (2020) Morphology control of WO_3 nanoplate film on W foil by oxalic acid for photocatalytic gaseous acetaldehyde degradation. J Photochem Photobiol A Chem 401. https://doi.org/10.1016/j.jphotochem.2020.112760
303. Sayama K, Hayashi H, Arai T, Yanagida M, Gunji T, Sugihara H (2010) Highly active WO_3 semiconductor photocatalyst prepared from amorphous peroxo-tungstic acid for the degradation of various organic compounds. Appl Catal B Environ 294(1–2). https://doi.org/10.1016/j.apcatb.2009.11.003
304. Chen J, Shan M, Shi X, Zhang S, Li J, Luan J (2022) $BiSnSbO_6$–TiO_2 composites enhance LED light-driven photocatalytic antibacterial activity (December). https://doi.org/10.1016/j.ceramint.2022.03.192. 2021
305. Katsumata K, Motoyoshi R, Matsushita N, Okada K (2013) Preparation of graphitic carbon nitride (g-C_3N_4)/WO_3 composites and enhanced visible-light-driven photodegradation of acetaldehyde gas. J Hazard Mater 260. https://doi.org/10.1016/j.jhazmat.2013.05.058
306. Jin Z, Murakami N, Tsubota T, Ohno T (2014) Complete oxidation of acetaldehyde over a composite photocatalyst of graphitic carbon nitride and tungsten(VI) oxide under visible-light irradiation. Appl Catal B Environ 150–151. https://doi.org/10.1016/j.apcatb.2013.12.048
307. Wang L, Xu X, Wu S, Cao F (2018) Nonstoichiometric tungsten oxide residing in a 3D nitrogen doped carbon matrix, a composite photocatalyst for oxygen vacancy induced VOC degradation and H_2 production. Catal Sci Technol 8(5). https://doi.org/10.1039/c7cy02572f
308. Kim H, Kim HN, Weon S, Moon GH, Kim JH, Choi W (2016) robust co-catalytic performance of nanodiamonds loaded on WO_3 for the decomposition of volatile organic compounds under visible light. ACS Catal 6(12):8350–8360. https://doi.org/10.1021/acscatal.6b02726
309. Xiao Q, Gao L (2013) One-step hydrothermal synthesis of C, W-codoped mesoporous TiO_2 with enhanced visible light photocatalytic activity. J Alloys Compd 551. https://doi.org/10.1016/j.jallcom.2012.10.040
310. Gondal MA, Suliman MA, Dastageer MA, Chuah GK, Basheer C, Yang D, Suwaiyan A (2016) Visible light photocatalytic degradation of herbicide (Atrazine) using surface plasmon resonance induced in mesoporous Ag-WO_3/SBA-15 composite. J Mol Catal A Chem 425. https://doi.org/10.1016/j.molcata.2016.10.015
311. Sheng C, Wang C, Wang H, Jin C, Sun Q, Li S (2017) Self-photodegradation of formaldehyde under visible-light by solid wood modified via nanostructured fe-doped WO_3 accompanied with superior dimensional stability. J Hazard Mater 328. https://doi.org/10.1016/j.jhazmat.2017.01.018
312. Li Y, Wang C, Zheng H, Wan F, Yu F, Zhang X, Liu Y (2017) Surface oxygen vacancies on WO_3 contributed to enhanced photothermo-synergistic effect. Appl Surf Sci 391. https://doi.org/10.1016/j.apsusc.2016.07.042
313. Rodríguez-González V, Hernández-Gordillo A (2019) Silver-based photocatalysts: a special class. https://doi.org/10.1007/978-3-030-10609-6_8
314. Wen C, Yin A, Dai WL (2014) Recent advances in silver-based heterogeneous catalysts for green chemistry processes. Appl Catal B. https://doi.org/10.1016/j.apcatb.2014.06.016
315. Li X, Xu P, Chen M, Zeng G, Wang D, Chen F, Tang W, Chen C, Zhang C, Tan X (2019) Application of silver phosphate-based photocatalysts: barriers and solutions. Chem Eng J. https://doi.org/10.1016/j.cej.2019.02.083
316. Huang K, Li C, Zheng Y, Wang L, Wang W, Meng X (2022) Recent advances on silver-based photocatalysis: photocorrosion inhibition, visible-light responsivity enhancement, and charges separation acceleration. Sep Purif Technol. https://doi.org/10.1016/j.seppur.2021.120194
317. Pham MT, Nguyen TMT, Bui DP, Wang YF, Tran HH, You SJ (2022) Enhancing quantum efficiency at Ag/g-C_3N_4 interfaces for rapid removal of nitric oxide under visible light. Sustain Chem Pharm 25. https://doi.org/10.1016/j.scp.2021.100596
318. Xue X, Gong X, Chen X, Chen BY (2021) A facile synthesis of Ag/Ag_2O@TiO_2 for toluene degradation under UV–visible light: effect of Ag formation by partial reduction of Ag_2O on photocatalyst stability. J Phys Chem Solids 150. https://doi.org/10.1016/j.jpcs.2020.109799

319. Li J, Yang X, Ma C, Lei Y, Cheng Z, Rui Z (2021) Selectively recombining the photoinduced charges in bandgap-broken Ag_3PO_4/$GdCrO_3$ with a plasmonic ag bridge for efficient photothermocatalytic VOCs degradation and CO_2 reduction. Appl Catal B Environ 291. https://doi.org/10.1016/j.apcatb.2021.120053
320. Li Z, Mao X, Liu Q, Song H, Ji Y, Xu D, Qiu B, Tian D, Wang J (2019) International journal of infectious diseases long-term effect of exposure to ambient air pollution on the risk of active tuberculosis 87:177–184.https://doi.org/10.1016/j.ijid.2019.07.027
321. Zhang W, Dong X, Liang Y, Sun Y, Dong F (2018) Ag/AgCl nanoparticles assembled on BiOCl/$Bi_{12}O_{17}Cl_2$ nanosheets: enhanced plasmonic visible light photocatalysis and in situ DRIFTS investigation. Appl Surf Sci 455. https://doi.org/10.1016/j.apsusc.2018.05.171
322. Guo J, Ouyang S, Li P, Zhang Y, Kako T, Ye J (2013) A new heterojunction Ag_3PO_4/Cr–$SrTiO_3$ photocatalyst towards efficient elimination of gaseous organic pollutants under visible light irradiation. Appl Catal B Environ 134–135. https://doi.org/10.1016/j.apcatb.2012.12.038
323. Deng X, Zhang D, Lu S, Bao T, Yu Z, Deng C (2021) Green synthesis of Ag/g-C_3N_4 composite materials as a catalyst for DBD plasma in degradation of ethyl acetate. Mater Sci Eng. B Solid-State Mater Adv Technol 272. https://doi.org/10.1016/j.mseb.2021.115321
324. Qi K, Liu SY, Zada A (2020) Graphitic carbon nitride, a polymer photocatalyst. J Taiwan Inst Chem Eng 109:111–123. https://doi.org/10.1016/j.jtice.2020.02.012
325. Jiang L, Yang J, Zhou S, Yu H, Liang J, Chu W, Li H, Wang H, Wu Z, Yuan X (2021) Strategies to extend near-infrared light harvest of polymer carbon nitride photocatalysts. Coord Chem Rev 439:213947. https://doi.org/10.1016/j.ccr.2021.213947
326. Liu J, Wang H, Antonietti M (2016) Graphitic carbon nitride “Reloaded”: emerging applications beyond (Photo)catalysis. Chem Soc Rev 45(8):2308–2326. https://doi.org/10.1039/c5cs00767d
327. https://creativecommons.org/licenses/by/3.0/
328. Lin H, Wu J, Zhou F, Zhao X, Lu P, Sun G, Song Y, Li Y, Liu X, Dai H (2023) Graphitic carbon nitride-based photocatalysts in the applications of environmental catalysis. J Environ Sci (China) 124:570–590. https://doi.org/10.1016/j.jes.2021.11.017
329. Wang W, Zhou C, Yang Y, Zeng G, Zhang C, Zhou Y, Yang J, Huang D, Wang H, Xiong W, Li X, Fu Y, Wang Z, He Q, Jia M, Luo H (2020) Carbon nitride based photocatalysts for solar photocatalytic disinfection, can we go further? Chem Eng J 2021(404):126540. https://doi.org/10.1016/j.cej.2020.126540
330. Jia J, Zhang X, Jiang C, Huang W, Wang Y (2020) Visible-light-driven nitrogen-doped carbon quantum dots decorated g-C_3N_4/Bi_2WO_6 Z-scheme composite with enhanced photocatalytic activity and mechanism insight. J Alloys Compd 835:155180. https://doi.org/10.1016/j.jallcom.2020.155180
331. Jiang Y, Fang S, Cao C, Hong E, Zeng L, Yang W, Huang L, Yang C (2022) Enhanced light harvesting and charge separation of carbon and oxygen co-doped carbon nitride as excellent photocatalyst for hydrogen evolution reaction. J Colloid Interface Sci 612:367–376. https://doi.org/10.1016/j.jcis.2021.12.077
332. Jo WK, Kumar S, Tonda S (2019) N-doped C dot/coal-layered double hydroxide/g-C_3N_4 hybrid composites for efficient and selective solar-driven conversion of CO_2 into CH_4. Compos Part B Eng 176(April):107212. https://doi.org/10.1016/j.compositesb.2019.107212
333. Feng H, Guo Q, Xu Y, Chen T, Zhou Y, Wang Y, Wang M, Shen D (2018) Surface nonpolarization of g-C_3N_4 by decoration with sensitized quantum dots for improved CO_2 photoreduction. Chemsuschem 11(24):4256–4261. https://doi.org/10.1002/cssc.201802065
334. Zhu J, Wei Y, Chen W, Zhao Z, Thomas A (2010) Graphitic carbon nitride as a metal-free catalyst for NO decomposition. Chem Commun 46(37):6965–6967. https://doi.org/10.1039/c0cc01432j
335. Wang H, He W, Dong X, Wang H, Dong F (2018) In situ FT-IR investigation on the reaction mechanism of visible light photocatalytic NO oxidation with defective g-C_3N_4. Sci Bull 63(2):117–125. https://doi.org/10.1016/j.scib.2017.12.013
336. Duan Y, Li X, Lv K, Zhao L, Liu Y (2019) Flower-like g-C_3N_4 assembly from holy nanosheets with nitrogen vacancies for efficient NO abatement. Appl Surf Sci 492(January):166–176. https://doi.org/10.1016/j.apsusc.2019.06.125

337. Wu X, Cheng J, Li X, Li Y, Lv K (2018) Enhanced visible photocatalytic oxidation of NO by repeated calcination of g-C_3N_4. Appl Surf Sci 2019(465):1037–1046. https://doi.org/10.1016/j.apsusc.2018.09.165
338. Li K, Cui W, Li J, Sun Y, Chu Y, Jiang G, Zhou Y, Zhang Y, Dong F (2019) Tuning the reaction pathway of photocatalytic NO oxidation process to control the secondary pollution on monodisperse Au nanoparticles@g-C_3N_4. Chem Eng J 378(April):122184. https://doi.org/10.1016/j.cej.2019.122184
339. Lu X, Chen F, Qian J, Fu M, Jiang Q, Zhang Q (2021) Facile fabrication of CeF_3/g-C_3N_4 heterojunction photocatalysts with upconversion properties for enhanced photocatalytic desulfurization performance. J Rare Earths 39(10):1204–1210. https://doi.org/10.1016/j.jre.2020.09.023
340. Yaghoot-Nezhad A, Moradi M, Rostami M, Danaee I, Khosravi-Nikou MR (2020) Dual Z-scheme CuO-ZnO@graphitic carbon nitride ternary nanocomposite with improved visible light-induced catalytic activity for ultrasound-assisted photocatalytic desulfurization. Energy Fuels 34(11):13588–13605. https://doi.org/10.1021/acs.energyfuels.0c02012
341. Kong L, Li X, Song P, Ma F (2020) Porous graphitic carbon nitride nanosheets for photocatalytic degradation of formaldehyde gas. Chem Phys Lett 2021(762):138132. https://doi.org/10.1016/j.cplett.2020.138132
342. Meng J, Wang X, Yang X, Hu A, Guo Y, Yang Y (2018) Enhanced gas-phase photocatalytic removal of aromatics over direct Z-scheme-dictated $H_3PW_{12}O_{40}$/g-C_3N_4 film-coated optical fibers. Appl Catal B Environ 2019(251):168–180. https://doi.org/10.1016/j.apcatb.2019.03.063
343. Pham TH, Jung SH, Kim TY (2021) Enhanced photodegradation of toxic volatile organic pollutants using Ni-doped graphitic carbon nitride under natural solar light. Sol Energy 224(May):18–26. https://doi.org/10.1016/j.solener.2021.05.087
344. Ding F, Chen P, Liu F, Chen L, Guo JK, Shen S, Zhang Q, Meng LH, Au CT, Yin SF (2019) Bi_2MoO_6/g-C_3N_4 of 0D/2D heterostructure as efficient photocatalyst for selective oxidation of aromatic alkanes. Appl Surf Sci 490(June):102–108. https://doi.org/10.1016/j.apsusc.2019.06.057
345. Zhang M, Cai S, Li J, Elimian EA, Chen J, Jia H (2020) Ternary multifunctional catalysts of polymeric carbon nitride coupled with Pt-embedded transition metal oxide to enhance light-driven photothermal catalytic degradation of VOCs. J Hazard Mater 2021(412):125266. https://doi.org/10.1016/j.jhazmat.2021.125266
346. Welderfael T, Yadav OP, Taddesse AM, Kaushal J (2013) Synthesis, characterization and photocatalytic activities of Ag-N-codoped ZnO nanoparticles for degradation of methyl red. Bull Chem Soc Ethiop 27(2). https://doi.org/10.4314/bcse.v27i2.7
347. Khosravi-Gandomani S, Yousefi R, Jamali-Sheini F, Huang NM (2014) Optical and electrical properties of P-type Ag-doped ZnO nanostructures. Ceram Int 40(6). https://doi.org/10.1016/j.ceramint.2013.12.145
348. Jongprateep O, Deedit P, Puranasamriddhi R, Meesombad K (2018) Synthesis of nanoparticulate Ti-doped ZnO by solution combustion technique. J Met Mater Miner 28(1). https://doi.org/10.14456/jmmm.2018.14
349. Georgekutty R, Seery MK, Pillai SC (2008) A highly efficient Ag-ZnO photocatalyst: synthesis, properties, and mechanism. J Phys Chem C 112(35). https://doi.org/10.1021/jp802729a
350. Abdelsamad AMA, Gad-Allah TA, Mahmoud FA, Badawy MI (2018) Enhanced photocatalytic degradation of textile wastewater using Ag/ZnO thin films. J Water Process Eng 25. https://doi.org/10.1016/j.jwpe.2018.07.002
351. Lam SM, Quek JA, Sin JC (2018) Mechanistic investigation of visible light responsive Ag/ZnO micro/nanoflowers for enhanced photocatalytic performance and antibacterial activity. J Photochem Photobiol A Chem. 353. https://doi.org/10.1016/j.jphotochem.2017.11.021
352. Jongprateep O, Meesombad K, Techapiesancharoenkij R, Surawathanawises K, Siwayaprahm P, Watthanarat P (2019) Influences of chemical composition, microstructure and bandgap energy on photocatalytic and antimicrobial activities of ZnO and Ag-doped ZnO by solution combustion technique. J Met Mater Miner 29(1)

353. Dhanalakshmi M, Saravanakumar K, Lakshmi Prabavathi S, Abinaya M, Muthuraj V (2018) Fabrication of novel surface plasmon resonance induced visible light driven iridium decorated SnO_2 nanorods for degradation of organic contaminants. J Alloys Compd 763. https://doi.org/10.1016/j.jallcom.2018.05.340
354. Sun C, Yang J, Xu M, Cui Y, Ren W, Zhang J, Zhao H, Liang B (2022) Recent intensification strategies of SnO_2-based photocatalysts: a review. Chem Eng J. https://doi.org/10.1016/j.cej.2021.131564
355. Li K, Gao S, Wang Q, Xu H, Wang Z, Huang B, Dai Y, Lu J (2015) In-situ-reduced synthesis of Ti^{3+} self-doped TiO_2/g-C_3N_4 heterojunctions with high photocatalytic performance under LED light irradiation. ACS Appl Mater Interfaces 7(17). https://doi.org/10.1021/am508505n
356. Fan CM, Peng Y, Zhu Q, Lin L, Wang RX, Xu AW (2013) Synproportionation reaction for the fabrication of Sn^{2+} self-doped SnO_{2-x} nanocrystals with tunable band structure and highly efficient visible light photocatalytic activity. J Phys Chem C 117(46). https://doi.org/10.1021/jp407296f
357. Sun M, Su Y, Du C, Zhao Q, Liu Z (2014) Self-doping for visible light photocatalytic purposes: construction of SiO_2/SnO_2/SnO_2:Sn^{2+} nanostructures with tunable optical and photocatalytic performance. RSC Adv 4(58). https://doi.org/10.1039/c4ra04356a
358. Chang YC, Yan CY, Wu RJ (2014) Preparation of Pt@SnO_2 core-shell nanoparticles for photocatalytic degradation of formaldehyde. J Chin Chem Soc 61(3). https://doi.org/10.1002/jccs.201300272
359. Lv T, Peng C, Zhu H, Xiao W (2018) Heterostructured Fe_2O_3@SnO_2 core-shell nanospindles for enhanced room-temperature HCHO oxidation. Appl Surf Sci 457:83–92. https://doi.org/10.1016/J.APSUSC.2018.06.254
360. Fu X, Wang J, Huang D, Meng S, Zhang Z, Li L, Miao T, Chen S (2016) Trace amount of SnO_2-decorated $ZnSn(OH)_6$ as highly efficient photocatalyst for decomposition of gaseous benzene: synthesis, photocatalytic activity, and the unrevealed synergistic effect between $ZnSn(OH)_6$ and SnO_2. ACS Catal 6(2). https://doi.org/10.1021/acscatal.5b02593
361. An T, Zhang M, Wang X, Sheng G, Fu J (2005) Photocatalytic degradation of gaseous trichloroethene using immobilized ZnO/SnO_2 coupled oxide in a flow-through photocatalytic reactor. J Chem Technol Biotechnol 80(3). https://doi.org/10.1002/jctb.1187

Photocatalytic Nitrogen Fixation on Semiconductor Materials: Fundamentals, Latest Advances, and Future Perspective

Merga Hailemariam Urgesa, Dwi Fortuna Anjusa Putra, Abdul Qadir, Uzair Ali Khan, Ting-Chen Huang, Yun Xuan Chiu, Jia Hung Lin, and Riski Titian Ginting

Abstract Photocatalysis has recently gained much scientific attention as a possible sustainable ammonia manufacture approach as an alternative to the regular Haber–Bosch process. One of the most critical problems for boosting conversion of solar to ammonia efficiency is the design of efficient photocatalysts. The bismuth-based photocatalytic for the reduction of nitrogen to ammonia has been the subject of extensive and promising study in recent years. Our study mainly summarized and discussed the recent advanced in photocatalytic nitrogen reduction to produce ammonia based on bismuth-containing semiconductors and related work. Photocatalysis using heterojunctions has recently surfaced as a possible solution to several environmental and energy issues, including nitrogen reduction to ammonia. The heterojunction photocatalysts offer the advantage of spatially separated photogenerated charge carrier and simultaneously preserving significant oxidation and reduction potentials of individual components, enabling light harvesting. The essential ideas of photocatalysis heterojunctions, the nitrogen reduction reaction mechanism, to ammonia production techniques, and the existing advantages in heterostructure and types heterostructures of photocatalysts for nitrogen reduction to ammonia are the main discussed in this chapter. Finally, key issues and promises for the specific topic of heterostructures photocatalytic nitrogen fixation are briefly discussed.

Keywords Nitrogen reduction · Photocatalysis · Semiconductor · Bismuth · Heterostructures

---

M. H. Urgesa (✉) · D. F. A. Putra · A. Qadir · U. A. Khan · T.-C. Huang · Y. X. Chiu · J. H. Lin
Department of Materials Science and Engineering, National Taiwan University of Science and Technology, Taipei 10672, Taiwan, ROC
e-mail: mergah2004@gmail.com

R. T. Ginting
Department of Electrical Engineering, Universitas Prima Indonesia, Medan, Indonesia

H. Abdullah (ed.), *Photocatalytic Activities for Environmental Remediation and Energy Conversion*, Green Energy and Technology,
https://doi.org/10.1007/978-981-19-6748-1_3

1 Introduction

With the tremendous increase of the human population and economic advancement in the recent decade, human civilization has had an insatiable and ferocious need for fossil fuels. Depletion of fossil fuels has become a worldwide challenge, including detrimental environmental conditions and energy crisis are the most critical issues affecting humanity. Organic micropollutants are becoming increasingly prevalent and persistent in water, thus presenting a new hazard to human health [1, 2]. Over the past few years, photocatalysis has been employed in wastewater treatment as a green and attractive means of removing different organic and inorganic impurities. Using zero or low-carbon energy to replace fossil fuels for achieving nearly zero emission is one of the most promising techniques for decreasing CO_2 emissions from industrial operations. Clean hydrogen generated from low- or zero-emission sources may assist the energy and industrial sectors achieve profound decarbonization [3].

Accordingly, photocatalytic water splitting in simultaneous H_2 and O_2 generation without a sacrificial agent is significant and encouraging. However, the main bottleneck of water oxidation process was exceedingly slow with slow kinetics due to the intricate four-electron transfer and enormous activation energy barrier. An increase in solar energy utilization efficiency and photocatalyst should have a relatively narrow band gap and suitable electronic band position to facilitate electron–hole-including redox reaction [3, 4]. To enable photo-oxidation processes, the conduction band must be more negative potential compared to the reduction potential of water. In contrast, the top of valence band must be more positive relative to the redox potential of O_2/H_2O. Despite extensive efforts to manufacture high-performing photocatalysts, the candidates that fulfill both criteria with high efficiency and stability are far from sufficient. An alternative approach is to investigate metal oxide semiconductors with a narrow bandgap and high photo-oxidation process, which is advantageous for photo-induced organic pollutant decomposition and water oxidation half-reaction while providing a tremendous amount of electron and photon sources to enhance H_2 release from water reduction half-reaction [5] In another way, we involve the photoelectrochemical system to drive and produce hydrogen and react it with nitrogen.

Natural nitrogen fixation, as one of the essential processes, turns solar energy into chemical energy in the phase of "fixed" nitrogen, with ammonia as a principal product necessary for animals, plants, and other living organisms. Therefore, switching to renewable energy is essential, particularly in light of the world's continued population increase. It is critical to developing artificial nitrogen fixation technology to balance the global nitrogen cycle, which has sparked scientists' attention since the eighteenth century. The Haber–Bosch process, which was initially employed in the early 1900s, is the most popular technique for commercial ammonia manufacturing. It is important to note that this process is energy-intensive at high temperatures (300–550 °C) and pressures (15–25 Mpa). In addition, the process is about approximately 1–2% of total energy in the world [1, 6].

Photocatalysis is a realistic and on-site-specific alternative in this sector, and it is a hot issue in chemistry and heterogeneous catalysis. In 1977, Schrauzer and Guth published the first work on photocatalytic nitrogen reduction using Ti-based nanostructured material. At the same time, TiO_2 was already placed as the typical photocatalyst due to cost-effectiveness and lack of toxicity. Nevertheless, TiO_2 is only effective in UV light, substantially hindering solar utilization efficiency. Recently, nitrogen fixation has been studied utilizing metal oxides, metal sulfides, g-C_3N_4, and bio-mimicking photocatalysts [7, 8]. However, the research in this field is still in its early phase, mainly due to poor solar to ammonia conversion efficiencies (rate of 100 mmol g^{-1} h^{-1}) [8]. It is still far from becoming economically feasible. In general, the main drawback is the main drawback of the poor quantum efficiency in photocatalytic processes, largely efficient visible light utilization, and highly recombining photo-generated electron–hole pair. Despite that, photocatalytic nitrogen reduction is an uphill reaction that involves the formation of high-energy molecules, activation, and breakdown of strongly nonpolar N≡N bond (941 kJ/mol) remains a barrier (N_2H and N_2H_2) [9]. These issues reveal that optimizing total photocatalytic output for adequate light-driven nitrogen fixation needs thorough engineering for light harvesting results to ensure the capacity of photocatalyst in reducing nitrogen [8]. In addition, it also needs to promote N_2 activation sites to improve the sluggish nitrogen excitation reaction mechanism.

Biomass carbon for energy storage and polycarboxylic acid as a water reduction for building material has shown tremendous research interest. To determine the photocatalyst performance, the band edge formed by p-orbital or sp-hybridization has been demonstrated to alter the physicochemical features of catalysts, resulting in greater phenomena in visible light irradiation [9, 10]. The CB edges may also be lowered and enhanced by the p-electrons inside the hybridization phase, having a substantially narrower bandgap over d-block semiconductors (e.g., TiO_2, ZnO, etc.). This occurrence is beneficial for visible-light absorption in semiconductors via optoelectronics and photocatalysis. Moreover, anisotropic space p and sp hybrid may lead in a highly dispersed band structure, lowering the effective quality and promoting charge separation and transfer inside the photocatalysis process [11, 12]—all of such advantages make p-block semiconductors attractive for photocatalysis using solar light. Bismuth-based photocatalysts, for instance, are chemically stable, have minimal toxicity, and are inexpensive. They also exhibit exciting photo activity owing to their high electrical conductivity and adequate intrinsic bandgaps. Although visible light may equally excite Bi^{5+}-based compounds, Bi^{3+}-based compounds have been explored extensively and have superior stability. The VB of these oxides is largely hybridized by 6s orbit of Bi and 2p orbit of O. Due to the introduction of Bi 6s orbital, which reduces the bandgap, the VB of such material is less than the oxide (2p orbital of O). Furthermore, such compounds are frequently layered, resulting in highly anisotropic structural, electrical, visual, and mechanical characteristics due to the strong intralayer bond formation and weak interlayer van der Waals interactions. Degradation, photocatalytic water split, and photocatalytic CO_2 conversion have all recently been explored with bismuth-based photocatalysts [13].

It is worth mentioning that photocatalytic N_2 fixing employing bismuth-containing systems has been widely researched. Bismuth molybdate has been demonstrated to boost photocatalytic NH_3 production during the hydrogenation process under moderate conditions. Furthermore, bismuth oxyhalides BiOX (X = if, Cl, Br) have gained a lot of research attention due to their layered structure with staggered [Bi_2O_2] sheet and two halogen atoms [13, 14]. The ability to engineer BiOX physicochemical properties towards N_2 adsorption and its activation converting by pH adjustment for ammonia production from N_2 reduction under solar illumination, phase transition, activation of facet-related internal electric field (IEF) intensity, and generation of surface oxygen vacancies allow for selectively exposed of various crystal facets. BiOX photocatalysts have provided novel and relevant discoveries [15].

Accordingly, this chapter discusses the current breakthroughs in the field of photocatalysts in nitrogen fixation to ammonia. To provide the most acceptable and trustworthy study, the thermodynamic and kinetic concepts of photocatalytic nitrogen fixation on bismuth-based photocatalysts, as well as the most widely utilized ammonia detection methodologies, are rigorously explored. The microscopic mechanism of photocatalytic N_2 reduction and the impact of material change on catalytic activity are studied. Furthermore, outlooks and perspectives on the doped photocatalysts for ammonia fixation are discussed. This chapter will give a valuable tutorial for photocatalytic fixation research and stimulate the rational design and manufacture of advanced bismuth-based photocatalysts for long-term nitrogen reduction.

1.1 Reaction Mechanisms of the Photocatalytic Nitrogen Fixation Process

This process may have proceeded one of two ways: nitrogen reduction reaction (NRR) or nitrogen oxidation reaction (NOR). Essentially, the photocatalytic N_2 reduction process is separated into multiple steps below.

During light irradiation, the conduction band (CB) excites electrons, leaving holes in the valence band (VB). Few electrons and protons are expelled into the systems. In the interim, additional photogenerated holes (h^+) oxidize water to create H^+ and O_2 (Eq. 1), and the N_2 production of NH_3 is induced by the reduction of hot electrons (Eq. 2). As a consequence, NH_3 is consequently under-regulated circumstances, conditions in which sunlight as a source of energy (Eq. 3).

$$2H_2O + 4h^+ \longrightarrow O_2 + 4H^+ \tag{1}$$

$$N_2 + 6H^+ + 6e^- \longrightarrow 2NH_3 \tag{2}$$

$$\text{Overall } 2N_2 + 6H_2O \longrightarrow 4NH_3 + 3O_2 \tag{3}$$

Figure 1a displays the potential of different photocatalytic nitrogen reduction processes at pH 14. Even though the total nitrogen reduction to ammonia process ($N_2 + 6H^+ + 6e^- / 2NH_3$) is preferable from a thermodynamic perspective for hydrogen evolution, several critical elemental bond cleavage potentials are very large. Since, initial hydrogenation of N_2 ($N_2 + H^+ + e^- / N_2H$) need a reduction potential of 3.2 V versus NHE. Another hurdle for photocatalytic ammonia synthesis is the conflict between nitrogen reduction and hydrogen evolution [3, 5, 12].

Full knowledge of a nitrogen fixation system requires active surface area, electronic structures, analysis of adsorbed intermediates, and product analyses. In this case, the impacts of nitrogen fixation are far more intricate than previously studied. Consequently, it is necessary to look at various response mechanisms [16]. In general, N_2-to-NH_3 conversion that includes dissociative and associative pathways as well as association of reaction intermediates are the major focus of nitrogen fixation,

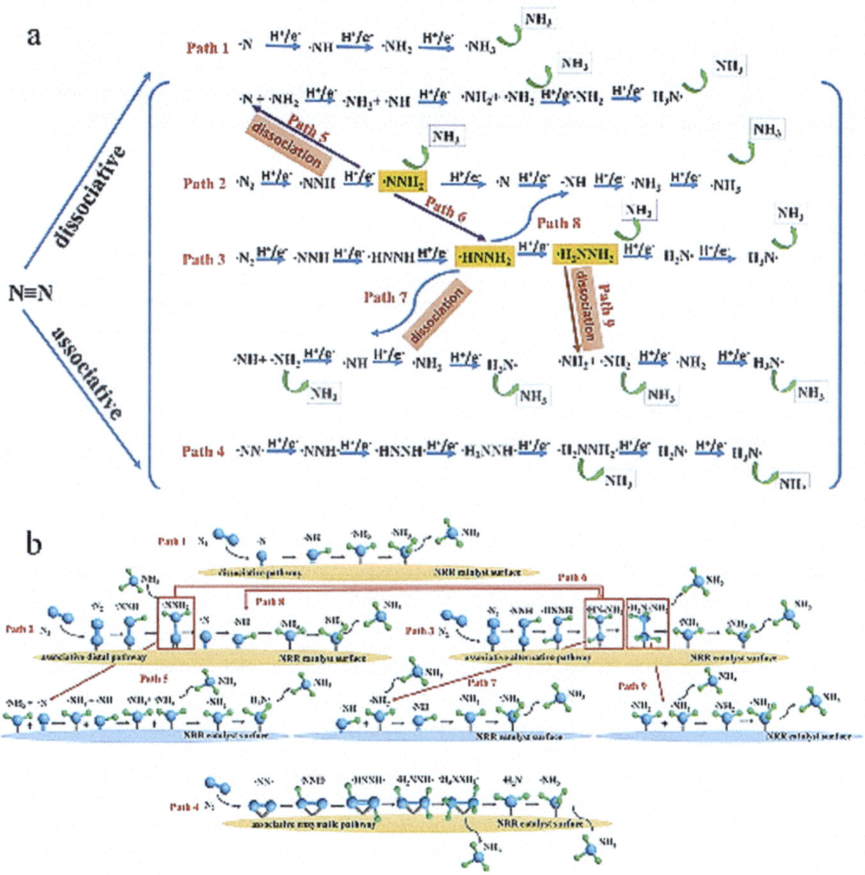

Fig. 1 **a** Different NH_3 production pathways and their intermediates and **b** their ball-and-stick representations (Reproduced with permission)

as shown in Fig. 1. For the dissociative approach (Fig. 1a, path 1), the triple-bond of the N_2 molecule must be engaged at the start of the process, resulting in two isolated atoms on the catalyst surface, which are protonated individually to form NH_3, similarly to the commercial NH_3 production technique. Because the N≡N triple bond may be broken at 9.79 eV (945 kJ mol^{-1}), this pathway has a higher reaction energy gap, and remain challenging to maintain on a catalyst surface under moderate condition [16, 17]. This procedure is complex for materials with layers of structure. The associative system, in contrast to the dissociative system, is similar to such chemical mechanism, including N_2 fixing enzymes: the adsorb N_2 molecules are prepared by the reaction progressively, connecting two atoms before the formation NH_3 molecules [18]. The associative approach sequence hydrogenation step may be separated into associative distal pathways and other associative paths 2–8, as shown in Fig. 1. In addition, determining the NRR routes requires the preliminary nitrogen adsorption configuration. There are two types of nitrogen adsorption systems in general: side–on and end–on. The associative distal (Fig. 1a, route 2) and alternative associative pathways (Fig. 1a, path 3) are reached through the side–on configuration, while the associative enzymatic pathway is accessed via the end–on configuration (Fig. 1a, path 4). The establishment of N_2 configurations was complex unless it utilizes sophisticated procedure in-situ FTIR methods. First-principle calculations lead to an additional information about adsorption configurations and processes. The catalyst adsorbent surface atom preferentially participates inside the hydrogenation process until the distant NH_3 molecule is formed. The second NH_3 molecule is created when another atom engages in the same hydrogenation process (route 2). As an alternative, two atoms are hydrogenated with a proton-coupled electron transfer (PCET), with the sequential release of two NH_3 molecules in the last step (route 3). Notably, the following hydrogenation of intermediate $*NNH_2$ was formed (where *represents the N_2 adsorption site) may proceed through either the remaining distal route or an alternate pathway to generate $*HN\text{-}NH_2$ (route 6). Subsequently, in route 3, the intermediate $*HN\text{-}NH_2$ might travel to the distal pathway and create *NH with an NH_3 molecule released (route 8). At potentials below 0.71 V, the intermediate $*NNH_2$, $*NHNH_2$, and intermediate hydrazine ($*NH_2NH_2$) were expected to separate and form multiple reaction pathways (route 5, 7, and 9). According to Back & Jung's calculations (Fig. 1a) Previously, Watt and Chrisp technique was used to identify hydrazine synthesis or compared the Gibbs values of *HN-NH or $*NH_2NH_2$, both of which may come from protonation of $*N_2H$ [18, 19]. In addition, the enzymatic technique (route 4) was based on a similar hydrogenation mechanism as the alternate pathway, with both atoms attached to a catalyst surface in an end-on coordinating arrangement, which is rare for heterogeneous catalysts. Li et al. have conducted another route such as Langmuir–Hinshelwood (LH) mechanism. Briefly, H_2 is first adsorbed towards the surface of a N_2 fragment. For particular transition-metal nitrides (TMNs) catalysts, a Mars-van Krevelen mechanism has been shown to be beneficial for electrolytic ammonia synthesis. In addition, the atoms in TMN lattice constantly combines with proton-electron pairs (PEPs) to generate NH_3 molecules, thus creating vacancies on the surface [19]. The voids were chemically adsorbed and activated N_2 molecules to aid the subsequent NRR. Overall, anticipating multi-electrons and the hydrogenation process is unfavorable because the various bonding mechanisms and reaction routes remain challenging. Accordingly, nine alternative reaction routes were discussed, as shown in Fig. 1, based on pioneering mechanistic research carried out using first-principal calculations.

Additionally, the dissociative mechanism as shown in Fig. 1b, N≡N triple bond cleavage needs very high energy input as the starting step during Haber–Bosch process. Two unique associative N_2 reduction mechanisms, namely, distal and alternating pathways, have been postulated, evoking a distinct intermediate. Two atoms are hydrogenated alternately in the possible path, releasing the first NH_3 and hydrazine-bound intermediate, as illustrated in Fig. 1b, after four hydrogenation stages. A single atom of N_2 is hydrogenated three times in the distal route to generate the first NH_3, and the other nitride -N is hydrogenated three times to produce another NH_3.

The thermodynamics of the photocatalytic nitrogen fixation process can be separated into two categories: whole reaction and half-reaction thermodynamics, which are independent of the catalyst, but also photon induction mechanisms that are dependent on the photocatalyst. Medford et al. conducted a comprehensive thermodynamic investigation of typical nitrogen-containing compounds utilizing explicit entropy and enthalpy to investigate the thermodynamics viability of different nitrogen-fixation procedures. Free energy may be estimated and used to predict the reaction starting potential provided the adsorption energy and vibrational energy of putative intermediate formed during nitrogen reduction are known. Based on data from the National Institute of Standards and Technology (NIST), Table 1 displays the thermodynamics of oxidative and reductive reactions at ambient temperature. In the photocatalytic reaction process, photocatalysts have several restrictions. To disrupt the N_2 bond cleavage on the photocatalyst surface, a reduction potential of 4.2 V vs NHE is required, which is relatively difficult to produce in most semiconductors. On the other hand, many electrical approaches are more practical [20, 21].

Table 1 Standard temperature and pressure, various nitrogen fixation thermodynamic potentials of various nitrogen fixation reactions (Reproduced with permission)

Reaction	ΔG (eV)
$N_2(g) + O_2 \leftrightarrow 2NO(g)$	1.81
$N_2(g) + 2O_2 \leftrightarrow 2NO_2(g)$	1.08
$N_2(g) + \frac{6}{5}H_2O(l) \leftrightarrow \frac{6}{5}NO(g) + \frac{4}{5}NH_3$	3.90
$N_2(g) + 3H_2O(l) \leftrightarrow 2NH_3(g) + \frac{3}{2}O_2$	7.03
$N_2(g) + H_2O(l) + \frac{5}{2}O_2 \leftrightarrow HNO_3(g)$	0.93
$N_2(g) + H_2O(l) + \frac{3}{2}O_2 \leftrightarrow HNO_2(g)$	1.55
$N_2(g) + 3H_2O(l) \leftrightarrow 2NH_2OH(g) + \frac{1}{2}O_2$	7.31

Based on a thermodynamic standpoint, it is straightforward to convert N_2 to a single-nitrogen molecule like ammonia or nitrate. However, it is challenging to convert a diazo compound to a mono-nitrogen compound. Although nitrogen fixation can be classed as a reduction or oxidation reaction, titanium-based photocatalysts can produce both nitrogen reduction products and oxidative products of nitrate under environmental conditions [22]. As a result, it is hardly understood whether distinct oxidation or a reduction reaction occurred. Redox product mutual conversion is time-consuming; therefore, the complex pathways of oxidation reaction between oxidation or a reduction reaction. Numerous researches have reported the oxidation of ammonia under environmental conditions. As the change of reduced products to oxidized products requires a driving force, photocatalysis under environmental circumstances may be a process of reducing product oxidation. However, compared to most nitrogen fixation processes, the product's reaction required only low voltage with a small amount of thermodynamic energy [23, 24].

In a protic solvent, the hydrogen evolution reaction (HER) is invariably a competitor of the nitrogen reduction process (NRR). Some interesting materials with strong nitrogen-fixation catalytic activity is also expected to have an *H binding energy close to the optimum for hydrogen evolution, which could make the competitive HER reaction easier. Accordingly, the selectivity between the HER and NRR must be considered during the development of metal cocatalysts for nitrogen reduction. The relationship between the catalytic performance of nitrogen reduction and the binding energy of intermediate was studied by Norskov et al. using density functional theory (DFT) on face-centered cubic (FCC) metals. The HER, for example, has a greater positive limiting potential than the NRR (211) [25].

Based on the (111) facet, the difference in limiting potentials between the HER and NRR is lesser than on the (211) facet. The metals had a limiting potential difference of more than 0.4 V between these two reactions, showing that HER uses selectivity (Fig. 2). The HER process should be slowed down over the catalysts while developing nitrogen reduction catalysts. The rate-limiting phase is when the metal-to-nitrogen is too weak in the reductive adsorption of N_2 to $*N_2H$. The hydrogenation of $*NH$ to $*NH_2$ or the development of $*NH_2$ to NH_3 determines the reaction rate when the metal-to-nitrogen is sufficiently strong. The best heterogeneous nitrogen reduction catalyst must have a high affinity for nitrogen and able to reduce nitrogen at a fast rate and a low affinity for intermediate intermediates.

2 Fundamentals of Nitrogen Fixation Electrocatalyst

Semiconductor photocatalysis has recently received considerable attention due to its potential applications in solving environmental issues and energy crisis. Researchers have already looked into homo/heterojunction, phase-junction, solid solution, surface engineering, and doping to improve photocatalytic performance. The preparation of semiconductor solid solution has aroused curiosity since it allows for perfect visible light absorption by altering the semiconductor band gap.

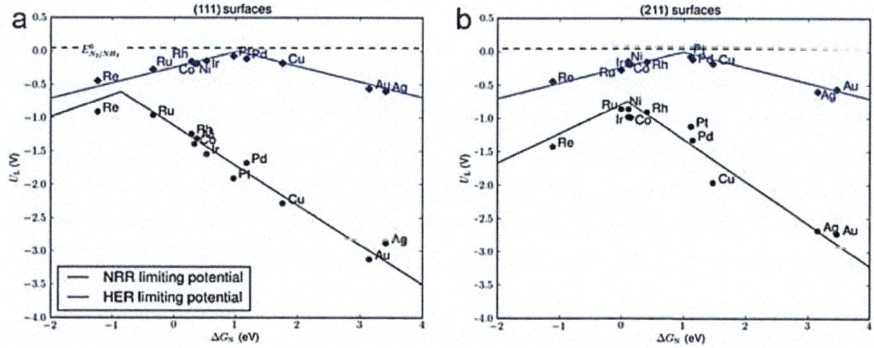

Fig. 2 HER and NRR limiting-potential volcanoes are compared. Individual metal points are indicated for (111) (**a**) and (211) (**b**) surfaces of HER and NRR overpotentials as a factor of the *N binding-energy parameter are displayed in blue and black, respectively [25]

Even though GaN nor ZnO is able to absorb visible light, the absorption edge of the $(Ga_{1-x}Zn_x)(N_{1-x}O_x)$ solid solution would be extended to longer wavelengths [24]. Bismuth oxyhalides have recently attracted much attention to distinctive layered catalyst structures and indirect transition bandgap properties. $[Bi_2O_2]^{2+}$ as the slabs on the alternate with double slabs of halogen ions in the tiered design. As a result, the layered structure of BiOX generates an internal electric field that helps to promote electron–hole pair separation in the [001] direction and significant photocatalytic activity. For example, BiOCl nanoplates exposed to [001] facets also get higher surface energy, and that also increased catalytic activity than BiOCl nanoplates exposed to [010] facets with lower surface energy only when exposed to ultraviolet (UV) light. Therefore, research has focused on developing a bismuth oxyhalide photocatalyst with exposed facets [001]. The photoactivity of single-crystal BiOCl nanosheets was approximately seven times higher than regular BiOI nanosheets for the breakdown of rhodamine B(RhB) [26, 27].

The Bi-based layered compounds with a square lattice structure are bismuth oxyhalides (BiOX, X = Cl, Br). The $[Bi_2O_2]^{2+}$ layer in the system polarizes the atoms, creating an internal electric field that efficiently separates and transmits charge. Bismuth oxyhalides are included as photocatalysts of visible light photocatalysts such as BiOBr and BiOI with band gap of 2.87 and 1.89 eV, respectively. However, the disadvantage of BiOBr is mainly due to its limited capacity to absorb visible light. Despite its bandgap being low enough to cover most of the visible light spectrum, BiOI capacity (redox) is adequately excellent, which is restricted owing to its high conduction band (CB) level. Sensitizing BiOBr photocatalyst onto the visible regime and suppressing redox potential, the bandgap might decrease CB level and enlarge to the valence band (VB) and *vice-versa.* To increase CB level and reduce the VB level, the bandgap may be widened to sensitize the BiOI photocatalyst redox capacity, thus lowering visible light absorption. Consequently, strong visible light absorption and appropriate redox capacity are mutually incompatible [28, 29].

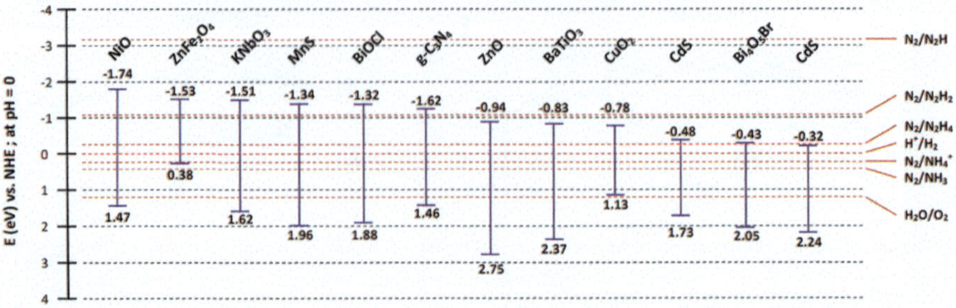

Fig. 3 Schematic diagram of nitrogen fixation photocatalyst material with strong photoactivity and the respective band edge positions (Reproduced with permission)

Photocatalytic nitrogen fixation, such as CO_2 reduction, pollutant degradation, and water splitting, needs the generation of highly active radicals and the dissociation of water molecules. Charge carriers travel to the surface without a wasteful recombination process, and spatially separated electrons and holes participate in reduction and oxidation reactions [15, 22]. Figure 3 shows the schematic representation of a typical photocatalytic process utilizing a single-component photocatalyst [15, 22].

Generally, photocatalyst efficiency is determined by considering reaction kinetics and thermodynamics. Photoexcited electrons convert N_2 molecules to NH_3 in the nitrogen fixation process, whereas photoexcited holes in the oxidation route can produce nitric acid (NO_3) [30]. As a result, photocatalysts for nitrogen fixation with enormous redox, i.e., with more negative conduction band on its potential, are preferred. Due to energy band alignment, Fig. 3 shows a typical list on the side of photocatalysts nitrogen fixation process.

In single-component photocatalysts, recombining before the electron is transported to the surface with redox processes has been a major issue for the charge carriers. The problem is exacerbated by the fact that fewer free electrons are available to drive the reduction of N_2 adsorbed in the molecules of photocatalytic N_2 fixation, resulting in slow kinetics. Furthermore, electron energy cannot be easily adjusted on a single component in the photocatalysts has to drive the energy-intensive N_2 reduction process. Therefore, bulk photocatalysts thus far have modest N_2 fixation activity (~100 mmol g^{-1} h^{-1}), much less than commercial relevance [13, 30]. Furthermore, most semiconductors have a large bandgap that is unable to photoexcited by visible light, limiting the effectiveness of photocatalysis. The UV light spectrum of TiO_2 is the most selectively photoactive of all the single-component photocatalysts. The ultraviolet spectrum of solar radiation is narrow, while visible light makes up most of the solar spectrum. Consequently, the photoactivity is limited in pristine TiO_2 and other wide bandgap semiconductors. Many modification procedures, such as the formation of heterojunctions, already address this issue with single-component photocatalysts. It is now the most alternative approach for improving spatial charge separation and visible light-harvesting with the formation of heterojunctions. Surface modification, doping, and defect engineering are examples of noble metal deposition [31].

Photocatalysis and other bismuth-based layered materials have emerged as promising and distinctive semiconductors for various applications. Because of the preferential hybridization between the Bi as its 6s orbitals with the O on to the 2p orbitals, most Bi-based can harvest visible light. Bi-based layered materials are also visible light photocatalysts because of the high stability of Bi^{3+} and the low cost of Bi-metal. Despite significant improvements in the NRR over the last decade, the results remain poor. It has recently been proposed that a Bi-based group metal with semiconducting properties could serve as a possible N_2 fixation catalyst [32, 33]. Bi-layered materials with unique and intriguing properties have made them attractive semiconductors for various applications, particularly in electronics. Less reactive metals may reduce HER by reductive N_2 adsorption to create N_2H^* without regarding to intermediate reactions' binding energy. In artificial NRR, Bi-based materials show remarkable effectiveness and selectivity [34, 25].

Two-dimensional Bi nanosheets synthesized by electrochemical reductions of bismuth oxoiodide (2D structure) are exceptionally selective in electrocatalytic NH_3 production. Bi-based NSs revealed a nanosheet-assembly micro flower structure in their natural state. Although inheriting the ultrathin nanosheet structure [34, 25], certain nanoflakes (10 nm) are scattered and alter the morphology, it can be seen that the Bi NSs surface appeared less smooth in Fig. 4b.

On the other hand, the Bi NSs outperformed the NRR performance on the electrocatalytic. The results were outstanding, where NH_3 output of 13.23 mmol/gh has been achieved, as shown in Fig. 3, RHE at –0.8 V, –1 V versus edging sites, and p-orbital electron delocalization was discovered. Additionally, the semiconducting attribute of restricting electron accessibility on the surface may improve fE (field electronic) at a sufficient degree of exposure, resulting in a more remarkable. The existence of inherent active edge sites in 2D Bi NSs makes a state of chemisorption along with activation of N_2 generate much easier by utilizing Bi as nanoparticles. As a result, it has been proposed that the metal Bi was an ideal morphological design. For N_2 activation, an electronic structure may be an effective method [18, 25].

A single crystal Bi NS was prepared using the same process as the Qiao previously described. A sheet-like morphology characterized the region, with a large surface area and several active zones. Catalytic properties of Bi NSs on carbon paper (Bi NSs/CP) were considered exceptional during the NRR [35]. Furthermore, the theoretical studies showed that exposed edge sites are more efficient than basal plane sites at adsorbing and activating N_2 molecules. Besides, it also revealed some enormous potential of Bi-based in the NRR materials. The discovery of Bi_2O_3, a Bi-based molecule with outstanding optical and electrical properties, has sparked interest in photocatalysis. However, due to its semiconducting properties, the limitation of charge transfer and ease of aggregation, which lacks the nanoscale state required to effectively limit N_2 adsorption and reactivity, thus Bi_2O_3 has received little interest for NRR. In addition, a substrate with a robust form is required to make higher conductivity of Bi_2O_3 NPs while also ensuring their dispersion [36]. In the other case, a unique free-standing catalyst (Bi_2O_3/functionalized exfoliated graphite, FEG) with the hybrid was prepared by attaching Bi_2O_3 with nanoplates form to the surface for FEG for NRR, leading to high nitrogen fixation with NH_3 output rate

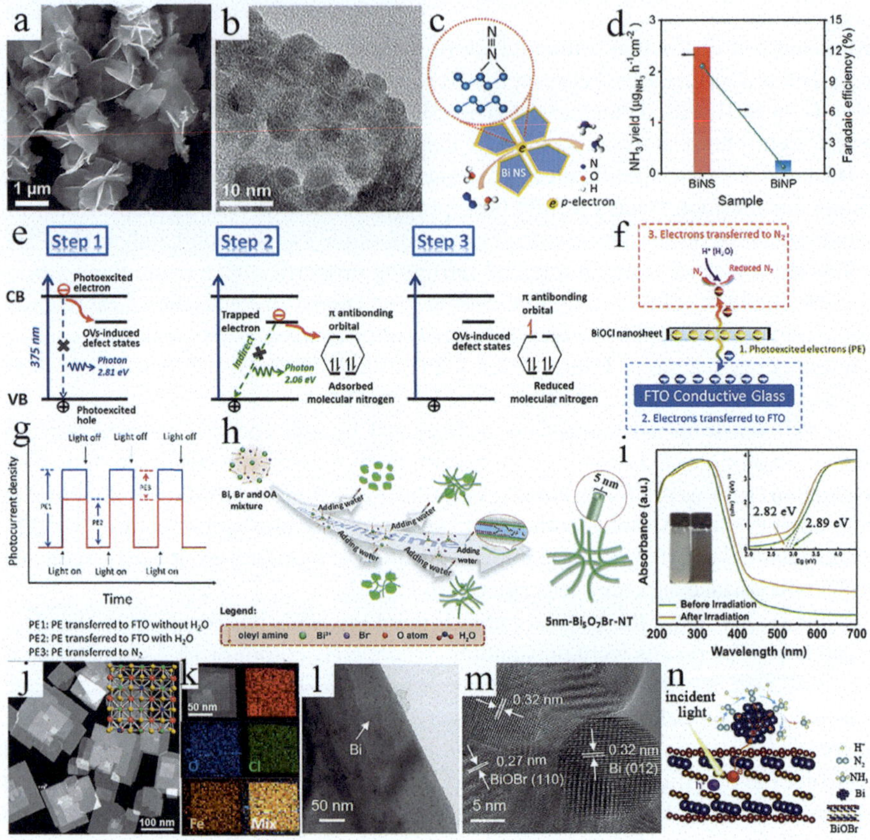

Fig. 4 **a** SEM and **b** HR-TEM pictures show the mosaic structure of Bi nanosheets. **c** NH_3 yields from Bi NSs and Bi NPs. **d** For the NRR, a schematic diagram of Bi NSs consists of three steps for the reduced molecular orbital. This image has been reproduced with permission. **e** An oxygen vacancy (OV)-induced approach for improving interfacial electron transport. **f** This micro-reaction system's electron transfer diagram and **g** this micro-reaction system electron transfer diagram of transient photocurrent response output curves. **h** Bi_5O_7Br NT and the UV visible light spectra (absorption) are formed before light irradiation and after **j** HAADF–STEM inset crystal structure, and **k** the mapping images of STEM–EDS of BiOCl NSs–Fe–5% **l** TEM with **m** HRTEM Bi/BiOBr. **n** Representation schematic of Bi/BiOBr composites for photocatalytic NRR

of 4.21 0.14 g^{-1} h^{-1} cm^2 and a FEG of 11.2% at –0.5 V versus RHE. The strong conductivity of FEG could compensate for Bi_2O_3 semiconducting nature and prevent aggregation.

There is also the issue of high selectivity but poor selectivity, as well as activity for present NRR catalysts and various reactive materials. The evolution of spots on the Bi [001]/[012] surface was examined. Theoretically, modeling catalytic NRR behavior, the immaculate (001) surface ability was slightly lower than that of the other surfaces. It was discovered that the catalytic ability has improved and Bi surfaces have been used to overcome several issues [37], including oxygen vacancy. NRR and a deficiency in Bi catalysts may boost HER activity. Extensive efforts have been devoted to increase the nitrogen fixation, however, due to protons preferentially coated these dopants, thus hindered the performance. The suitable catalyst for NRR applications was not derived using transition metal or Ca doping techniques.

Furthermore, the B dopants achieved the best results in increasing both NRR activity and selectivity [38]. Researchers will be able to better understand the NRR process in Bi catalysts due to these new defect engineering results, which will lead to new techniques to prepare more efficient Bi catalysts. The unique structure of BiOX aids in the production of active vacancies, which speeds up the catalytic process, while the indirect bandgap prevents charge recombination. Finally, in recent years, this kind of material has become more important in photocatalytic NRR [37–39].

On the other hand, the BiOCl-010-OV NRR route took a different strategy, with N_2 adsorbed side-on on its surface. This work elucidated proton transport on interfacial redox reactions by developing a micro-reaction system. In a non-protic environment, photoexcited electrons (PE_1) from BiOCl NSs could ordinarily travel from the ITO glass. When H_2O replaces the electrolyte, the competing photoexcited electron adsorbed N_2 and produces an obstruction that prevents photoexcited electrons (PE_2) from being transmitted to the ITO glass. Furthermore, based on the varying proportionate H_2O, transient photocurrent response levels indicated that the difference between photoexcited electrons (PE_3) from BiOCl to the adsorbed N_2 mirrored that of PE_1 and PE_2, as shown in Fig. 4g. According to the findings, the N_2 on OVs improves the adsorbed of N_2 by Lewis basicity, facilitating N_2 fixation and making the subsequent proton attack more favorable. Bi_2MoO_6 NSs that could possibly be rich in OV_S were created to have improved nitrogen-creating capabilities by visible light, inspired by this OV method [40]. On the other hand, the OVs on the BiOX surface are easily oxidized, resulting in a rapid drop in NRR activity.

To address this issue, researchers developed an ultrafine Bi_5O_7Br nanotube (Br_5O_7Br NT) with abundant and long-lasting OVs for better solar-driven NRR performance using a water-assisted self-assembled approach. After light irradiation, the band edge was red-shifted, as illustrated in Fig. 4i, and predicted at 2.89 and 2.82 eV [41, 42]. Meanwhile, the matching dispersion color shifted from light yellow to dark grey after irradiation. The dark grey dispersion will fade to a light golden tone. When the light is turned off, light-induced OVs are replenished by collecting O atoms from water or oxygen. As a result of the reversible color change outcomes, the as-obtained catalysts can generate OV sustainably and reversibly by light control. The doping technique has also resulted in a successful breakthrough in increased photocatalytic activity by securely separating the electron and hole of the catalysts. The photocatalytic NRR with OVs and doped Fe (BiOCl NSs–Fe) was greatly assisted by BiOCl nanosheets. When Fe doping reached 5%, the BiOCl NSs–Fe showed a strong NH_3 production of 1022 mmol g^{-1} h^{-1} under full-spectrum light irradiation, although the corresponding AQE was 1.8% at 420 nm [43].

Figure 4j and k exhibit the BiOCl NSs–Fe morphology that was dominated by square forms as we can see on the figures, with Fe components evenly distributed throughout the NSs. In addition to these approaches, Bi nanoparticles (NPs) were coated on BiOBr NSs as semiconducting support to hybridize the Schottky–junction photocatalyst based on Bi/BiOBr, where Bi NPs were uniformly loaded on the BiOBr NSs as shown in Fig. 4l and m. The interfacial charge transmission can be assisted by the intimate interface contacts [44]. As a result, the composite with the 2.0 wt % Bi loading of metal ratio produced 65 times the quantity of NH_3 compared to bare BiOBr. In addition, Fig. 4n demonstrated the surface adsorption process of N_2 with Bi nanoparticles, which corresponded to an earlier finding that N_2 could preferentially be adsorbed by BiOBr–OVs by the interaction of partially Bi that reduced its cations with an end-on bound on its structure. Finally, metal Bi plays a key role in managing focused photogenerated electron inflow and N_2 activation by NRR activation, serving as a paradigm for optimizing the potential of the Bi-metal hybrids semiconductor [45].

Bi-based layered materials, especially BiOX, have been predominantly used in photocatalytic systems, which can be useful for energy band fluctuation and the development of effective heterostructures. Future studies should focus on other crucial areas, such as electrocatalytic NRR. Furthermore, as enhancing the stability and photo-corrosive character of Bi-based photocatalysts is still a work in progress, further research is required to completely understand their photostability in photocatalysis [45, 46]. However, recent discoveries and properties of Bi-based layered materials lead to a bright future in nitrogen fixation.

To summarize, under the ambient condition, the critical challenges in the diverse layer are structured based on the material systems that could produce NH_3 yield and with the NRR selectivity. Several materials, namely, MXene, TMDs (transition-metal dischalcogenides), COFs (covalent organic frameworks), MOFs (metal–organic frameworks), LDHs (layered double hydroxides), and Bi-based compounds have low bandgap but high yields. Meanwhile, due to their lower cost, metal-free materials are more likely than metal-containing materials to meet the requirements of long-term practical usage [47]. NRR cognition is still in early state, despite the recent release of a huge amount of theoretical research on layered materials. Metal-free materials (such as graphene, CN, BP, and B-based materials) have high Es, but low yields at low applied potentials due to nonmetallic characteristics. MXenes, TMDs, COFs, MOFs, LDHs, and Bi-based compounds, on the other hand, have low fees but high yields. Meanwhile, due to their lower cost, metal-free materials are more likely than metal-containing materials to meet the requirements of long-term practical usage [48].

2.1 Properties of Nitrogen Molecules

The natural N_2 conversion is comparable to artificial N_2 conversion. In the case of nodular bacteria, it has been demonstrated that they can generate 50–150 kg of air/acre, hence a characteristic decentralized with an industrially inappropriate source of N_2. Nitrogenase, a metalloenzyme that converts N_2 to ammonia under suitable conditions than the Haber–Bosch process in industrial systems, this is used to convert N_2 to ammonia in biological systems. Some microorganisms require

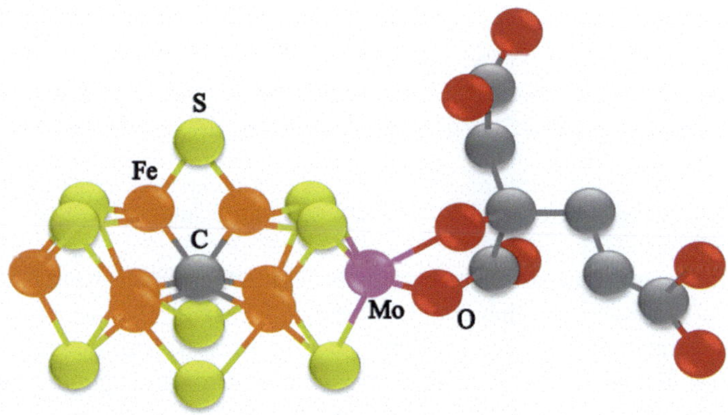

Fig. 5 Schematic illustration of mitrogenase active site and its co-factor, which contains Mo and Fe (Reproduced with permission)

nitrogenase enzymes to survive [2, 3]. In nitrogenase active site, based on the co-factor composed of iron, molybdenum, and sulfur atoms based on interstitial on the carbon atom was found as depicted in Fig. 5. The chemical energy required for this action is provided by hydrolyzed-ATP, which is known as adenosine-5-triphosphate. The electrons are delivered rapidly, Fe-based as the component of the enzyme that accomplishes hydrolyzed—ATP. According to Eq. (4), the entire reaction needs the transmission of eight electrons [13, 49]:

$$N_2 + 8e^- + 16ATP + 8H^+ \longrightarrow 2NH_3 + H_2 + 16ADP + 16P \quad (4)$$

Increased N_2 Lewis acid/base activity by activating t at the co-factor overcomes the difficulties of cleaving the N_2 triple bond, which has the enthalpy value about 945 kJ/mol or 10.0 eV. Protonation and reduction become easier as a result of electron donation to the metal center to the antibonding orbitals of p* in the N_2 [50].

Artificial photosynthesis, in particular, has become a widely debated topic due to the replacement of biological stages with chemical pathways directly or indirectly, which mimics the processes involved in natural photosynthesis (although it still is less efficient than natural photosynthesis). The artificial N_2 conversion can also be accomplished via a photocatalytic approach.

The nitrogen reduction by direct light capture is more thermodynamically better than water splitting [19, 49]. Since the 1940s, active photocatalysts for N_2 photoreduction have been identified as TiO_2, Fe_2O_3, and ZnO_2. These compounds are abundant in desert soils and can remove up to 107 tons of nitrogen annually. These catalysts could theoretically produce ammonia from air and water (Eq. 5):

$$N_2(g) + 3H_2O(g) \longrightarrow 2NH_3(g) + 1.5O_2 \quad \Delta H = 633.2\ kJmol^{-1} \quad (5)$$

At pH = 0, ammonia in aqueous will be divided by two sections that make the solution become ammonium (Eqs. 6 and 7).

$$3H_2O(1) \longrightarrow 1.5O_2(g) + 6H^+(aq) + 6e^- \; E^0_{SHE} = 1.23\ V \tag{6}$$

$$N_2(g) + 8H^+(aq) + 6e^- \longrightarrow 2NHH_4^+(aq) \; E^0_{SHE} = 0.274\ V \tag{7}$$

As a result, photocatalyst semiconductors with approximate bandgap 1 eV or more are suitable band gap that could theoretically execute the function. It is also worth mentioning that hydrogen evolution (0 V at pH = 0) and nitrogen reduction have identical potentials. The reduction of one molecule of N_2 in photocatalysis and photoelectrochemistry, for example, needs the transfer of six electrons, leading to rate-determining phases, favoring the formation of H_2. Catalysts will be required to overcome the charge buildup activation barrier. Furthermore, the O_2 evolution process typically requires large overpotentials (OER). This phenomenon is similar to multi-electron transfer-based CO_2 reduction techniques [50, 51]. Therefore, nitrogen reduction and CO_2 reduction compete with hydrogen evolution processes, posing a significant stumbling block for ammonia selectivity. As limited solubility of nitrogen in the water, the most widely used reaction medium, limits the ability to obtain high ammonia selectivity [11] and the solubility of nitrogen in water. Besides, the most commonly used reaction medium causes limitation ability to achieve high ammonia selectivity.

Schnauzer found hydrazine with the concept of a progressive reduction of N_2H_2 and N_2H_4. Even though the first chemical reaction is similarly high endothermic, the free energy per electron is 2.5 eV lower than TiO_2 bandgap [11]. Water vapor, as well as sacrificial chemicals, have been proven to increase activity under specific circumstances. The optimal reaction temperature is between 80 and 120 °C. The transfer of an electron and a proton to N_2 initiates the reduction of adsorbed nitrogen, resulting in the HN = N* species. The proton or electron transfer occurs sequentially at one nitrogen, ammonia released while leaving the other reaction, or the transfer occurs alternatingly at both N_2, ammonia released while leaving one in place (Fig. 5). The associative and dissociative mechanisms for N_2 hydrogenation are distinguished by the sequence in which the N = N bond is broken [26].

2.2 *Nitrogen Fixation to Ammonia via Photocatalysis*

Photocatalysts for N_2 reduction to NH_3 must be capable of adsorbing and activating highly inert N_2 molecules, allowing for a moderate thermodynamically uphill NRR, according to the conditions described earlier. Kinetic constraints imposed by reactant (N_2) diffusion to active sites and electron transfer mechanisms presently limit NRR reaction rates. When choosing a photocatalyst for NRR, the inherent features of the semiconductor (e.g., band gap, VB and CB levels, stability, morphology) and how it is given for the photoreaction must be considered. In this section, we divided the photocatalysts in NRR into inorganic photocatalysts, biomimetic photocatalysts, and other emergent photocatalysts. Inorganic photocatalysts are classified by the kind of inorganic material utilized, which includes titanium dioxide-based materials, bismuth oxyhalides materials, and carbonaceous materials.

Capturing solar energy to manufacture fertilizers and fuels is challenging and time-consuming [52]. The green ammonia produced by nitrogen photo fixation has recently attracted researchers' interest. After biological N_2 fixation and lightning discharges, abiotic photo fixation of dinitrogen in soils and sands has been proposed as the third most important source of natural nitrogen fixation. Solar-driven N_2 fixation is essential knowledge to understand and control the nitrogen cycle. Even though early research concentrated mostly on titania-based catalysts, in which the photocatalytic N_2 fixation has progressed substantially in recent years.

Lightning bolts and volcanic eruptions only produce a small percentage of the oxidatively fixed nitrogen because direct nitrogen oxidation is a kinetic and thermodynamically expensive process. In addition, no living creature has ever created an enzyme capable of oxidative nitrogen-fixing, proving that N_2 oxidation is impossible. Thus, nitrogen-fixing is always performed through the nitrogen reduction process, whether in nature or industry (NRR). By using photocatalysis, N_2 molecules can be captured and activated on the photocatalyst surface, which generates electrons and protons in the electrolyte that react with them to form NH_3 and other nitrates by using photogenerated electrons and protons [52, 53]. Because of the competing hydrogen evolution process, the limited selectivity of existing photocatalytic NRR is a critical issue (HER). Based on the NHE reference scale, the classical redox by its equilibrium potentials for the HER and NRR are 0 V and +0.092 V, respectively. The demonstration of the HER, which uses a two-electron transfer mechanism, is more thermodynamically advantageous than the NRR, which uses a six-electron transfer mechanism. kinetically unfavorable. The huge energy difference between the HOMO (s-2p) and antibonding LUMO p* 2p and the weak proton affinity of dinitrogen molecules make NRR. Selectively choosing electrolytes that restrict proton transfer while enhancing N_2 solubility would improve total NRR efficiency significantly. Since NRR only contains the photocatalyst reduction component, scavenging agents like methanol and EDTA are frequently utilized in its reaction fluid to prevent electrons from interacting with the hole, at the end which can minimize charge carrier recombination [16, 54].

The NRR process of heterogeneous photocatalytic can be assumed to be similar to that of a nitrogenase enzyme, except that it does not produce H_2:

$$N_2 + 6e^- + 6H^+ \longrightarrow 2NH_3 \quad E^0 = +0.55\text{v Vs NHE} \tag{8}$$

The proton-coupled electron transfer (PCET) transfer mechanism is used in the first step of the NRR process when N_2H formed on the catalyst surface based on Eq. (9)

$$N_2 + H^+ + e^- \longrightarrow N_2H \quad E^0 = -3.2\text{v Vs NHE} \tag{9}$$

Equation (10) demonstrates that producing N_2H requires significant negative reduction potential. Other intermediates produced in the NRR include hydrazine (N_2H_4) and diazene (N_2H_2), which is due to the participation of four and two-electron transfer, respectively, according to equations below:

$$N_2 + H^+ + 4e^- \longrightarrow N_2H_4(g);\quad E^0 = -0.36v \text{ Vs NHE} \tag{10}$$

$$N_2 + 2H^+ + 2e^- \longrightarrow N_2H_2(g);\quad E^0 = -1.10v \text{ Vs NHE} \tag{11}$$

On a catalytic surface that is heterogeneous,

The mechanism of the NRR can currently be classified into two types: dissociative and associative processes. As seen in Fig. 6, the triple N≡N bond of the N_2 molecule is first broken, followed by the hydrogenation process, which leads to the production of NH_3 molecules. In the (HB) Haber–Bosch process, where the interaction between H atoms occurs after the cleavage of N_2 and H_2 molecules (Fig. 6a), the dissociative mechanism is thought to be active. Unlike the dissociative mechanism, the associative mechanism begins with N_2 hydrogenation, breaking N≡N bonds and forming NH_3. Furthermore, the associative mechanism has two hydrogenation channels: distal and alternating (Fig. 6b and c). The other possible route is that protons are added to nitrogen atoms far from the catalytic surface, but in the alternate pathway, two nitrogen atoms are protonated before converting to NH_3 [19, 52]. Since photon does not have the energy to break the highly stable N≡N bonds, the photooxidation process must rely on an associative mechanism, similar to biological nitrogen fixation (BNF) operates when nitrogenase enzymes are present. In summary, trapping N_2 molecules at the photocatalyst surface, extending light-harvesting to the visible spectrum, and spatial separation of photogenerated charge carriers are all essential factors in improving the overall photocatalytic NRR efficacy.

2.3 NH_3 Quantification/detection Using Various Methods

Due to catalytic efficacy is weak, the production rate of NH_3 is relatively low ppm/ppb, and some species will interfere in the measurement. However, detecting and quantifying NH_3 generated from N_2 fixation remains challenging [5, 12, 55]. This area is heavily governed by international standardized organizations such as the American Public Health Association (APHA), the American Society of Testing Materials (ASTM), and the United States Environmental Protection Agency (EPA) (Fig. 7).

The presence of specific ions, the chemical makeup of the substance, the pH of the medium, the type of sacrificial agent used, and the solvent have a role in the accuracy of detection procedure. The nano/micromolar level of NH_3 produced by the N_2 photofixation mechanism was involved in the measurement [55]. As a result, repeating or applying multiple testing procedures is critical to obtain a consistent result or a

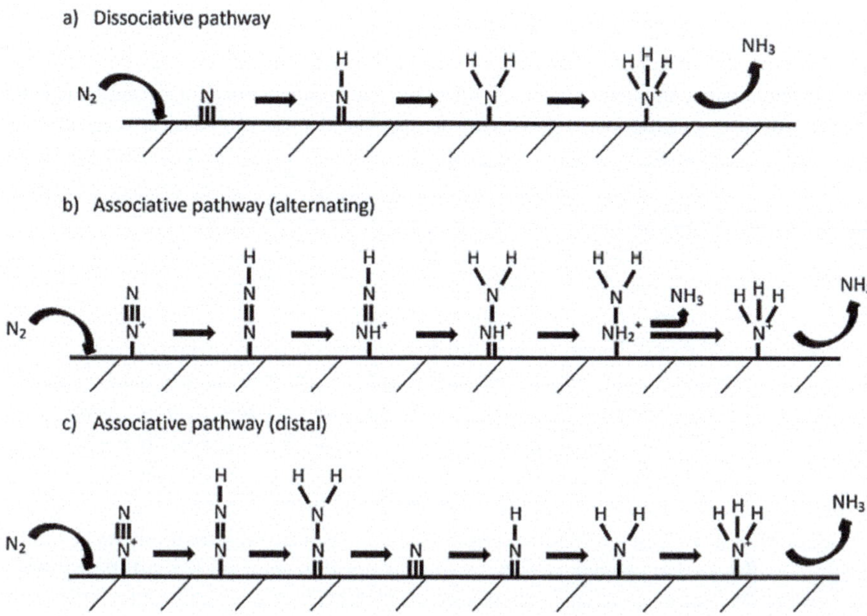

Fig. 6 Theoretically hypothesized nitrogen-fixing pathway, the three nitrogen-fixing mechanisms are the dissociative route, the alternate associative pathway, and the distal associative pathway (Reprinted with permission)

Fig. 7 Several ammonia detection techniques (Reproduced with permission)

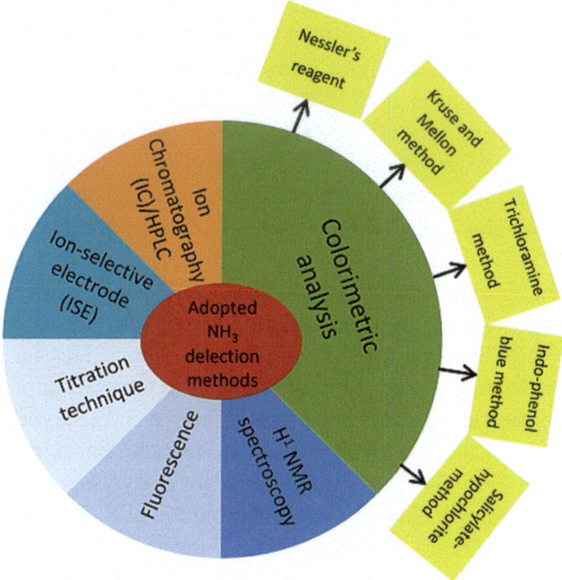

minor data error. For example, NH_x-containing photocatalysts and sacrificial agents, as well as their secondary products, interact with the generated NH_3, resulting in an incorrect result; thus, these issues must be addressed to obtain accurate results [56]. To improve production, sacrificial reagents (hole scavengers such as methanol, humid acid, ascorbic acid, ethanol, isopropanol, and others) or their oxidized product are used. Nevertheless, the combined synthesis of ammonia on rare occasions results in inconsistencies [57].

Furthermore, chemical species containing N-atoms used in photocatalyst synthesis (i.e. HMT, oley-amine, TAA, etc.) may be leached or dissolved into the experimental solution, affecting the quantification result. A control experiment in the presence of Ar or $^{15}N_2$ is carried out to prevent these phenomena [57, 58]. In addition, a unique catalyst was prepared (i.e. g-C_3N_4, boron nitride, etc.) with the catalyst containing N-atoms. The NH_3 yield is incorrect when nitrogen (amine/nitro functionalized MOF, etc.) is evaluated for N_2-fixation. Accordingly, if the solvents used in the fixation experiment contain a trace of NH_3 from another source, such as ambient ammonia, the aqueous medium in which the N_2 photoreduction occurs may be contaminated with an overestimation of NH_3 concentration after analysis [17, 25]. Isotopic leveling, also known as ^{15} N testing, is another method for validating NH_3 synthesis by N_2 photofixation, with the result ($^{15}NH_3$) being assessed by H^1-NMR methods. Enzymatic analysis can also determine ammonia production, bypassing quantification and detection issues.

In summary, in the 6.8–7.05 ppm range of H1-NMR spectra, $^{14}NH^{4+}$ generated from ^{14}N exhibits a triplet with a J-coupling of 52 Hz, whereas ^{15}N exhibits a doublet with a J-coupling of 73 Hz. The experiment and NMR analysis for the ^{15}N isotope leveling market is completed using the readily available $(^{15}NH_4)_2SO_4$ reagent. The H1-NMR spectra show both ^{14}N and ^{15}N signals. However, the ^{15}N isotope has a higher intensity than ^{14}N [44, 59, 60]. It is difficult to find an effective quantification/testing process for NH_3 concentration at ppm and ppb standards that consider the effects of interferants. This problem is due to a lack of database or knowledge on the interfering variables that affect NH_3 detection at nano/micromolar concentrations. This collection covers the most common NH_3 detection/quantification procedures, including Nessler's reagent method, indophenol blue method, SE, and chromatography method, while considering pH, presence, sacrificial agents, the solvent employed, and N-atom containing catalyst. The approach is more reproducible and helpful than spectroscopic processes such as Nessler's and indo-phenol blue, but it is also more expensive and complex in terms of apparatus [61, 62]. Conversely, the colorimetric approach is a cost-effective and straightforward determination method.

The colorimetric approach is favored because of high selectivity/sensitivity, higher trustworthiness, and high efficiency/assured durability/compatibility [7, 57]. According to Nessler's reagent (solution of K_2HgI_4 and KOH), NH_3 forms a reddish-brown colored complex with iodine and mercury in an alkaline medium (in the absence of interferants, Eq. (12)) that has an optical absorbance of about 420 nm and is directly proportional to the concentration of ammonia in the solution. Rochelle salt is sometimes added to the reagent during testing to reduce the impact of interfering ions [62, 63].

$$2[HgI_4]^{2-} + NH_3 + 3OH^- \longrightarrow Hg_2ONH_2I + 7I^- + 2H_2O \quad (12)$$

There are several factors to consider for N_2 detection. The Nessler's chemical is only stable for three weeks, so it should be used instantly; (i) if the reagent has to be prepared, ultrapure water or an NH_3-free solvent should be used; (ii) the reagent must be handled with extreme care because of the presence of toxic mercury; and (iii) data should be obtained within 10–30 min after mixing the NH_3-containing solution with Nessler's chemical [64] Furthermore, specific ions that increase the turbidity of the medium or N_2 fixation side products/intermediates (urea, hydrazine, glycine, amine, etc.) can affect the detection procedure [65, 66]. As shown in Eq. (13) (Berthelot reaction), an alkaline environment leads to ammonia reacting with phenol and hypochlorite to produce a blue product (indophenol blue).

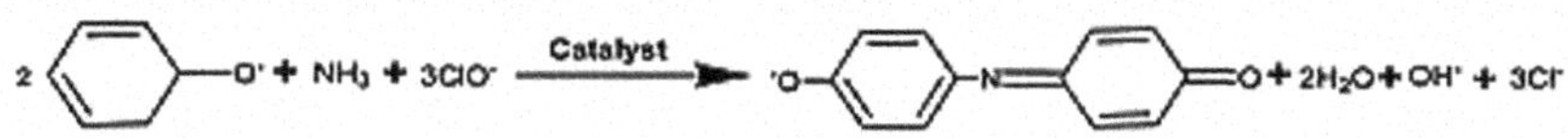

(13)

In this experiment, the color intensifier sodium nitroprusside and the pH stabilizer citrate buffer were utilized, as well as indophenol and Nessler's reagent detection techniques, which yield reliable and acceptable findings at lower NH_3 concentrations, i.e., within 0.5 g/L, according to Zhang et al. Above this range, however, Nessler's procedures are dependable and effective, while the indophenol process fails to yield precise data, resulting in erroneous/overestimated NH_3 concentration [66]. A change in pH also impacts the detecting process, resulting in a more accurate result. Nessler's method works across a wide pH range, from acid to alkaline to neutral (in an acidic medium). The role of ammonia, iodine, and mercury in acidic media could cause this issue. Various pH solution causes the color of the solution changes from yellow (in neutral) to green (in acidic), altering the measuring technique. Since sodium hypochlorite is unstable, this method could mislead ammonia concentration in the acidic solution. Consequently, indophenol blue techniques can only be employed with a low error rate in alkaline and neutral media, but Nessler's reagent can be used in all three media [45].

Furthermore, the presence of specific ions affects detection processes. In Nessler's approach [22, 47], for example, ions with a wavelength of 420 nm that chemically reacts with the reagent to produce interference may lead to an error. Similarly, sacrificial agents, the oxidized product generated during the reaction, and their fraction in solution cause inaccuracies in concentration estimation using the colorimetric approach in photocatalytic N_2 fixation. The hue solution color is highly affected by a variety of agents and intermediates (methanol, ethanol, DMF, DMSO, acetone, formaldehyde, formic acid, etc.). The hue looks pink in formaldehyde and light yellow in the presence of formic acid when employing the indophenol blue technique. Consequently, the concentrations determined by Nessler or the indophenol blue approach are erroneous or exaggerated. Neither spectroscopic analytical approach is optimal for ammonia quantification in organic sacrificial chemicals; nonetheless, the calormetric method is favored when measurements are done under controlled conditions and in the presence of organic sacrificial reagents (as shown in Fig. 8).

Furthermore, owing to its large detection range of 30 ppb to 1000 ppm, good selectivity, and low sensitivity to interfering compounds, the ion-selected (ISE) detection approach is frequently employed [22, 68]. However, the deficient concentrations and instrumental drift make maintaining the operating state challenging. ISE offers two key benefits over colorimetric methods: it has an extensive detection range (0.03–1400 NH_3-N mg L^{-1}) and is not affected by turbidity or color. Finally, since columns

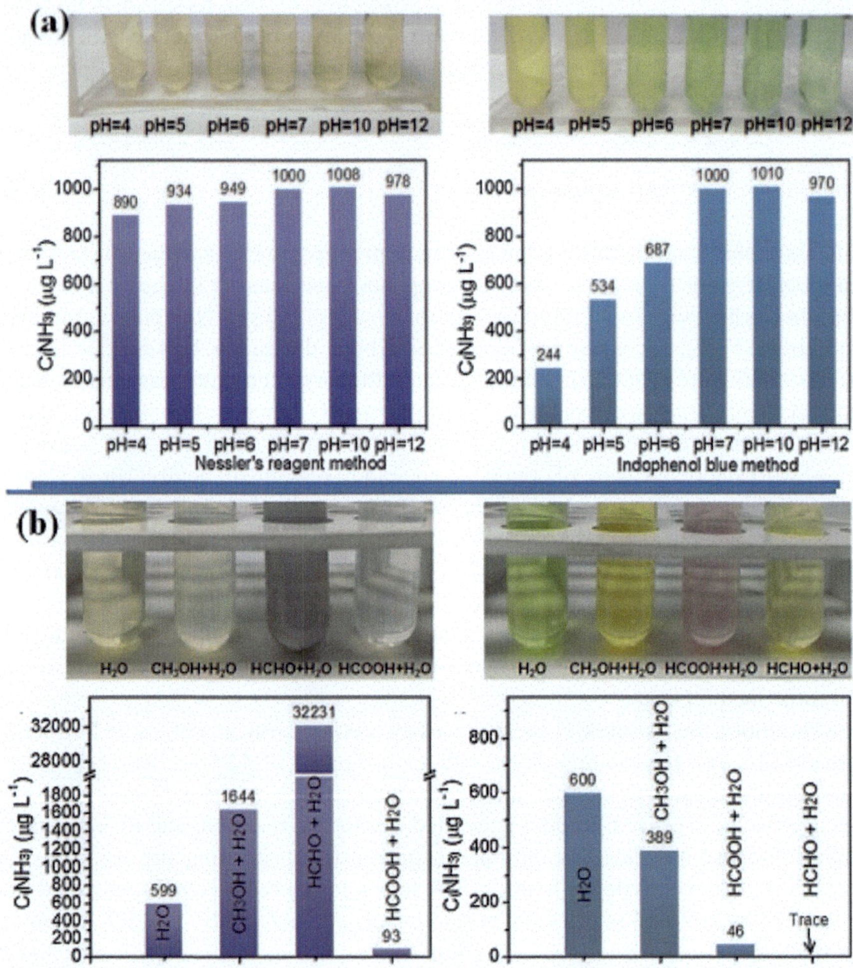

Fig. 8 **a** Photographs of various ammonia concentrations calculated by each method; **b** ammonia concentrations corresponding with photographs of NH_3 calculated by Nessler's reagent method with the indophenol blue method for checking various oxidizing reagents to check the presence [67] (Reprinted with permission)

separate the substrate components, the chromatography approach is more reliable. Numerous worldwide organizations and scientific groups supported this approach for NH_3 measurement. Adding a cation-exchange column and a conductive detector to atomize the apparatus expands the range of NH_3 estimation while decreasing analysis time, resulting in reproducible, precise, and accurate findings [18, 41]. Figure 8 demonstrates a practical approach for detecting ammonia in specific conditions.

3 Research Works

Bismuth (Bi)-containing materials for photocatalytic have garnered vast attention in recent years owing to extraordinary optical properties (lower band gap enables to absorb wide solar spectrum) and robust photocatalytic performance for carbon dioxide reduction, organic pollutant degradation, water splitting, and nitrogen fixation to ammonia. Bi with oxy-halides like $BiOI_{1-x}Br_x$ a sort of +5, +6, +7 treble composite semiconductors that are usually designated by the basic formulation [$BilO_mX_n$], where X represents iodide, chloride, and bromide [30]. It has a layered structure and consists of two photocatalytic responses namely internal electric field (IEF) and a superficial potential (Vo) [69]. These topographies promote the formation and departure of photo-generated charge carriers, as well as activation of reactants [70]. In 2018, Xin Y. et al. effectively prepared Bi ox-bromide (BiOBr) nanosheets with abundant oxygen vacancy on the unprotected (001) surfaces (Fig. 9a), which confirmed the outstanding role of photocatalytic N_2 reduction (104.2 mol/gh of NH_3 under visible light illumination) without any additional sacrificial agents or co-catalysts (Fig. 9b). The bond length of triple bond in nitrogen absorbed on oxygen vacancy site was 1.133 against 1.078 for molecular nitrogen in the gas phase), according to theoretical forecasts, the lengthening of the triple bond in bond providing a sturdy indication for the activation of nitrogen (Fig. 9c and d). Interfacial electron transfer mechanism was strategically designed based on extensive lab experiments, where a large number of oxygen vacancies on the nanosheet shallow band of BiOBr formed a defect state between the conduction band and valency band. Therefore, it can be dynamic trapping excited electrons from the conduction band of BiOBr, resulting in straight recombination of photoexcited electron and hole formation. To complete N_2 reduction activation, the electrons might be injected into the vacancy of antibonding orbitals and adsorbed by nitrogen molecules. In the presence of oxygen vacancy (OV), reducing nitrogen to ammonia provided a low-energy pathway for ammonia production.

Kaiyue et al. investigate the $Bi_{12}O_{17}Br_2$, and Nan et al. determined BiOCl with oxygen vacancy on the remarkably uncovered (010) and (001) surface to get a more profound knowledge of nitrogen reduction to ammonia on diverse facets (Fig. 10a); nitrogen molecules adsorbed on the surface of OV was transformed into ammonia via numerous minimum-energy chemical processes including consecutive proton-assisted electron transmission step [71]. The density functional theory simulations and in-situ R where both N_2 adsorption mode and the energy levels for

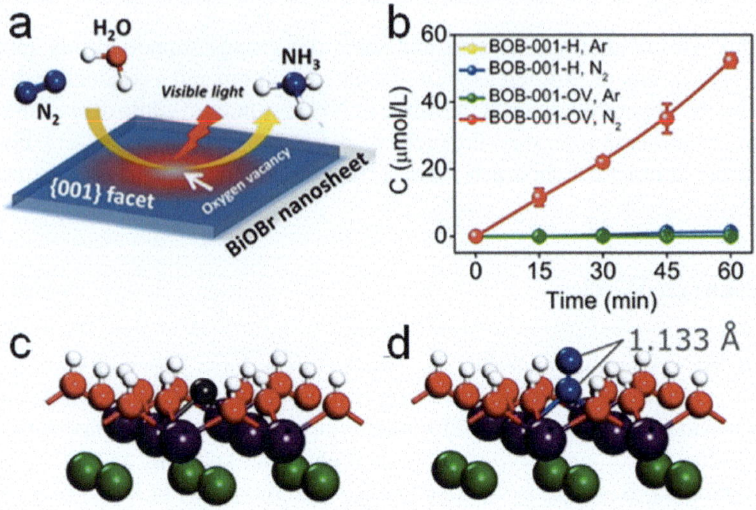

Fig. 9 **a** Illustration of nitrogen fixation on BiOBr-(001)—oxygen vacancy, **b** the rate of ammonia generation when exposed to visible light, **c** BiOBr (001) shallow with oxygen vacancies from the side, **d** a theoretical calculation of nitrogen adsorption and initiation on a BiOBr (001) exterior oxygen vacancies. Purple compasses represent bismuth atoms, green, red, and white spheres represent Br, O, and H atoms, respectively, meanwhile blue spheres and grey for oxygen vacancy [65, 68] (Reprinted with permission)

activation site are altered by the BiOCl surfaces (Fig. 10b). Nitrogen adsorbed on the oxygen vacancy places of the BiOCl (001) facet with end-on direction style, allowing for an as-symmetric distal route to NH_3. The vacancy oxygen on the BiOCl plane (010) surface, on the other hand, preferred nitrogen adsorption through a side bridging coordination geometry, subsequent in mostly N_2H_4 production via an irregular electron and proton allocation way (Fig. 10c). The superficial investigation highlighted how various semiconductor surfaces consists of oxygen vacancy may impact nitrogen fixation development, providing the outline for upcoming research toward more well-organized ammonia synthesis photocatalysts. Bandgap structures may be developed to increase the activation of nitrogen in the calculation of oxygen vacancy. The methods of calcination and hydrolysis may be utilized to generate oxygen vacancy exposed on Bi_5O_7I nanosheets with surfaces of plane (001) and (100) facets. The better-quality NH_3 production rate over Bi_5O_7I with plane (001) surfaces (BiOI) plane (001) was clarified by a negative conduction band (1.45 eV) than its complement, leading to more favorable electrons with greater energy to drive nitrogen fixation for ammonia creation under simulated light irradiation. Other Bi-based compounds, such as BiO, [34] Bi_2MoO_6, [25], and $Bi_2O_2CO_3$ [35], have been exploited as photocatalysts for NH_3 production on the compound of bismuth oxyhalides. Under the simulated sunlight process, BiO quantum dots with an average size of 2–5 nm under the simulated sunlight process confirmed catalytic activity for nitrogen reduction reaction [34]. The activation sites for nitrogen reduction were determined by low valence surface Bi^{2+} species. Bi_2MoO_6 photocatalysts have also been proven effective in producing ammonia from atmospheric air. The (100) edge exposed coordinatively unsaturated Mo atoms in Mo–O polyhedral contributed to the Bi_2MoO_6 crystals inherent N_2 activation capability. The solvothermal synthesis procedure followed by a hydrogenation treatment was employed to obtain altered crystal structure and the flawed engineering in the mesoporous Bi_2MoO_6 basis [70, 72].

The two-dimensional Bi-oxyhalides and similar Bi-based photocatalysts are used as model materials to investigate the effect of oxygen vacancy, surface engineering, and band gap structure alteration on photocatalytic nitrogen reduction to ammonia [73]. The investigation was mainly based on computational theoretical to understand

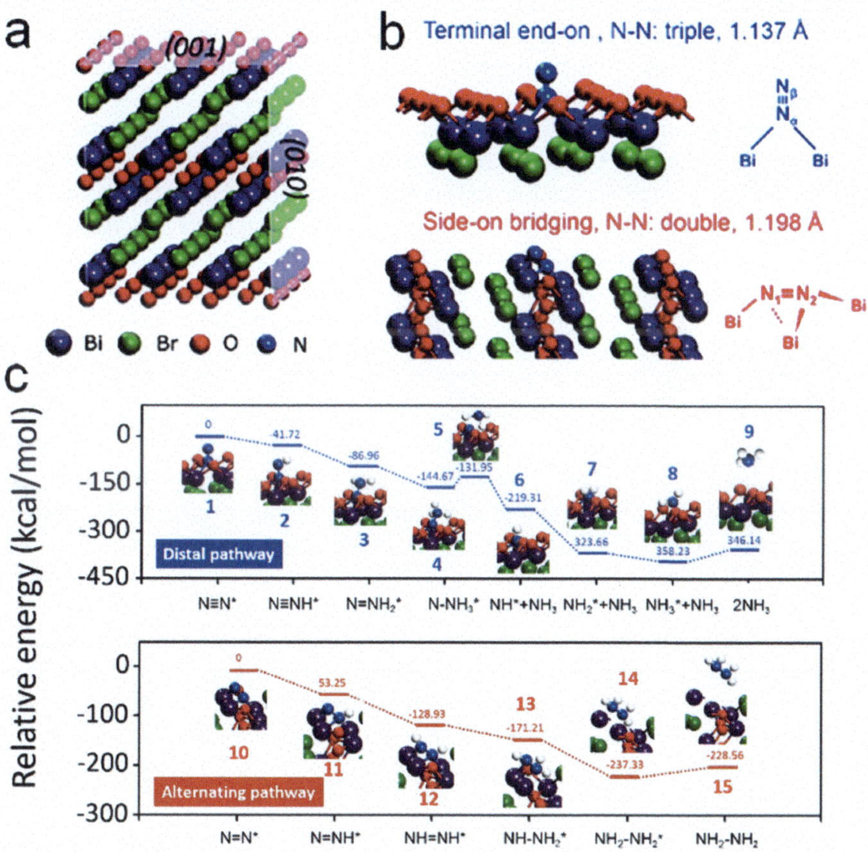

Fig. 10 **a** BiOCl structure with surfaces (001), (002), and (010) facets. **b** The side on bridging adsorption mode of nitrogen on (010) facets of BiOCl and the terminal end on adsorption mode of nitrogen on (001) facets, **c** the distal and alternate pathways for nitrogen fixation are mediated by oxygen vacancy on the (001) and (010) facets. This image has been reproduced with permission

the facet-dependent nitrogen adsorption mechanism as well as the ammonia production process (distal or alternating paths), paving the way for the upcoming growth of better 2D photocatalysts.

3.1 Nitrogen Fixation Using Different Materials

Photocatalysts for nitrogen fixation are now being researched using bismuth-based materials, g-C_3N_4 grounded resources, and metal supplies with SPR. Additionally, different types of semiconductors displayed considerable promise in this application. TiO_2 is a good photocatalyst with a track record for photocatalytic N_2 reduction. Besides, there have been many recent breakthroughs in photocatalytic N_2 fixation action. Zhang et al. produced a tunable TiO_2 by heating TiO_2 with $NaBH_4$ at optimal temperatures (310–360 °C) in an Ar gas. As the temperature increased, the color of the TiO_2 samples deteriorated, as shown in Fig. 11.

The maximal ammonia output of the sample obtained at 340 °C was 324.86 mol/h.g, which is 3.85 times higher than pristine TiO_2. The oxygen vacancies can enhance charge separation and nitrogen adsorption capacity, resulting in augmented photocatalytic activity. The heterojunction design may increase TiO_2 ability to fix nitrogen photochemically. For example, Rong et al. discovered that $TiO_2/ZnFe_2O_4$ heterojunction photocatalysts with a Z-scheme required great advanced photocatalytic performances than single TiO_2 or $ZnFe_2O_4$ [44]. The Z-scheme heterojunction can increase electron–hole pair separation and transport, leading to high photocatalytic activity [73, 74].

Other types of new photocatalysts for N_2 reduction have been promoted and have the possibility to be developed. Luo et al., for example, Fe-$SrMoO_4$ with considerably increased nitrogen fixation photocatalytic activity. The catalytic performance can be enhanced due to lowered band gap, advanced light pre-occupation, and newly produced active centers Fe-Mo [75]. Using heterojunctions is a highly efficient way of improving the parting and broadcasting of photogenerated electron–hole pairs. A CeO_2/FeS_2 heterojunction and a $CeCO_3OH$/g-C_3N_4/CeO_2 heterojunction have been characterized and potential as photocatalysts for N_2 reduction. Engineering defect is a popular way of altering catalysts to produce high-performance N_2 reduction. Sun et al. developed sulfur V-rich O-MoS_2 modified CdS nanorods with a photochemical NH_3 production yield of 8.2 mmol/l.g.h. It was discovered by introducing oxygen into the S-vacancy enhances conductivity and reduces the bandgap for N_2 fixation, which is essential in high photocatalytic performance [73, 74].

3.2 Reaction Mechanism in Yielding NH_3.

Understanding the basic N_2 fixation process is significantly important since the better we understand the reaction process, the more effectively we can work to advance N_2

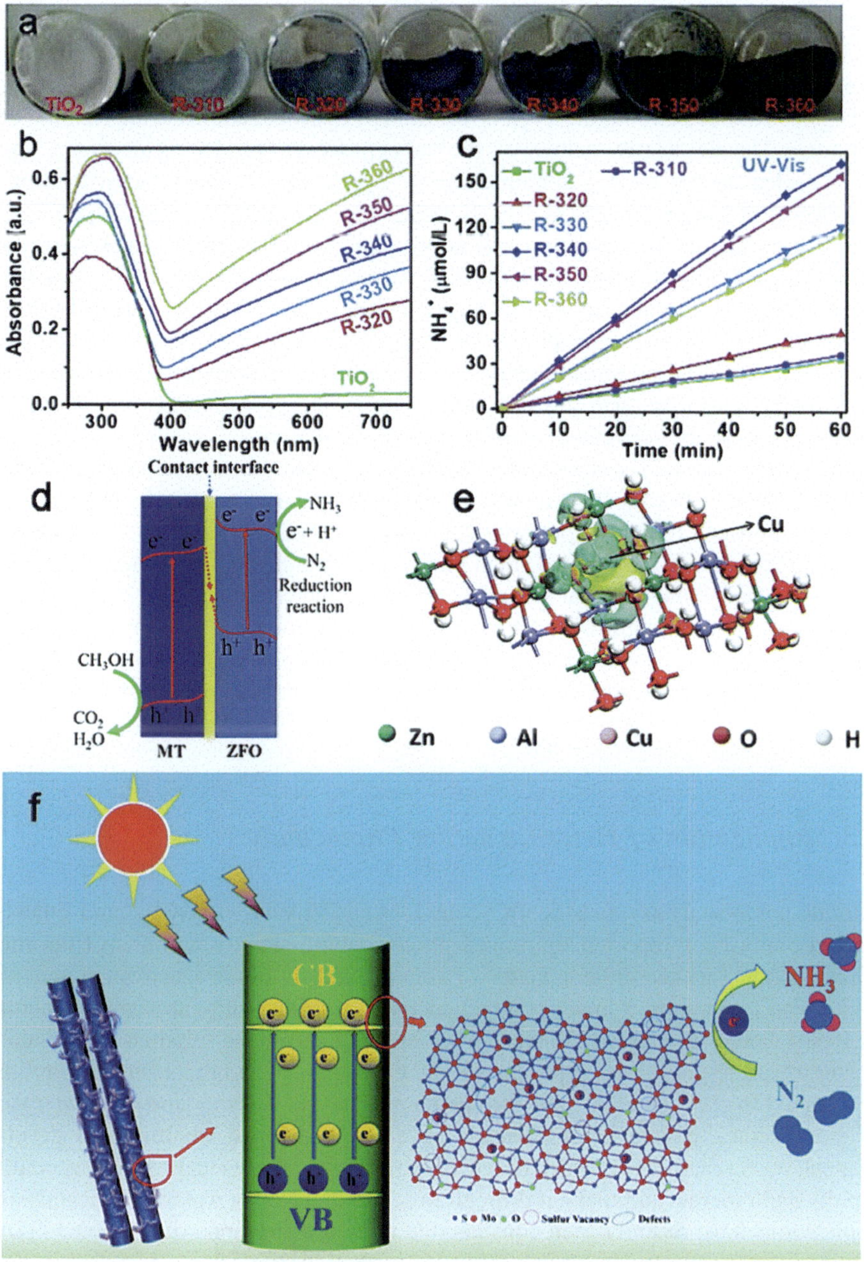

Fig. 11 **a** Photograph image of TiO_2 samples, **b** DRS UV–Vis, **c** photocatalytic performance for NH_3 yield, **d** nitrogen fixation mechanism of the $TiO_2/ZnFe_2O_4$ heterojunction photocatalyst using the Z-scheme, **e** the structure of ZnAl-LDH and charge density pattern on its surfaces, **f** illustration of N_2 reduction mechanisms and charge separation on MoS_2/CdS merged photocatalysts [47]

reduction competence. There are well-known catalytic nitrogen fixation methods, each involving four pathways: distal associative, dissociative, alternative associative, and enzymatic. The dissociative pathway uses high energy to break triple-bond nitrogen on molecules completely; it is followed by hydrogenation and the creation of ammonia. The dissociative pathway is the Haber process of producing ammonia. The $N{\equiv}N$ is cleaved step by step in both associative pathways, although the hydrogenation occurs in various routes [75, 76]. In the cases of the distal associative pathway, the occurrences of hydrogenation first at N_2 atom farthest from the catalyst. After the nitrogen atom is hydrogenated, the NH_3 molecule will be released. The alternative approach is in which two N_2 atoms are hydrogenated at the same time. For the enzymatic pathway, hydrogenation is typically observed in the reaction transported out by nitrogenases and other catalysts. N_2 is adsorbed on nitrogenases/catalysts at the atom edges, rather than just one side, as previous approaches. In this process, each nitrogen atom undergoes synchronous hydrogenation.

In this process, each nitrogen atom undergoes synchronous hydrogenation. The surface hydrogenation mechanism was proposed by Ling et al. Surface hydrogenation as a new nitrogen reduction pathway can occur on the superficial of noble-metal-based metal catalysts. The initial stage in this mechanism is the low-energy conversion of H^+ to $^*H$, and thought to be a possible rate-determining step. N_2 molecule will then combine straight with $^*H$ to produce higher-energy $^*N_2H_2$. When $^*N_2H_2$ interacts with H^+ and an electron, NH_3 is produced. Increasing the concentration of alkali-metal cations, which operate likewise to H^+ in an N_2, could be an excellent strategy to accelerate the nitrogen fixation mechanism [77].

3.3 Application of Heterostructure Photocatalysts

Various photocatalysts, such as TiO_2, ZnO, WO_3, $ZnWO_4$, Bi_2WO_6, and others, have recently been reported. Improved photocatalytic activity will arise from the connection of various semiconductors. Excited electrons of the sensitizer semiconductor can replace by a lower conduction band of wide band-gap semiconductor in type I heterostructure. However, holes are trapped in the sensitized semiconductor, resulting in improved photocatalytic efficiency and a broader light sorption spectrum [76–79]. Due to their strong photocatalytic activity and relative ease of manufacture, powder photocatalysts have received much attention and development by researchers. Powder-based photocatalysts dispersed in solution could rapidly agglomerate and make it difficult to separate from the water. Photocatalysis applications have recently seen considerable influence from one-dimensional (1D) semiconductor nanostructures with a type I core/shell structure formed on a substrate. Photocatalytic characteristics of photocatalysts with a powder type I heterostructure and 1D type I core/shell structures produced by substrates based on widely utilized TiO_2 and ZnO material groups are discussed in this section. Other photocatalysts, such as $ZnWO_4$, Bi_2WO_6, and a-Fe_2O_3 that have lately been extensively researched are also discussed. Some type I structures with high photocatalytic performance are particularly emphasized [80].

Due to their strong photocatalytic activities and relative ease of productions, powdered photocatalysts are widely employed. To create type I heterostructure, a broader bandgap semiconductor is commonly covered with a small bandgap semiconductor. In response, the efficiency of photoinduced charge, change of integrity and solar energy semiconductor have greatly improved. Heterostructure morphology, geometry, surface texture, and sensitizer particle size affect charge transfer and photocatalytic performance [81, 82].

BiOX crystals for structural engineering have attracted much attention for point defect management and precise facet tailoring. The existence of point defects in BiOX materials, such as lattice vacancies, causes photo-generated charge carrier entrapment and its visible light sensitivity performance. Thermal conditions or the protective cover of a reducing atmosphere like Ar typically introduce point defects like oxygen or metal vacancies, resulting in optimum electronic band levels. These defects provide charge transport and the color of BiOCl shifts to black due to the formation of oxygen-containing functional groups, contributing to the imperfection state located below the conduction band. The new state acts as electron traps, allowing for the separation of electron/hole pairs and increased catalytic properties. Therefore, the photocatalyst band structure has been fine-tuned to make it a noticeable photocatalyst that converts solar energy more efficiently. Charge carriers can indeed employ these holes as a trap, according to Zhao et al. By exposing white BiOCl powder to UV light while covered by Ar gas, oxygen vacancies in BiOCl nanosheets were produced. Since the synergistic impact of the 3D/2D $BiO_{1-X}Br$/BiOCl heterostructures and the insertion, figure on the $BiO_{1-X}Br$/BiOCl showed considerably greater photocatalytic activity than pure $BiO_{1-X}Br$ and BiOCl [54] (Fig. 12).

4 Heterojunction Photocatalyst Principles

Building a heterojunction is a useful technique to increase the photocatalytic activity of bismuth oxyhalide. The formation of heterojunction at the intersection of various semiconductor materials could provide the separation of photogenerated charges in a photocatalytic system. The sorts of heterojunctions offered include traditional direct Z-scheme heterojunctions and 2D/3D heterojunctions. Because of its optimum band edge positioning, which allows for effective charge extraction, the typical semiconductor heterojunction (class I) is the most explored so far. Peng et al. employed $BiOBr/Bi(OHC_2O_4)_2H_2O$ to generate $Bi_{12}O_{17}Br_2/Bi_{24}O_{31}Br_{10}$ type-II heterostructure. $Bi_{12}O_{17}Br_2/Bi_{24}O_{31}Br_{10}$ demonstrated exceptional electrochemical properties for the destruction of RhB and phenol due to the rapid electron–hole dissociation in the type-II heterojunction. In semiconductor heterostructure systems, bismuth oxyhalide nanoparticles may perform three significant functions. Bismuth oxyhalides exhibit a wide range of band redox couples, making them simple to match in the

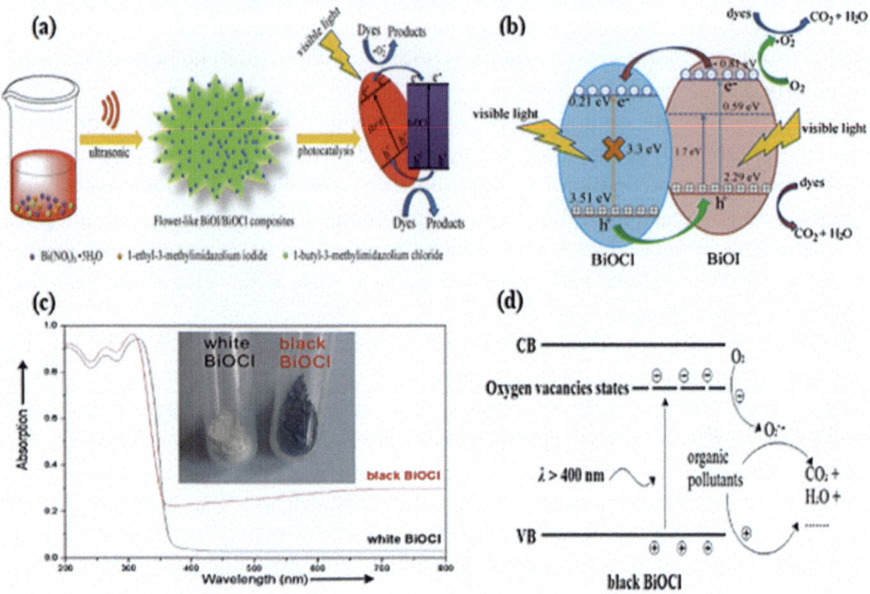

Fig. 12 **a** Photocatalytic reaction mechanism over BiOI/BiOCl heterostructure; **b** flower-like BiOI/BiOCl heterostructure production schematic; **c** ultraviolet–visible DRS spectra for as-prepared BiOCl specimens; **d** schematic diagram of bandgap and photocatalytic reaction mechanism over black BiOCl (Reproduced with permission [54])

energy level of various semiconductors and acting as a driving factor for charge-directed separation [18, 83]. To absorb a broader solar spectrum, narrow bandgap bismuth oxyhalide semiconductors (BiOI and BiOBr) can be utilized to photosensitize other semiconductors. BiOI was mixed with TiO_2. For instance, p-n nanostructures were formed using a chemical inertness bath approach. The chemical composition and photocatalytic efficiency range of TiO_2/BiOI and ZnO/BiOI heterostructures may be precisely changed by adjusting the molar ratio of Bi/Ti/Zn. BiOI has a greater negative conduction and valence band level than TiO_2 and ZnO, which might lead to efficient photogenerated electron transmission between the catalysts, thereby prolonging the lifetime of photogenerated charges and lowering the possibility of recombination (Fig. 13). Lastly, because bismuth oxyhalides have a wide structural variety, carbon-based nanoparticles may readily be coupled with them, allowing for higher charge separation & transfer. Due to the apparent greater relationship between bismuth oxyhalide and graphene, BiOCl/graphene, BiOBr/graphene, and BiOI/graphene have already been employed [18, 84, 85]. Graphene may also be utilized as a cocatalyst to extract photogenerated electrons. Consequently, electron

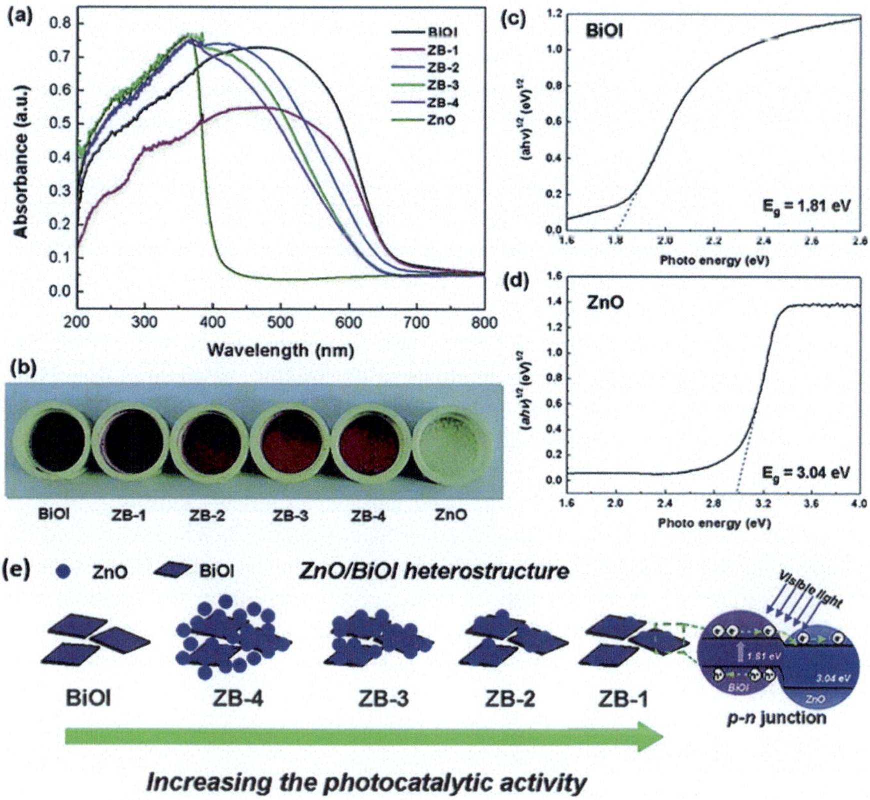

Fig. 13 **a** Ultraviolet diffuse reflectance (DRS) spectroscopy, **b** powder colors of pure BiOI, pure ZnO, and ZnO/BiOI heterostructures. Bandgap energy of **c** BiOI and **d** ZnO, **e** diagram representation of various Bi/Zn molar ratios in ZnO/BiOI heterostructures. The American Chemical Society [67] granted permission for this reprint

transport between bismuth oxyhalides to graphene efficiently suppresses electron–hole recombination, lengthening carrier lifespan, and boosts photocatalytic effectiveness. The heterojunction approach, more frequently utilized in solar photocatalytic activity and photoelectrocatalysis, was rarely employed to make binary or multi-component composite photocatalytic systems [67].

4.1 The Photocatalytic System with p-n Heterojunction

Although the category-I heterojunction capacity efficiently divides charge couples in space, the increase in charged particle isolation across the heterojunction is insufficient to address fast electron–hole replication on the semiconductors. In this regard, it has been proposed to develop a p-n heterojunction photocatalyst that can provide an additional electric field to speed up electron–hole pair transfer and increase photocatalytic performance. The electron–hole pair separation of p-n heterojunction is better than that of a type-I heterojunction, which can boost photocatalytic activity significantly. The p-n heterojunction photocatalyst was formed by combining p-type and n-type semiconductor materials [27, 67, 86]. The Fermi level of a p-type diode position at VB, whereas the Fermi level of an n-type semiconductor is close to the CB. The holes will move from the p-type to the n-type semiconductor before light irradiation, leaving a negatively charged species. Until the Fermi level system reaches equilibrium, thus hinders the electron–hole pairs diffusion. Finally, an electric field was formed at the interface between two kinds of semiconductors. The photoinduced electron–hole pairs will be separated under the effect of the electric field. Due to the presence of NH/NH_2 groups and individuals in its structure as electron donors, g-C_3N_4 is principally an n-type semiconductor material. As a result, constructing a p-n heterojunction can help g-photocatalytic C_3N_4 operate better [28, 29].

CuS is a common p-type type of transistor for photocatalytic activity, with favorable properties such as a narrow band gap and excellent stability [29]. CuS was hybridized with g-C_3N_4 to produce p-n heterojunctions, which extended the composite shorter exposure range by establishing more interfaces for efficient transfer of photogenerated electron–hole pairs to reduce recombination, as well as increasing the range of optical absorption range of photosensitizer. Using an insitu synthesis approach, Cai group [30] developed a g-C_3N_4/CuS p-n heterostructures composite. The increased photocatalysis efficiency might be attributed to the g-C_3N_4/CuS heterojunction interface. Both CuS and g-C_3N_4 may excite charge carriers when exposed to visible light. Therefore, the VB of g-C_3N_4 became less positive than that of CuS, photogenerated hole on the VB of CuS could readily transport to the VB of g-C_3N_4. CuS does have a lower CB energy than g-C_3N_4. Therefore, photoinduced electrons on CB of g-C_3N_4 might diffuse to CuS CB. Ma group [69] produced $Cu_{1.96}S$/g-C_3N_4 p-n heterojunctions by altering the thermodynamic efficiency factor of $Cu_{2-x}S$. Through connecting 3–10 nm nanoparticles toward the surface of g-C_3N_4 wrinkled layers, 0D/2D p-n heterojunctions were formed, which can supply extra photocatalytic reaction centers and ease of charge transfer. The $Cu_{1.96}S$/g-C_3N_4 nanocomposites had a considerably decreased PL intensity, indicating that the p-n heterojunction effectively prevented charge pair combination.

4.2 Photocatalytic Systems Based on the Z-scheme

Because of their smart charge mechanism, Z-scheme-based photocatalysts provide remarkable photocatalytic activity by jointly delivering high-redox potential and visible light harvesting. As a result, the drawback of total redox potential loss may be overcome when switching to type-II heterojunctions. The theoretical efficiency of 40% against 30% with single-component photocatalysts is substantially stronger charge recombination reduction and higher photoactivity in the electrolysis of water. Furthermore, certain Z-scheme heterojunctions for nitrogen fixation were investigated. Feng et al. [30] proposed all Z-scheme photocatalysts devices based on Schiff base chemistry and Ga_2O_3/g-C_3N_4 functionalized with 3,4-dihydroxybenzaldehyde. After exposure to light irradiation, nitrification reached up to 112.5 mmol L^{-1} g^{-1}, much higher than single-component g-C_3N_4. The aromatic ring 3,4-dihydroxybenzaldehyde acted as an electron mediator at the Ga_2O_3 and g-C_3N_4 interface, while a strong oxidizing potential assured the generation of hydroxyl radicals. Nitrogen fixation was performed by oxidizing methanol to CO_2 and its chemical intermediates, then reducing N_2 using CO_2 radicals. It is generally established that adding defects improves photoactivity by creating significant charge carriers engaging in the surface reaction through electron–hole pairs. The incorporation of nitrogen vacancy in the crystalline structure of the photocatalyst might be useful for trapping and activating N_2 molecules due to the selectivity of the nitrogen reduction process. Guang et al. developed a g-C_3N_4/Ag_2CO_3 photocatalysts system based on the Z-scheme mechanism, in which nitrogen vacancies were created by ammonium dissolving in solution. While N_2 and O_2 contents were proportionately equal, the catalyst generated more N_2 at 11 mg L^{-1} h^{-1} g. Nitrogen protonation followed a two-path method. Initially, the nitrogen vacancies were captured, nitrogen molecules activated, and afterward, the trapped molecules were reduced with hydrogen ions to form ammonium ions. Oxygen molecules were decreased to generate H_2O_2, which was further reduced to form OH radicals in the second reaction pathway. These radicals interacted with methanol to form CO_2 radicals, further reducing N_2 molecules to NH_3. The overall favorability of charge transport between O_2 molecules and the catalyst was connected to the high rate of N_2 photooxidation in a mixed environment. Metal–organic frameworks (MOFs) have recently received much interest for photocatalytic activity due to their higher porosity and surface area, allowing the reaction occurs with many active sites. When utilized as a single-component photocatalyst, MOFs exhibit weak photocatalytic activity. To boost their photoactivity, MOFs are routinely coupled with other photoactive chemicals. Ding et al. demonstrated the Z-scheme properties by using MOF-74 for manufacturing nitrogen defect-containing g-C_3N_4 [69–71]. The pure nano-MOF-74 had no visible light photoactivity, whereas the thin-film g-C_3N_4 showed poor photoactivity. The photoactivity increased threefold when nano-MOF-74 and thin-g-C_3N_4 were utilized in tandem production, demonstrating that MOFs offered a source of highly photosensitive sites for the photocatalytic process. Finally, when nitrogen defects were introduced to the nano-MOF-74 and thin-film supported g-C_3N_4, the photoactivity of pristine g-C_3N_4 was enhanced two-fold.

The significant porosity and surface area of MOFs cause the high photoactivity of 2.32 mmol g^{-1} h^{-1}. Moreover, the activation and decrease of binding energies for the shattering of the N≡N bond came from the trapping of nitrogen molecules via nitrogen vacancies. Liang et al. prepared a $W_{18}O_{49}$/g-C_3N_4 photocatalyst capable of generating a 2.6 mg L^{-1} h^{-1} N_2 photooxidation rate under visible-light irradiation. Surface electrons oscillated coherently due to the oxygen vacancies in $W_{18}O_{49}$, leading to a greater photocatalytic NH_3 production rate. The $W_{18}O_{39}$ component allowed light harvesting over the visible light spectrum and many electron injections to recombine with holes in g-C_3N_4. Over 40 h of operation, the catalyst demonstrated good photostability [33, 34].

To optimize the cocatalyst function, the lattice and electronic structure of the cocatalyst and the semiconductor photocatalyst should be similar and have a suitable band energy level. Xiao et al. synthesized $(BiO)_2CO_3$ using gold nanoparticles and found that it had high catalytic activity. The photocatalyst with a shorter response area and the increased rate is formed when a wide bandgap semiconductor is substituted with a narrow energy band gap semiconductor or two types of narrow bandgap semiconductors are connected. As a consequence, building a heterojunction might help bismuth-based photocatalysts work better. Each of these heterostructures will be addressed in depth in this section. In addition, cocatalysts can increase photocatalytic nitrogen fixation efficiency by providing active surface area, trapping charge carriers, decreasing the activation energy of surface processes, and regulating photogenerated electron and hole recombination. The relationship between surface plasmon resonance stimulation and resonance photons can considerably boost solar energy usage [83]. With the presence of plasmon resonance on the Ag surface, the ultra-thin Ag-Bi_2O_3 sheet forms visual traps when irradiated, which prevents charged particle replication but also induces light-induced oxygen. Photogenerated electrons from gold nanoparticles are transferred directly to the semiconductor bandgap, assisting charge separation. The multilayer ultra-thin structure of the Ag-modified d-Bi_2O_3 photocatalyst allows it to be used in many applications (Figs. 14 and 15).

To address BiOCl low response to visible light, Guo et al. employed solvothermal treatment to generate 2D/2D $ZnIn_2S_4$/BiOCl heterojunctions [58]. The substrate 2D/2D feather requires the carrier mobility of $ZnIn_2S_4$. Zhou et al. used an in situ bismuth reduction process to construct oriented Bi_2WO_6, a highly orientated lattice structure, using the metal Bi as a lattice junction. This oriented structure clearly offers a tight-knit interface and transfer channels for dissociating photogenerated carriers, leading to adequate nitrogen fixation. Charge separation and photocatalysts degradation efficiency can be improved using type I heterojunctions. In the literature, its heterojunction photocatalyst has been described. The design and fabrication of type I heterostructure catalysts can significantly improve the catalyst efficiency in reducing nitrogen. For instance, the heterostructure may improve the spacing of photogenerated electron–hole pairs [18]. Gao and colleagues synthesized AgCl/Bi_2O_3 nanosheets with a consistent thickness of around 2.7 nm using a cost-effective hydrodynamic regulated precipitation approach [68]. The photocatalytic nitrogen-fixing activity of the AgCl/Bi_2O_3 combination was enhanced. This activity was achieved by forming a p-n heterojunction structure of AgCl/Bi_2O_3 that provided enough electrons to reduce N_2. Additionally, the high surface area of ultrathin 2D nanosheets increases active sites on catalytic surfaces and two-dimensional anisotropic, which minimizes charge migration distance. Performance of photocatalysis related to surface defects and disorganized surface structure could be improved

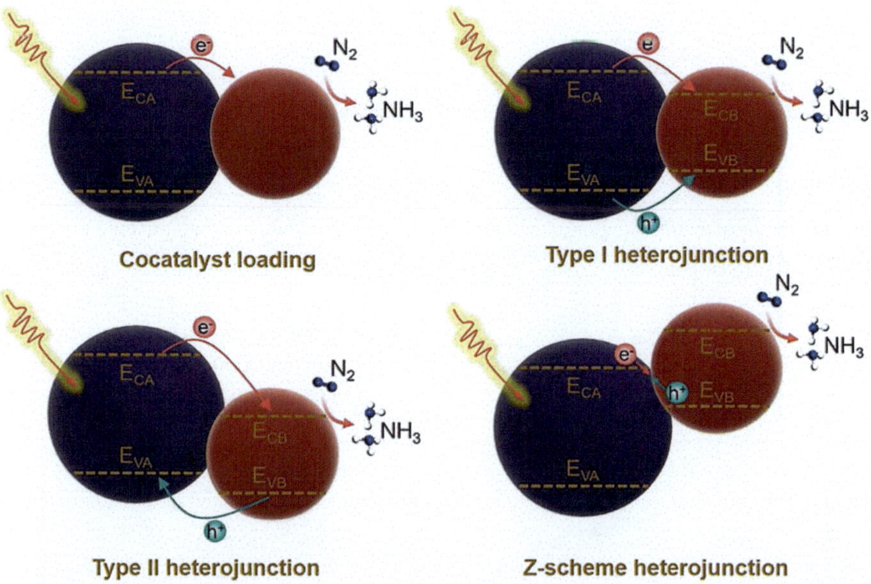

Fig. 14 Schematic illustration of various heterostructure buildings for photocatalysts nitrogen reduction to ammonia absorption edges and large interfacial sections demonstrates effective photocatalytic performance for creating NH^{4+} when exposed to visible light. With permission from Ref. [57]

(Fig. 15a). For a 3% $AgCl/Bi_2O_3$ sample, TEM images revealed a 2D ultrathin structure, including 2D ultrathin nanosheets (Fig. 15b). Biocompatibility, minimal toxicity, and poor water solubility are only a few of the advantages of graphene nanocrystals. For the first time, combined graphene nanoparticles and Bi_2WO_6 to make a heterostructure for photocatalytic nitrogen fixation [62]. The composite surpasses pure graphene nanocrystals and Bi_2WO_6 in catalytic performance by 33.8 and 8.88 times, respectively.

HRTEM images of $AgCl/Bi_2O_3$ indicated AgCl particles scattered throughout the surfaces of thick Bi_2O_3 sheets indicated in squares with a length of roughly 250 nm range (Fig. 15c). The synthesized materials exhibited a substantially greater ability for photocatalysts nitrogen fixation, as demonstrated in Fig. 15d and e, the generation of NH_4^+ concentrations increased linearly as the reaction time increased. The gathering with 3% $AgCl/Bi_2O_3$ exhibited a faster rate of photocatalytic fixing and a higher concentration of NH_4^+ (Fig. 15d). Nitrogen gas molecules are particularly simple to adsorbed and activation because of the vast number of oxygen-containing functional groups on the surface of the OV-BiOBr nanosheets. This function group might potentially be used as an electronic trap to detect ions in N_2 fixation operations. Using an ion-exchange method, Xu et al. produced a $Bi_2S_3/BiOBr$ hybrid structure [70]. The hybrid structure offers increased light capturing capacity and photogenerated carrier propagation due to the small band gap of Bi_2S_3 and the staggered bandgap of the $Bi_2S_3/BiOBr$ hybrid group. Xue and colleagues [88] used a simple

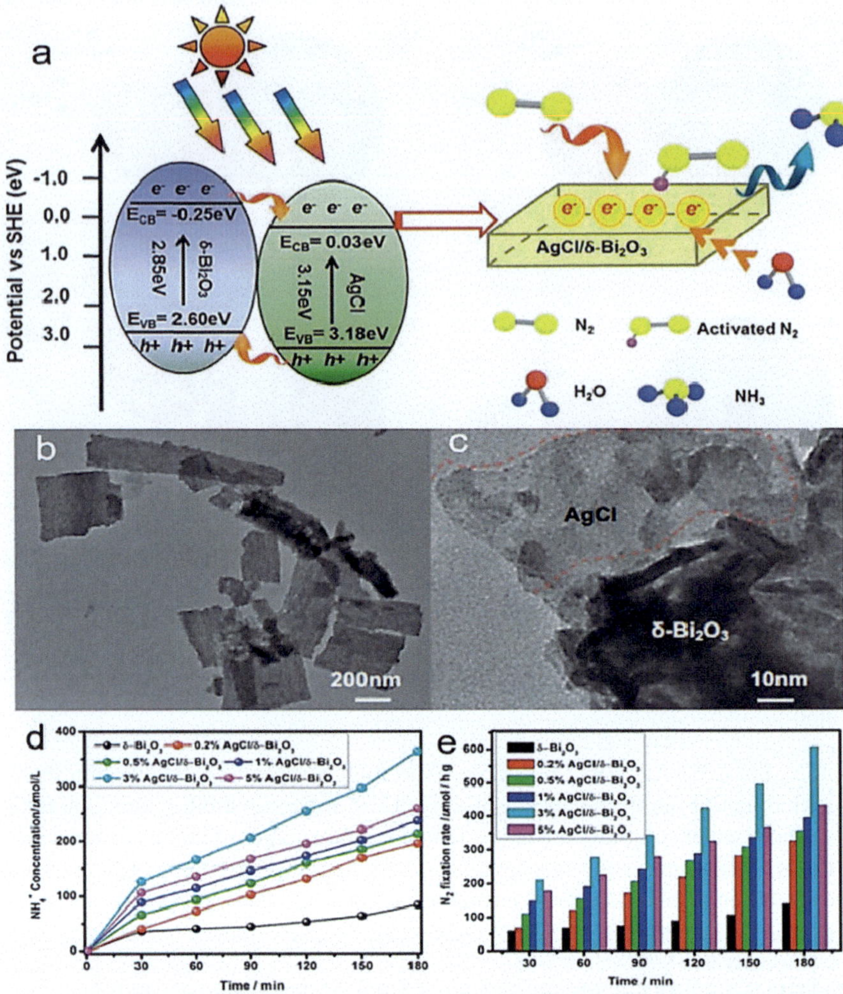

Fig. 15 **a** Illustration of a potential photocatalytic reaction, **b** TEM and **c** HRTEM of 3% AgCl/d-Bi_2O_3, **d** the NH_4^+ in liquid concentration, and **e** the efficiencies of photocatalytic nitrogen fixation on various catalysts as a function of time. With permission from Ref. [68]

solution technology to make heterojunctions with O vacancies out of n-type Bi_2MoO_6 nanorods and p-type BiOBr nanosheets (Bi_2MoO_6/OV-BiOBr). The interfacial of n-type Bi_2MoO_6 nanotubes is protected with ultra-thin p-type OV-BiOBr nanosheets, yielding a nanostructured phase. Moreover, the Bi_2MoO_6/OV-BiOBr heterojunction hierarchy, which has a larger specific surface area, can provide more active nitrogen fixation sites. A simple hydrothermal and photo-deposition technique resulting in Ag/AgI-Bi_2O_3 photocatalyst was obtained with a type II heterojunction [72]. The nanocomposite photocatalytic nitrogen fixation performance in waters is 420 mmol without the use of a sacrificial agent. Whenever Ag/AgI nanoparticles were added to Bi_2O_3, it caused morphological changes, increased oxygen vacancy concentration, and formed plasmon-sensitized heterostructures, resulting in improved light absorption and carrier separation. The photocatalytic performance of Ag/AgI-Bi_2O_3 is owing to its high N_2 adsorption and activation. Artificial heterogeneous Z-scheme photocatalyst systems that replicate natural photosynthetic can transcend the limits of single-component photocatalysts, allowing for a broad absorption range and long-term stability of photocatalysts. [73, 84]. The Ag/AgI-Bi_2O_3 photocatalyst was made into a type II heterojunction [72]. The nanocomposite photocatalytic nitrogen fixation performance in water is 420 mmol without the use of a sacrificial agent. In addition, when Ag/AgI nanoparticles were added to Bi_2O_3, it caused morphological alterations, increased oxygen vacancy concentration, and the formation of plasmon-sensitized heterostructures, resulting in improved light absorption and carrier separation. The photocatalytic performance of Ag/AgI-Bi_2O_3 is owing to its high N_2 adsorption and activation. Artificial heterogeneous Z-scheme photocatalyst systems that replicate natural photosynthesis can transcend the limits of single-component photocatalysts, allowing for a wide absorption range and long-term stability of photocatalysts [84]. Chen et al. created an Ag/AgBr/$Bi_4O_5Br_2$ nanocomposite for photocatalytic nitrogen fixation using hydrothermal and ion exchange techniques. [73] (Fig. 16)

The composite material photocatalyst reduction activity is impressive. The NH_3 generation of the β-Bi_2O_3/BiOCl composite is roughly 25 times of BiOCl when irradiated with Xenon light. Photocurrent and electrochemical impedance spectroscopy measurement show that producing β-Bi_2O_3/BiOCl heterojunctions reduces photogenerated electron and hole combination [75].

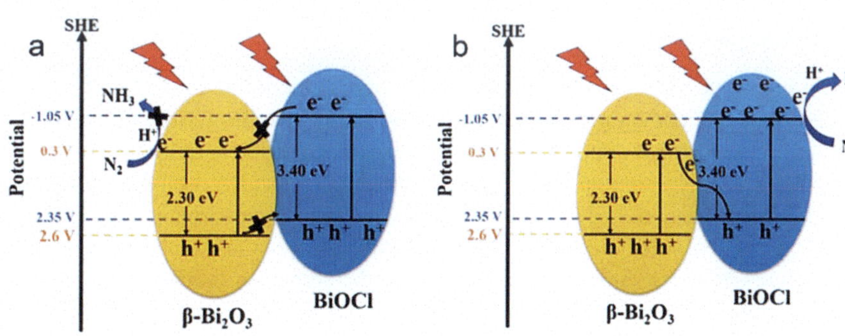

Fig. 16 Photocatalysts reductions over the b-Bi_2O_3/BiOCl heterojunction based on Type II **a** and Z-scheme **b** mechanisms. With authorization from Ref. [74] and has been reproduced. Copyright: Springer, 2019

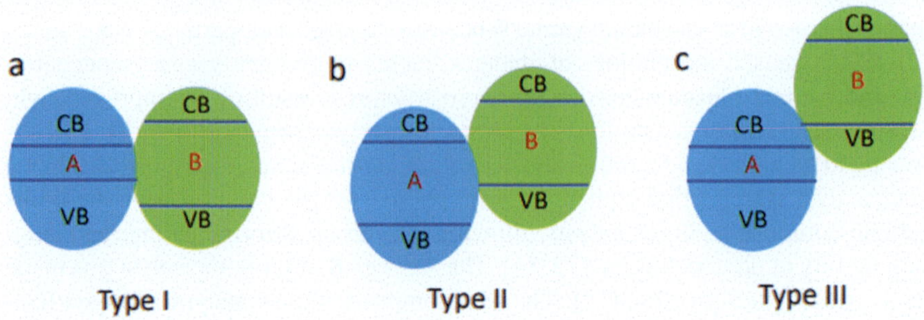

Fig. 17 Schematic diagram of the three most common forms of heterojunctions. **a** type I, **b** type II, and **c** type III semiconductors, where A and B are denoted as two distinct semiconductors (Reproduced with permission)

4.3 Heterojunction Photocatalysis System

Photocatalysis efficiency is heavily influenced by the separation and distribution of photoactivated electron/hole pairs. To achieve high photocatalytic effectiveness, the generated electron/hole pairs must be successfully separated and transported to active sites rapidly enough to initiate the proper reactions [89]. Recombination of charge at the surfaces and space charge region should be prevented. A possible solution is to utilize semiconductor heterostructures to induce charge separation for better photocatalytic activity [90]. Figure 17 demonstrates that semiconductor heterostructures can be categorized into three primary types based on band alignment characteristics: type I straddling heterojunction, type II staggered heterojunction, and type II broken gap heterojunction [18].

Type I heterojuctions

In heterojunctions with type-I semiconductors, one semiconductor (CB and VB) is restricted inside the energy gap of other semiconductors, resulting in band alignment. The main property of produced charges and the photocatalytic activity is influenced by the VB and CB alignment. Both charge carriers are localized on one semiconductor; this heterojunction does not affect photocatalytic performances [31]. Due to potential reduction differences, photogenerated electrons are projected to travel from $SrZrO_3$ CB to the $SrTiO_3$ CB. Due to the difference in oxidation potentials, h^+ generated in the VB of $SrZrO_3$ will migrate to the valence band of $SrTiO_3$. As a result, electrons and holes will concentrate, creating considerable recombination in the $SrTiO_3$ semiconductor [91].

A heterojunction is the meddling among the two portions of a typical semiconductor with a rough band structure that produces interfacial band alignments. Based on the different energy levels of both CB and VB of two semiconductor materials, the product of heterojunctions can be categorized into two types: type I (straddling alignment) and type II (staggered alignment), as shown in Fig. 18. In the type I heterojunction, the VB and CB of semiconductor 1 are lower and higher than those

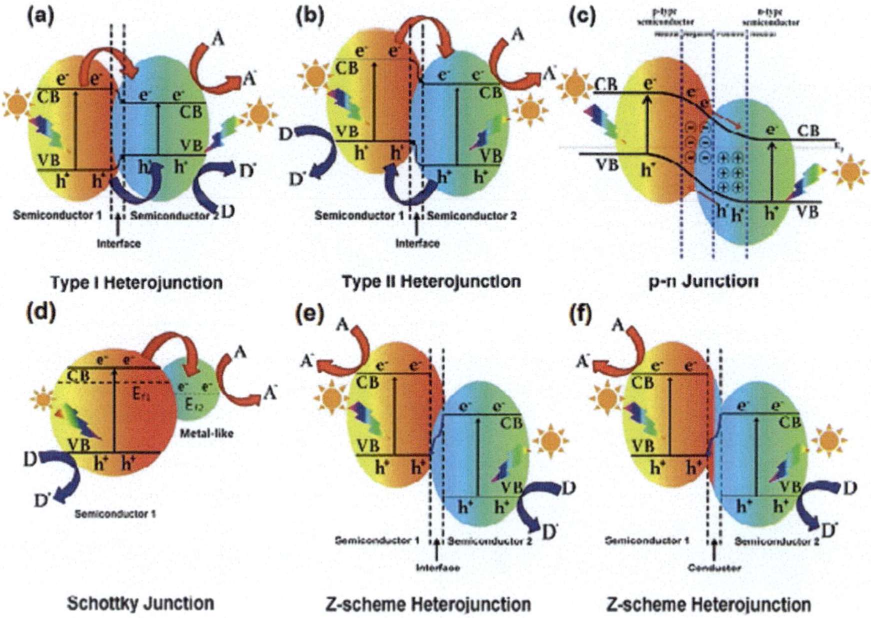

Fig. 18 Schematic illustration of different types of heterojunction nanocomposite: **a** heterojunction type I, **b** heterojunction type II, **c** p-n heterojunction, **d** Schottky junction, **e** Z-scheme heterojunction (without an electron mediator), and (f) indirect Z-scheme heterojunction (Reprinted with permission from Refs. [86, 91])

of semiconductor 2 (Fig. 18a). As light induces photogenerated holes and photoinduced electrons increase the energy, they may fold away from VB of semiconductor 1 to VB of semiconductor 2. In contrast, photoinduced electrons may travel from CB of semiconductor 1 to CB of semiconductor 2 [91].

A low bandgap of 2.5 eV with a unique electrical structure on the tin sulfide (SnS_2) performs well under visible light [86, 91]. SnS_2 has a little photocatalytic performance due to a high recombination rate of photogenerated electron–hole pairs. Hybridizing SnS_2 with g-C_3N_4 to create a heterojunction interface to improve SnS_2 photocatalytic performance has been a tremendously exciting research topic. The diverse SnS_2 morphologies, such as nanosheets, nanoparticles, and the 3D flower-like nanostructures were combined to create SnS_2/g-C_3N_4 hybrid heterojunctions. Figure 19a depicts the SnS_2 nanoparticles that were spread fully on g-C_3N_4 superficial. SnS_2 nanosheets (3D flower-like) were uniformly distributed on the g-C_3N_4 layer, as shown in Fig. 19b and c. The diagrams displayed the ability to improve simulated visible light absorption when compared with pure g-C_3N_4. It was attributed to the interaction between g-C_3N_4 and SnS_2 in the composites, based on the visual characteristics of various SnS_2 morphologies (Fig. 19d).

Zhang et al. [86] developed a hydrogen photocatalytic evolution 2D MoS_2/g-C_3N_4 heterojunction. Planar development of MoS_2 across the surface of g-C_3N_4 should be

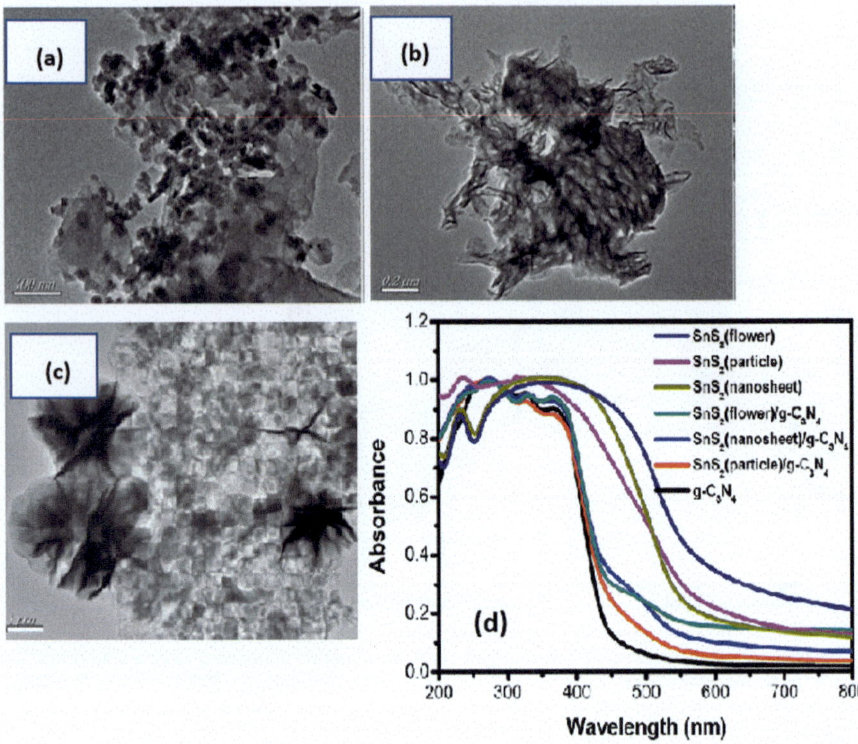

Fig. 19 **a** Synthesized SnS_2 nanoparticles/g-C_3N_4 composites, **b** SnS_2 nanosheets/g-C_3N_4 composites, and **c** 3D-flower-like SnS_2/g-C_3N_4 composites, **d** DRS of numerous morphologies of SnS_2, g-C_3N_4, and SnS_2/g-C_3N_4 composites (Reprinted with permission from Ref. [92])

simple since MoS_2 and g-C_3N_4 have similar layered structures. The photocatalytic activity of 0.5 wt% MoS_2/g-C_3N_4 composites was higher than that of the 0.5wt% Pt/ g-C_3N_4 composites. Improved hydrogen photocatalytic activity of MoS_2/g-C_3N_4 composite heterojunctions was attributed to the adjustment of the MoS_2 co-catalyst, which resulted in excellent performances of permitting charge carrier for hydrogen evolution.

Heterojunction Type II

Heterojunction type II includes its band structure among the two semiconductors. It may help to upsurge the spatial separation of electrons and holes and decrease recombination. If semiconductors 1 and 2 in Fig. 17 with similar band potentials are intimately coupled, they will generate a specific heterostructure type. In a type II heterostructure, the photogenerated holes in the VB of one semiconductor will drift to another if the valency band level of the semiconductor is subordinate to others under light irradiation. If the conduction band level of a semiconductor is greater than that of another semiconductor, photogenerated electrons in CB of one semiconductor may migrate to the CB of another semiconductor, causing charge carriers to be separated spatially. Finally, the internal field may help with charge carrier separation and migration. The chance of electron–hole pair recombination is diminished, and the lifespan of electrons is enhanced, which can be evaluated using transient spectroscopic methods. When only one semiconductor can be affected, the other semiconductor may operate as an electron/hole acceptor. The best band positions for competent separation of photogenerated electron–hole pairs are found on type II heterojunctions. For photocatalysis, the fabrication of Type II alignment is particularly desired [16, 86].

Due to its high oxidation in the photochemical reaction, non-toxicity, and affordable cost, TiO_2 photocatalyst is the most widely inspected. The high recombination ratios of photoinduced electron–hole pairs and TiO_2 limited reactivity to visible light have obstructed its use in photocatalysis. The development of a type-I heterostructure by linking TiO_2 with a narrow bandgap semiconductor upsurges photocatalytic action and expands the light absorption spectrum of TiO_2. This subject has involved much attention. Cds/TiO_2, CdSe/TiO_2, WS_2/TiO_2, and Bi_2S_3/TiO_2 are examples of TiO_2 type I heterostructures. A narrow bandgap chalcogenide is frequently exploited as a photosensitizer in the production of TiO_2-founded heterostructures and has attracted much attention. For example, in a CdS/TiO_2 heterostructure, CDs with a band gap of 2.25 eV may be excited by visible light to create electron–hole pairs, but TiO_2 will only operate when exposed to UV light. Since CDS has an advanced CB position than TiO_2, excited electrons on CdS may inject into toTiO_2 while the hole stays in CdS, established in Fig. 20 [24, 86, 91].

The charge separation efficiency of visible light-generated electron transfer from CdS to TiO_2 could be improved while the spectrum responsiveness was also enhanced. Liu and his colleagues used micro-emulsion mediated solvothermal hydrolysis followed by acid peptization to make CdS/TiO_2 semiconductors nanoparticles. On visible light irradiations (1 > 420 nm), the activity of photocatalytic

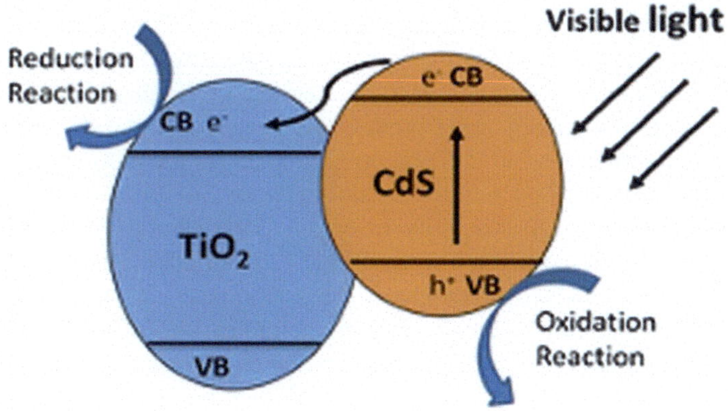

Fig. 20 Illustration of charge transfer in CdS/TiO_2 heterostructure (Reprinted with permission from Ref. [24])

of CdS/TiO_2 photocatalyst was validated using MB degradation. To make nano-sized CdS-sensitized TiO_2 nanocrystals, Yu and colleagues used a microemulsion-mediated solvothermal method. The morphology of CdS/TiO_2 is shown in Fig. 21a and b. When exposed to visible light, CdS/TiO_2 nanocrystals are more effective than pure TiO_2 at decomposing methylene blue (Fig. 21c). Electron paramagnetic resonance (EPR) was used to confirm Ti^{3+} formation on TiO_2 during irradiation, exhibiting effective photogenerated electron transfer from the CdS conduction band to the TiO_2 conduction band. Using $CdSe/TiO_2$ heterostructures, a photocatalyst was built using an ultrasound-driven solution method. $CdSe/TiO_2$ heterostructures displayed better photocatalytic activity than pure TiO_2 in the degradation of 4-chlorophenol when exposed to light (1 > 400 nm) [31, 93]. The quantum size effect of CdSe increases charge separation efficiency and photocatalytic performances of the $CdSe/TiO_2$ heterostructures.

In a linked semiconductor system, quantum confinement effects provide an acceptable variation for the energy levels of the CB and VB borders, making MoS_2 and WS_2 suitable for photo-sensitization of light. MoS_2/TiO_2 and WS_2/TiO_2 heterostructures have advanced photocatalytic activity in MB (methylene blue) and 4-chlorophenol degradation than pure TiO_2 when exposed to visible light ($\lambda > 400$ nm). Other narrow bandgap semiconductors, such as Ag_2O, $CuAlO_2$, $LaVO_4$, $AgIn_5S_8$, and V_2O_5 could be combined with TiO_2 to form a type I heterostructure, increasing the activity of photocatalyst and broadening TiO_2 absorption spectrum. Ternary hybrid catalysts can be made by combining type-I heterostructures with another material. For example, noble metals are frequently combined with TiO_2 type I heterostructures to improve photocatalytic activity in hydrogen generation. The development of the potential gradient at the interface between CdS and TiO_2 with Pt is an important factor in regulating the hydrogen generation efficiency of $CdS/TiO_2/Pt$ [31, 90].

Even though TiO_2 is the most frequently used photocatalyst, ZnO looks to be a potential substitute. ZnO is moderately priced, and its photo-degradation process is identical to TiO_2 [18]. Furthermore, some studies indicate that ZnO is more effective than TiO_2 at photo-catalytically destroying certain hues in water. Link to a narrow band gap semiconductor, such as C_3N_4, Zhang et al. [26] ZnSe, Zhang et al. [86] In_2O_3, Zhang et al. [27], is a practical approach to increase charge separation efficiency and solar light consumption of ZnO. Lee and colleagues employed simple solution-based surface modification processes to generate three distinct types of ZnSe/ZnO heterostructures [40, 86]. Figure 22a displays the morphology of three different kinds of ZnSe/ZnO heterostructures. Those ZnSe/ZnO heterostructures showed greater photocatalytic activity in organic degradation than pristine ZnO structures under visible-light illumination (Fig. 22b). In addition, the influence of ZnO and exposed crystal faces also induced the performance. Yu. and his colleagues employed a coprecipitation technique to manufacture In_2O_3/ZnO hetero nanostructures [27]. The type-I heterostructure causes a better charge separation and transport. Various compositions and annealing temperatures alter the photocatalytic activities. When annealing at 800 C with a Zn/In the ratio of 1 to 1, the greatest photocatalytic activity was observed, it is worth mentioning that, owing to photo deterioration, ZnO has limited stability during the extended photocatalytic periods [28, 91].

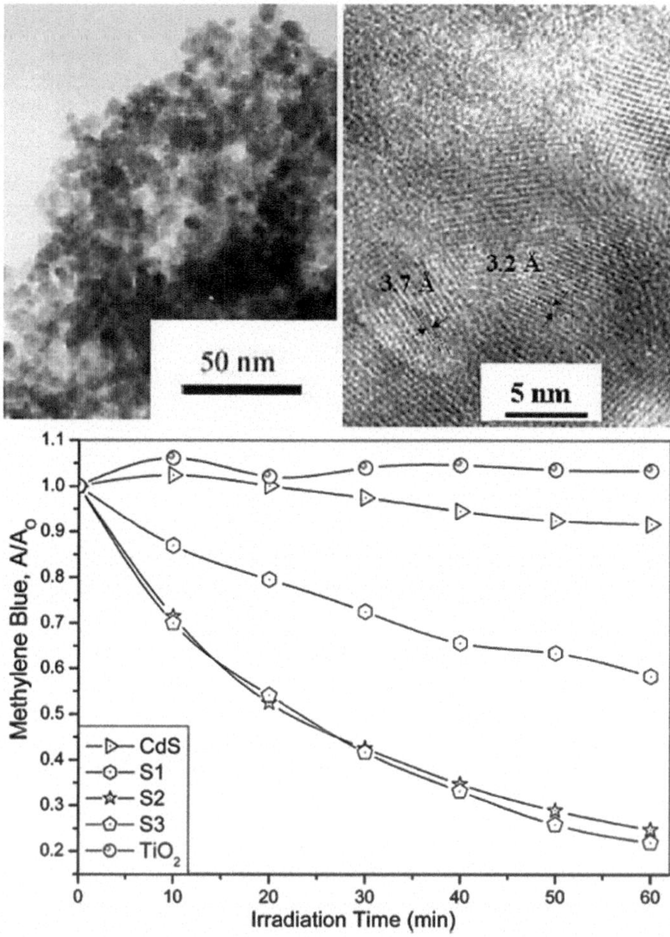

Fig. 21 **a**, **b** TEM and HRTEM images of CdS/TiO_2 with **c** MB degradation under visible-light irradiation at 660 nm (Reprinted with permission)

Zhu and colleagues produced a type-II graphite-like C_3N_4/ZnO heterostructure. When exposed to UV–visible light irradiation, ZnO was significantly increased by coupling with g-C_3N_4 (Fig. 22a and b). In addition, photo corrosion of ZnO in the g-C_3N_4/ZnO heterostructure crystal was successfully inhibited. However, as the VB position of ZnO is lower than that of C_3N_4, photogenerated holes on the ZnO surfaces would instantly transfer to g-C_3N_4, resulting in efficient photo corrosion inhibition. The photo activity of the g-C_3N_4/ZnO heterostructure is induced by the injection of excited electrons from the conduction band of g-C_3N_4 to the conduction band of ZnO.

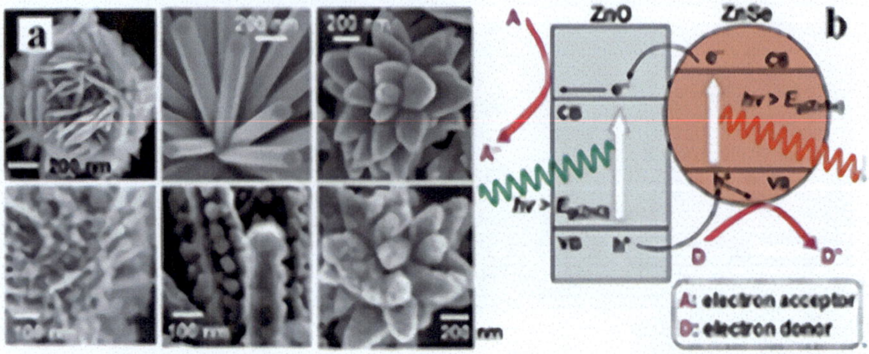

Fig. 22 **a** SEM images of ZnO/ZnSe and **b** energy band structure of ZnO/ZnSe (Reprinted with permission)

Photocatalytic Z-Scheme Heterojunction System

Although the above heterojunction may boost hole electron separation efficiency, the redox ability of composite may be degraded. To solve these challenges, Bard et al. [86] contributed the pioneering effort of designing a Z-scheme photocatalytic device in 1979. This perfect charge transfer is inspired by the photosynthesis process of green plants in nature, and it can effectively separate charge carriers and recollect strong and high redox abilities simultaneously. These Z-scheme systems are applied to the liquid phase by redox mediators [24, 91]. As a consequence, all solid-state Z-scheme photocatalytic devices [33] are projected to be created either with mediator electron (Fig. 23). Howsever, Yao et al. [67] and Zhange et al. [69] introduced a potential type of mediator-free deals in this domain in 2013. Photogenerated electrons on the lower level conduction band of semiconductor II may quickly migrate to the valence band of semiconductor I and recombine with holes. A direct Z-scheme transmission mechanism is a term for this charge transfer. In recent years, substantial research has been explored the Z-unique scheme structure [90].

All solid states of the Z-schemed photocatalytic device are more extensively done in liquid, solid, and gas–solid schemes than in the liquid Z-scheme [35]. They have been commonly employed as electron mediators in Z-scheme systems due to their strong electrical conductivity. Raizada et al. [92] successfully built Z-scheme Cd/Au/C_3N_4 ternary photocatalytic system based on these advantages based on hollow C_3N_4 nanospheres (HCNS). The C, N, S, Cd, and Au were homogeneously distributed in the spheres based on the elemental mapping of CdS/Au/HCNS photocatalysts (Fig. 23a). A significant emission peak was found in the photoluminance spectra of HCNS (Fig. 23b). The strong PL was slaked immediately after integrating CdS and Au, representing that the charge recombination was significantly decreased in the photocatalytic CdS/Au/HCNS. The efficient unintended Z-scheme heterojunction greatly improved couple separation competence. Figure 23c indicates a notable increase in photocurrent for CdS-Au-HCNS, which was approximately three times bigger than naked HCNS. The excellent mobility of charge carriers facilitated this

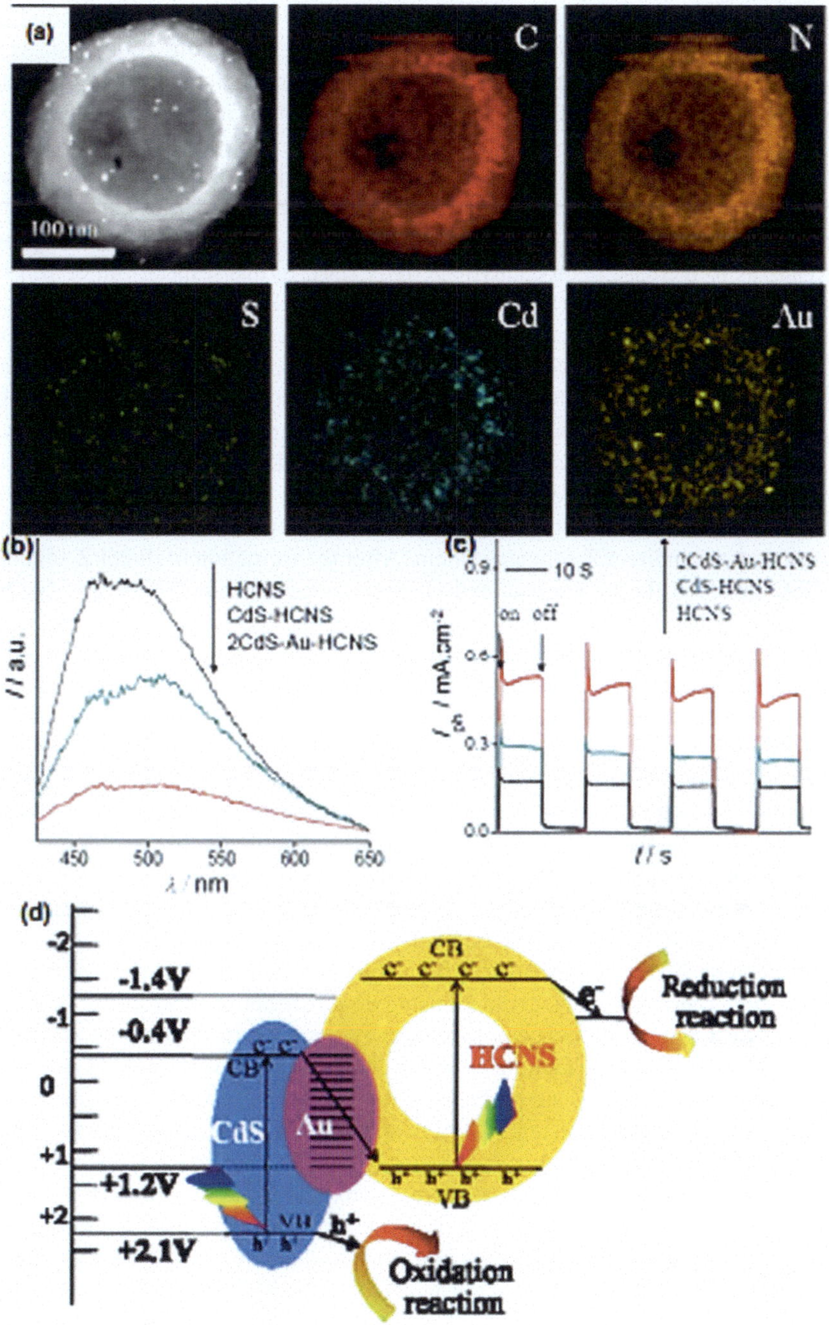

Fig. 23 **a** Elemental mapping, **b** photoluminescence, **c** photocurrent response, and **d** Z-scheme electron transfer of CdS/Au/HCNS nanocomposites (Reprinted with permission)

because of their high electrical conductivity. Because visible light could excite CdS and HCNS, photoelectron on the conduction band of CdS soon migrated Au. After that, the electrons in HCNS recombined with the photo-generator hole. Consequently, the electron induced in the HCNS conduction band and hole in the CdS valence band displayed significant reducing and oxidizing capabilities, respectively [31, 89].

As previously mentioned, using an electron mediator between two semiconductors to create a direct Z-scheme could dramatically enhance charge transfer. In addition to the direct Z-scheme, constructing a conventional Z-scheme arrangement allows mechanisms to increase photocatalytic activity [24]. In this respect, Shi et al. [90] developed C_3N_4 nanosheets (NS)/$CuInS_2$ (GSC) direct Z-scheme system composite photocatalysts. GCC had a considerably higher photocurrent density than pristine $CuInS_2$ (CIS) and g-C_3N_4 NS, indicating an increased charge separation efficiency. At whatever light intensity, the GSC had the lowest reaction mechanism, indicating that electrons in the GSC were more easily transferred to photons in the electrolyte to contribute to hydrogen evolution than in the pure CIS and C_3N_4. Astonishingly, there was a sixfold rise in the lifetime of photocarriers in GSC. It suggests that the direct Z-scheme arrangement works well in separating photoinduced electron–hole couples than the indirect Z-scheme system. After the formation of direct Z-scheme, the GSC involved the photogenerated electron transfer in the CB of g-C_3N_4 NS to the valency band of CIS. Because of the Z-scheme mechanism, more electrons in the CIS survived, creating more hydrogen molecules while retroactively. As a result, this research provides a solid foundation and novel insights into using a Z-scheme technique to build highly efficient g-C_3N_4-based metal sulfide heterojunction photocatalysts.

As the above explanation shows, the type II and Z-schemes have a similar energy band topologies with differing charge transfer routes. Consequently, how discrimination between the two forms of heterojunctions has become a hot topic in scientific inquiry. We may assume that the hybrid photocatalysts follow a Z-scheme when there is a mediator in the heterojunction. Nonetheless, if there is no mediator, a straight Z-scheme and type-II heterojunction are likely to occur. Consequently, more complete solutions are necessary to support the charge transfer procedure. The mechanism of electron–hole transport can be explored using double beam photo-acoustic (DB-PA) spectroscopy. Ohno group [34] employed DB-PA analysis to validate the Z-scheme route of WO_3/g-C_3N_4 instead of the conventional type II heterojunction. Aside from that, Z-scheme mechanisms are perhaps postulated by smearing a systematic chemical probe to validate the existence of $*O_2$ and *OH radicals concerning the redox potential in the reaction average.

Even though nitrogen molecule is the most ubiquitous gas in the atmosphere, the bioavailability of the fixed N_2 from living organisms is lower [91]. Consequently, technological solutions integrating green and sustainable energy sources are in tremendous demand. Photocatalytic nitrogen reduction, which employs never-ending solar energy, is a green and cost-effective process for converting nitrogen to ammonia. Cao et al. described aromatic ring form 3, 4-dihydroxybenzaldehyde (DBD) as an electron mediator in a Ga_2O_3-DBD/G-C_3N_4 as Z-scheme heterojunctions for photocatalytic N_2 reduction utilizing solar light. Due to the development of the conduction interface between Ga_2O_3/DBD and C_3N_4, the as-tailored nanocomposites were confirmed to exhibit superior photocarrier separation, improved simulated light absorption, and high redox capacity. In agreement with experimental findings, the hypothesized charge migration mechanism was the Z-scheme employing DBD cycling rings as electron mediators instead of type II, as shown in Fig. 24b. The radical CO_2 was validated are the main active species responsible for nitrogen reduction using the spin trapping electron paramagnetic resonance approach (Fig. 24c).

Furthermore, the photo-assisted fixation of nitrogen to ammonia was driven using the simple production of radical CO_2 ($E^0 = 1.8$ V) radicals with passable potential. In Fig. 24d shows the photocurrent measurements for 2.4% Ga_2O_3/C_3N_4 and 2.4% Ga_2O_3-DBD/C_3N_4. It can be demonstrated that the photocurrent response of 2.4%

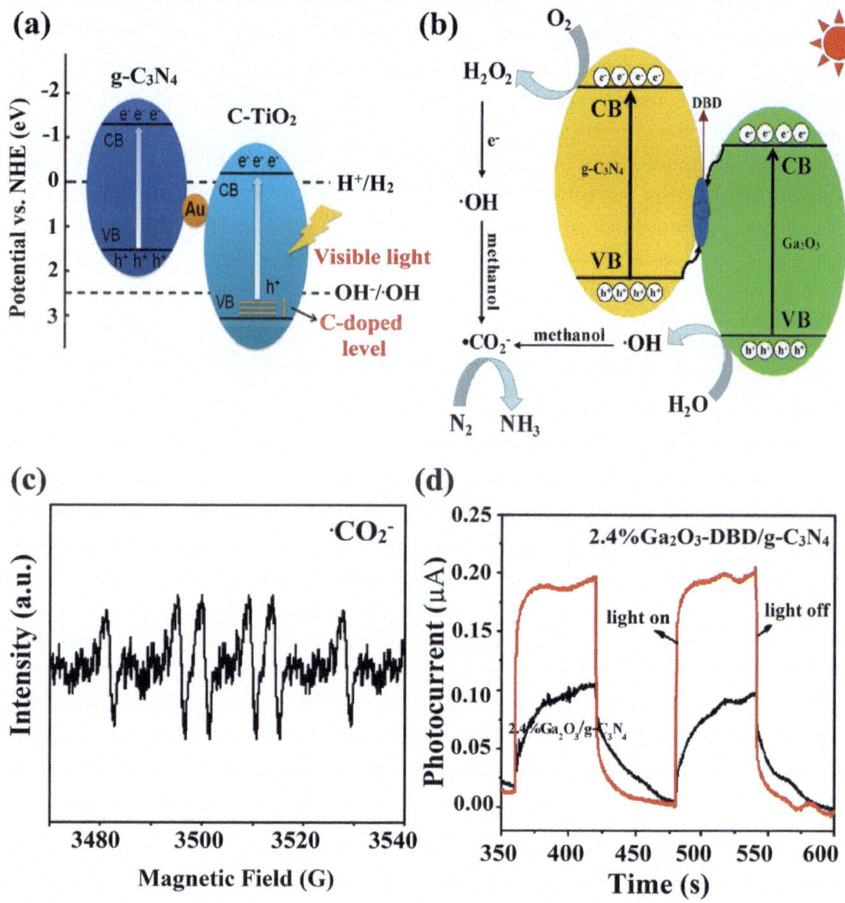

Fig. 24 **a** Hydrogen generation yield, **b** nitrogen reduction with light, **c** CO_2 in aqueous methanol solution in DMPO spin trapping, and **d** photocurrent density of Ga_2O_3/C_3N_4 systems (Reprinted with permission)

of Ga_2O_3-DBD/C_3N_4 is much larger than that of 2.4% Ga_2O_3/C_3N_4. It improved space charge separation by the usage of DBD cycling rings as an electron mediator in the Z-scheme. High potential photogenerated conduction band electrons of C_3N_4 converted O_2 to H_2O_2. However, the hole of Ga_2O_3 in VB with appropriate redox potential to form OH radical from H_2O over the Z-scheme charge migration pathway. Furthermore, OH radicals reacted with CH_3OH to generate CO_2, enabling N_2 to be reduced to NH_3.

5 Summary of Present Issues, Possibilities, and Prospects

Over the past century, a significant human population rise has resulted in various environmental and energy-related anxieties, the most serious of which is the unsustainable generation of fixed nitrogen supply. The Haber–Bosch process has been used for over a century and is being used to create ammonia. Photocatalytic nitrogen fixation, which transforms atmospheric nitrogen to NH_3 and other nitrogenous compounds under moderate circumstances, has aroused much interest in this field. However, current photocatalytic systems for nitrogen reduction to ammonia have lower conversion efficiency due to numerous critical factors, including hostile light harvesting, slow capture and activation of nitrogen, photogenerated charge separation, and low selectivity of the nitrogen reduction (NRR). Consequently, developing an adequate reduction nitrogen system based on photocatalytic requires a logical material design for advanced conversion efficiency. The heterojunction photocatalytic material based on type-I, II, and Z-scheme charge transmission channels has confirmed good conversion efficiencies, allowing special charge separating and light harvesting.

Furthermore, anionic vacancy in photocatalytic functions as activation and trapping sites for nitrogen molecules boosts the selectivity of the reduction of nitrogen process. Because of the confined superficial plasmon resonance (SPR), further changes, such as the addition of precious metals, may dramatically increase photocatalytic activity. As a consequence of the modification, nitrogen fixation rates have risen dramatically. To produce a viable photocatalytic reduction of nitrogen, cultivating the selectivity for nitrogen fixation using SPR concept can be a great solution.

Photocatalyst engineering, shape, composition, and surface characteristics that favor atom binding over H atom binding are key problems that must be overcome to suppress the HER and increase the NRR. Furthermore, devising a consistent technique for detecting NH_3 production rate needs particular care since the most often reported spectroscopic procedures, which use indophenol and Nessler reagent, are prone to producing inaccurate data. Consequently, we suggest researchers use the recommended experimental approach for estimating N_2 with reliability criteria. Finally, a complete mechanical analysis of photocatalytic nitrogen reduction processes is essential to understand, which leads to poor nitrogen reduction performance.

Particular emphasis should be paid to experimental and theoretical nitrogen reduction systems based on the biological nitrogenous enzyme containing Fe, Mo, and

S components. These methods have revealed that they can create NH_3 in natural conditions. Also, by altering the chemical and physical composite of photocatalytic materials, these material chemical and physical characteristics required for the PCET reaction may be avoided. The literature is combed for reports on heterojunction-based nitrogen fixation, methods, and criteria for performance assessment. Even though present nitrogen fixation by photocatalytic performance is still far from commercial feasibility, progress in this field seems to have a good accidental of success. Hopefully, this review will provide an overview of the literature for presently investigated heterojunction-based nitrogen reduction to ammonia through photocatalytic systems.

References

1. Lu Z, Saji SE, Langley J, Lin Y, Xie Z, Yang K, Bao L, Sun Y, Zhang S, Ng YH, Song L, Cox N, Yin Z (2021) Selective N_2/H_2O adsorption ointo 2D amphiphilic amorphous photocatalysts for ambient gas-phase nitrogen fixation. Appl Catal B: Environ 294:120240
2. Liu X, Luo Y, Ling C, Shi Y, Zhan G, Li H, Gu H, Wei K, Guo F, Ai Z, Zhang L (2022) Rare earth La single atoms supported MoO_{3-x} for efficient photocatalytic nitrogen fixation. Appl Catal B: Environ 301:120766
3. Chena X, Zhang X, Lia Y-H, Qia M-Y, Lia J-Y, Tang Z-R, Zhou Z, Xua Y-J (2021) Transition metal doping BiOBr nanosheets with oxygen vacancy and exposed {102} facets for visible light nitrogen fixation, Appl Catal B: Environ 281:119516
4. Mohebinia M, Wu C, Yang G, Dai S, Hakimian A, Tong T, Ghasemi H, Wang Z, Wang D, Ren Z, Bao J (2021) Ultrathin bismuth oxyiodide nanosheets for photocatalytic ammonia generation from nitrogen and water under visible to near-infrared Light. Mater Today Phys 16:100293
5. Li X, Sun X, Zhang L, Suna S, Wang W (2018) Efficient photocatalytic fixation of N_2 by KOH-treated g-C_3N_4. J Mater Chem A 6:3005
6. Sun S, An Q, Wang W, Zhang L, Liu J, Goddard WA (2017) Efficient photocatalytic reduction of dinitrogen to ammonia on bismuth monoxide quantum dots. Mater Chem A 5:201
7. Yang J (2018) Progress of metal oxide (sulfide)-based photocatalytic materials for reducing nitrogen to ammonia. Hindawi J Chem 2018, Article ID 3286782, 8
8. Ren W, Mei Z, Zheng S, Li S, Zhu Y, Zheng J, Lin Y, Chen H, Gu M, Pan F (2020) Wavelength-dependent solar N_2 fixation iinto ammonia and nitrate in pure water. AAAS Res 2020, Article ID 3750314, 12
9. Hao D, Ren J, Wang Y, Arandiyan H, Garbrecht M, Bai X, Shon HK, Wei W, Ni B-J (2021) A green synthesis of Ru modified g-C_3N_4 nanosheets for enhanced photocatalytic ammonia synthesis. AAAS Energy Mater Adv 2021, Article ID 9761263, 12
10. Bian S, Wen M, Wang J, Yang N, Chu PK, Yu X-F (2020) Edge-rich black phosphorus for photocatalytic nitrogen fixation. J Phys Chem Lett 11:1052–1058
11. Vu M-H, Sakar M, Do T-O (2018) Insights into the recent progress and advanced materials for photocatalytic nitrogen fixation for ammonia (NH_3) production. MDPI, Catals 8:621
12. Sridharan K, Shenoy S, Girish Kumar S, Terashima C, Fujishima A, Pitchaimuthu S (2021) Advanced two-dimensional heterojunction photocatalysts of stoichiometric and non-stoichiometric bismuth oxyhalides with graphitic carbon nitride for sustainable energy and environmental applications. Catalysts 11:426
13. Conte F, Pellegatta V, Tripodi A, Ramis G, Rossetti I (2021) Photo-oxidation of ammonia to molecular nitrogen in water under UV Vis and Sunlight Irradiation. Catalysts 11:975
14. Han Q, Jiao H, Xiong L, Tang J (2021) Progress and challenges in photocatalytic ammonia synthesis. Mater Adv 2:564
15. Zhang S, He Z, Li X, Zhang J, Zang Q, Wang S (2020) Building heterogeneous nanostructures for photocatalytic ammonia decomposition. Nanoscale Adv 2:3610

16. Sun S, An Q, Wang W, Zhang L, Liu J, Goddard WA (2017) Efficient photocatalytic reduction of dinitrogen to ammonia on bismuth monoxide quantum dots. J Mater Chem A 5:201
17. Ali H, Masar M, Guler AC, Urbanek M, Machovsky M, Kuritka I. Heterojunction-based photocatalytic nitrogen fixation: principles and current progress. https://doi.org/10.1039/d1na00565k
18. Li P, Zhou Z, Wang Q, Guo M, Chen S, Low J, Long R, Liu W, Ding P, Wu Y, Xiong Y (2020) Visible-light-driven nitrogen fixation catalyzed by Bi_5O_7Br nanostructures: enhanced performance by oxygen vacancies. J Am Chem Soc 142:12430−12439
19. Li H, Shang J, Ai Z, Zhang L (2015) Efficient visible light nitrogen fixation with biobr nanosheets of oxygen vacancies on the exposed 001 facets. J Am Chem Soc 137:6393–6399
20. Liu Q, Yuan J, Gan Z, Liu C, Li J, Liang Y, Chen R (2021) Photocatalytic N_2 reduction: uncertainties in the determination of ammonia production. ACS Sustain Chem Eng 9:560–568
21. Hao Q, Liu C, Jia G, Wang Y, Arandiyan H, Weia W, Ni B-J (2020) Catalytic reduction of nitrogen to produce ammonia by bismuth-based catalysts: state of the art and future prospects. Mater Horiz 7:1014
22. Vu M-H, Nguyen C-C, Do T-O (2020) Synergistic effect of Fe doping and plasmonic Au nanoparticles on $W_{18}O_{49}$ nanorods for enhancing photoelectrochemical nitrogen reduction. ACS Sustain Chem Eng 8:12321−12330
23. Yaoa X, Zhang W, Huanga J, Dua Z, Honga X, Chenc X, Hua X, Wang X (2020) Enhanced photocatalytic nitrogen fixation of Ag/B-doped g-C_3N_4 nanosheets by one-step in-situ decomposition-thermal polymerization method. Appl Catal A 601:117647
24. Sun B, Liang Z, Qian Y, Xu X, Han Y, Tian J (2020) Sulfur vacancy-rich O-doped 1T-MoS2 nanosheets for exceptional photocatalytic nitrogen fixation over CdS. ACS Appl Mater Interf 12:7257−7269
25. Yu X, Qiu H, Wang Z, Wang B, Meng Q, Sun S, Tang Y, Zhao K (2021) Constructing the Z-scheme TiO_2/Au/BiOI nanocomposite for enhanced photocatalytic nitrogen fixation. Appl Surf Sci 556:149785
26. Zhang K, Ai Z, Huang M, Shi D, Shao Y, Hao X, Zhang B, Wua Y (2021) Type II cuprous oxide/graphitic carbon nitride p-n heterojunctions for enhanced photocatalytic nitrogen fixation. J Catal 395:273–281
27. Zhang Y, Di J, Ding P, Zhao J, Gu K, Chen X, Yan C, Yin S, Xia J, Li H (2019) Ultrathin g-C_3N_4 with enriched surface carbon vacancies enables highly efficient photocatalytic nitrogen fixation. J Colloid Interf Sci 553:530–539
28. Xue Y, Guo Y, Liang Z, Cui H, Tian J (2019) Porous g-C_3N_4 with nitrogen defects and cyano groups for excellent photocatalytic nitrogen fixation without co-catalysts. J Colloid Interf Sci 556:206–213
29. Fei T, Yu L, Liu Z, Song Y, Xu F, Mo Z, Liu C, Deng J, Ji H, Cheng M, Lei Y, Xu H, Li H (2019) Graphene quantum dots modified flower like Bi_2WO_6 for enhanced photocatalytic nitrogen fixation. J Colloid Interf Sci 557:498–505
30. Feng Y, Zhang Z, Zhao K, Lin S, Li H, Gao X (2021) Photocatalytic nitrogen fixation: oxygen vacancy modified novel micro-nanosheet structure $Bi_2O_2CO_3$ with band gap engineering. J Colloid Interf Sci 583:499–509
31. Zhang W, Xing P, Zhang J, Chen L, Yang J, Hu X, Zhao L, Wu Y, He Y (2021) Facile preparation of novel nickel sulfide modified $KNbO_3$ heterojunction composite and its enhanced performance in photocatalytic nitrogen fixation. J Colloid Interf Sci 590:548–560
32. Song M, Wang L, Li J, Sun D, Guan R, Zhai H, Gao X, Li X, Zhao Z, Sun Z (2021) Defect density modulation of La_2TiO_5: an effective method to suppress electron-hole recombination and improve photocatalytic nitrogen fixation. J Colloid Interf Sci 602:748–755
33. Hu X, Zhang W, Yong Y, Xu Y, Wang X, Yao X (2020) One-step synthesis of iodine-doped g-C_3N_4 with enhanced photocatalytic nitrogen fixation performance. Appl Surf Sci 510:145413
34. Kong Y, Lv C, Zhang C, Chen G (2020) Cyano group modified g-C_3N_4: Molten salt method achievement and promoted photocatalytic nitrogen fixation activity. Appl Surf Sci 515:146009
35. Qiu P, Xu C, Zhou N, Chen H, Jiang F (2018) Metal-free black phosphorus nanosheets-decorated graphitic carbon nitride nanosheets with CeP bonds for excellent photocatalytic nitrogen fixation. Appl Catal B: Environ 221:27–35

36. Xiao C, Zhang L, Wang K, Wang H, Zhou Y, Wang W (2018) A new approach to enhance photocatalytic nitrogen fixation performance via phosphate-bridge: a case study of $SiW_{12}/K\text{-}C_3N_4$. Appl Catal B 239:260–267
37. Xu H, Wang Y, Dong X, Zheng N, Ma H, Zhang X (2019) Fabrication of In_2O_3/In_2S_3 microsphere heterostructures for efficient and stable photocatalytic nitrogen fixation. Appl Catal B: Environ 257:117932
38. Shen Z-K, Cheng M, Yuan Y-J, Pei L, Zhong J, Guan J, Li X, Li Z-J, Bao L, Zhang X, Yu Z-T, Zou Z (2021) Identifying the role of interface chemical bonds in activating charge transfer for enhanced photocatalytic nitrogen fixation of Ni_2P black phosphorus photocatalysts. Appl Catal B: Environ 295:120274
39. Lianga C, Niub H-Y, Guoa H, Niua C-G, Huangc D-W, Yanga Y-Y, Liua H-Y, Shaoa B-B, Fenga H-P (2020) Insight iinto photocatalytic nitrogen fixation on graphitic carbon nitride: defect-dopant strategy of nitrogen defect and boron dopant. Chem Eng J 396:125395
40. Zhang G, Sewell CD, Zhang P, Mi H, Lin Z (2020) Nanostructured photocatalysts for nitrogen fixation. Nano Energy 71:104645
41. Xin Y, Wang S, Yuan H, Hou T, Zhu W, Liu Y, Yao Y, Zhang E, Liang S, Wang L (2021) Atomic-level insights iinto the activation of nitrogen via hydrogen-bond interaction toward nitrogen photofixation. Chem. Elsevier Inc. 7:2118–2136
42. Xu C, Qiu P, Li L, Chen H, Jiang F, Wang X (2018) Bismuth subcarbonate with designer defects for broad-spectrum photocatalytic nitrogen fixation. ACS Appl Mater Interf 10:25321−25328
43. Liu Q-Y, Wang H-D, Tang R, Cheng Q, Yuan Y-J (2021) Rutile TiO_2 nanoparticles with oxygen vacancy for photocatalytic nitrogen fixation. ACS Appl Nano Mater 4:8674–8679
44. Medford AJ, Hatzell MC (2017) Photon-driven nitrogen fixation: current progress, thermodynamic considerations, and future outlook. ACS Catal 7:2624−2643
45. Shi R, Zhao Y, Waterhouse GIN, Shuai Zhang, and Tierui Zhang, Defect engineering in photocatalytic nitrogen fixation. ACS Catal 9:9739−9750
46. Shang S, Xiong W, Yang C, Johannessen B, Liu R, Hsu H-Y, Gu Q, Leung MKH, Shang J (2021) Atomically dispersed iron metal site in a porphyrin-based metal−organic framework for photocatalytic nitrogen fixation. ACS Nano 15:9670−9678
47. Li J, Wang D, Guan R, Zhang Y, Zhao Z, Zhai H, Sun Z (2020) Vacancy-enabled mesoporous TiO_2 modulated by nickel doping with enhanced photocatalytic nitrogen fixation performance. ACS Sustain Chem Eng 8:18258–18265
48. Comer BM, Medford AJ (2018) Analysis of photocatalytic nitrogen fixation on Rutile $TiO_2(110)$. ACS Sustain Chem Eng 6:4648–4660
49. Gao X, Wen Y, Qu D, An L, Luan S, Jiang W, Zong X, Liu X, Sun Z (2018) Interference effect of alcohol on Nessler's reagent in photocatalytic nitrogen fixation. ACS Sustain Chem Eng 6:5342−5348
50. Mou H, Wang J, Zhang D, Yu D, Chen W, Wang D, Mu T (2019) A one-step deep eutectic solvent assisted synthesis of carbon nitride/metal oxide composites for photocatalytic nitrogen fixation. J Mater Chem A 7:5719
51. Mao C, Wang J, Zou Y, Li H, Zhan G, Li J, Zhao J, Zhang L (2019) Anion (O, N, C, and S) vacancies promoted photocatalytic nitrogen fixation. Green Chem 21:2852
52. Zhang G, Yang X, He C, Zhang P, Mi H (2020) Constructing a tunable defect structure in TiO_2 for photocatalytic nitrogen fixation. J Mater Chem A 8:334
53. Huang Y, Zhang N, Wu Z, Xie X (2020) Artificial nitrogen fixation over bismuth-based photocatalysts: fundamentals and future perspectives. J Mater Chem A 8:4978
54. Hao D, Liu C, Xu X, Kianinia M, Aharonovich I, Bai X, Liu X, Chen Z, Wei W, Jiae G, Ni B-J (2020) Surface defect-abundant one-dimensional graphitic carbon nitride nanorods boost photocatalytic nitrogen fixation. New J Chem 44:20651
55. Gao W, Li X, Zhang X, Su S, Luo S, Huang R, Jing Y, Luo M (2021) Photocatalytic nitrogen fixation of metal–organic frameworks (MOFs) excited by ultraviolet light: insights into the nitrogen fixation mechanism of missing metal cluster or linker defects. Nanoscale 13:7801
56. Wei Y, Jiang W, Liu Y, Bai X, Hao D, Ni B-J. Recent advances in photocatalytic nitrogen fixation and beyond. https://doi.org/10.1039/d2nr00198e

57. Yan Y, Yang H, Yi Z, Xian T (2019) $NaBH_4$-reduction induced evolution of Bi nanoparticles from BiOCl nanoplates and construction of promising Bi@BiOCl hybrid photocatalysts. Catalysts 9:795. https://doi.org/10.3390/catal9100795
58. Xie T, Hu J, Yang J, Liu C, Xu L, Wang J, Peng Y, Liu S, Yin X, Lu Y (2019) Visible-light-driven photocatalytic activity of magnetic BiOBr/SrFe12O19 nanosheets. Nanomaterials 9:735
59. Bu T-A, Hao Y-C, Gao W-Y, Su X, Chen L-W, Zhang N, Yin A-X (2019) Promoting photocatalytic nitrogen fixation with alkali metal cations and plasmonic nanocrystals. Nanoscale 11:10072
60. Lee J, Tan L-L, Chai S-P (2021) Heterojunction photocatalysts for artificial nitrogen fixation: fundamentals, latest advances and future perspectives. Nanoscale 13:7011
61. Lianga C, Niub H-Y, Guoa H, Niua C-G, Yanga Y-Y, Liu H-Y, Tanga W-W, Fenga H-P (2021) Efficient photocatalytic nitrogen fixation to ammonia over bismuth monoxide quantum dots-modified defective ultrathin graphitic carbon nitride. Chem Eng J 406:126868
62. Zeng L, Zhe F, Wang Y, Zhang Q, Zhao X, Hub X, Wu Y, He Y (2019) Preparation of interstitial carbon doped BiOI for enhanced performance in photocatalytic nitrogen fixation and methyl orange degradation. J Colloid Interf Sci 539:563–574
63. Li X, He C, Zuo S, Yan X, Dai D, Zhang Y, Yao C (2019) Photocatalytic nitrogen fixation over fluoride/attapulgite nanocomposite: effect of upconversion and fluorine vacancy. Solar Energy 191:251–262
64. Ye L, Han C, Ma Z, Leng Y, Li J, Ji X, Bi D, Xie H, Huang Z (2017) Ni_2P loading on $Cd_{0.5}Zn_{0.5}S$ solid solution for exceptional photocatalytic nitrogen fixation under visible light. Chem Eng J 307:311–318
65. Gao K, Zhang C, Zhang Y, Zhou X, Gu S, Zhang K, Wang X, Song X (2022) Oxygen vacancy engineering of novel ultrathin $Bi_{12}O_{17}Br_2$ nanosheets for boosting photocatalytic N_2 reduction. J Colloid Interf Sci 614:12–23
66. Ren X, Wu K, Qin Z, Zhao X, Yang H (2019) The construction of type II heterojunction of Bi_2WO_6/BiOBr photocatalyst with improved photocatalytic performance. J Alloys Comp 788:102e109
67. Yao X, Zhang W, Huang J, Du Z, Hong X, Chen X, Hu X, Wang X (2020) Enhanced photocatalytic nitrogen fixation of Ag/B-doped g-C_3N_4 nanosheets by one-step in-situ decomposition-thermal polymerization method. Appl Catal A, Gen 601:117647
68. Kong XY, Ng B-J, Tan KH, Chen X, Wang H, Mohamed AR, Chai S-P (2018) Simultaneous generation of oxygen vacancies on ultrathin BiOBr nanosheets during visible-light-driven CO_2 photoreduction evoked superior activity and long-term stability. Catal Today 314:20–27
69. Zhang W, Xing P, Zhang J, Chen L, Yang J, Hu X, Zhao L, Wub Y, He Y (2021) Facile preparation of novel nickel sulfide modified $KNbO_3$ heterojunction composite and its enhanced performance in photocatalytic nitrogen fixation. J Colloid Interf Sci 590:548–560
70. Mao C, Cheng H, Tian H, Li H, Wen-Jing Xiao HuXu, Zhao J, Zhang L (2018) Visible light driven selective oxidation of amines to imines with BiOCl: Does oxygen vacancy concentration matter? Appl Catal B 228:87–96
71. Liang C, Niu H-Y, Guo H, Niu C-G, Yang Y-Y, Liu H-Y, Tang W-W, Feng H-P (2021) Efficient photocatalytic nitrogen fixation to ammonia over bismuth monoxide quantum dots-modified defective ultrathin graphitic carbon nitride. Chem Eng J 406:126868
72. Xiao C, Zhang L, Wang K, Wang H, Zhou Y, Wang W (2018) A new approach to enhance photocatalytic nitrogen fixation performance via phosphate-bridge: a case study of SiW_{12}/K-C_3N_4. Appl Catal B: Environ 239:260–267
73. Lina Q, Li Y-H, Qi M-Y, Li J-Y, Tang Z-R, Anpo M, Yamada YMA, Xu Y-J (2020) Photoredox dual reaction for selective alcohol oxidation and hydrogen evolution over nickel surface-modified $ZnIn_2S_4$. Appl Catal B: Environ 271:118946
74. Wang G, Huo T, Deng Q, Yu F, Xia Y, Li H, Hou W (2022) Surface-layer bromine doping enhanced generation of surface oxygen vacancies in bismuth molybdate for efficient photocatalytic nitrogen fixation. Appl Catal B: Environ 310:121319
75. Sun B, Qian Y, Liang Z, Guo Y, Xue Y, Tian J, Cui H (2019) Oxygen vacancy-rich BiO_2-x ultra-thin nanosheet for efficient full-spectrum responsive photocatalytic oxygen evolution from water splitting. Sol Energy Mater Sol Cells 195:309–317

76. Baia Y, Baia H, Qu K, Wang F, Guan P, Xu D, Fan W, Shi W (2019) In-situ approach to fabricate BiOI photocathode with oxygen vacancies: understanding the N_2 reduced behavior in photoelectrochemical system. Chem Eng J 362:349–356
77. Hou J, Jiang T, Wang X, Zhang G, Zou J-J, Cao C (2021) Variable dimensional structure and interface design of g-C_3N_4/BiOI composites with oxygen vacancy for improving visible-light photocatalytic properties. J Clean Prod 287:125072
78. Xiong J, Song P, Di J, Li H (2020) Atomic-level active sites steering in ultrathin photocatalysts to trigger high efficiency nitrogen fixation. Chem Eng J 402:126208
79. Li G, Yang W, Gao S, Shen Q, Xue J, Chen K, Li Q (2021) Creation of rich oxygen vacancies in bismuth molybdate nanosheets to boost the photocatalytic nitrogen fixation performance under visible light illumination. Chem Eng J 404:127115
80. Xue X, Chen H, Xiong Y, Chen R, Jiang M, Fu G, Xi Z, Zhang XL, Ma J, Fang W, Jin Z (2021) Near-infrared-responsive photo-driven nitrogen fixation enabled by oxygen vacancies and sulfur doping in black TiO_2−xSy nanoplatelets. ACS Appl Mater Interf 13:4975−4983
81. Xue X, Chen R, Chen H, Hu Y, Ding Q, Liu Z, Ma L, Zhu G, Zhang W, Yu Q, Liu J, Ma J, Jin Z (2018) Oxygen vacancy engineering promoted photocatalytic ammonia synthesis on ultrathin two-dimensional bismuth oxybromide nanosheets. Nano Lett 18:7372−7377
82. Liu Y, Su Y, Quan X, Fan X, Chen S, Yu H, Zhao H, Zhang Y, Zhao J (2018) Facile ammonia synthesis from electrocatalytic N_2 reduction under ambient conditions on N-doped porous carbon. ACS Catal 8:1186−1191
83. Xu Y, Fu H, Zhao L, Jian L, Liang Q, Xiao X (2021) Insight iinto facet-dependent photocatalytic H_2O_2 production on BiOCl nanosheets. New J Chem 45:3335
84. Hirakawa H, Hashimoto M, Shiraishi Y, Hirai T (2017) Photocatalytic conversion of nitrogen to ammonia with water on surface oxygen vacancies of titanium dioxide. J Am Chem Soc 139:10929–10936
85. Zhang N, Jalil A, Wu D, Chen S, Liu Y, Gao C, Ye W, Qi Z, Ju H, Wang C, Wu X, Song L, Zhu J, Xiong Y (2018) Refining defect states in $W_{18}O_{49}$ by Mo doping: a strategy for tuning N_2 activation towards solar driven nitrogen fixation. J Am Chem Soc 140:9434−9443
86. Zhang K, Ai Z, Huang M, Shi D, Shao Y, Hao X, Zhang B, Wu Y (2021) Type II cuprous oxide/graphitic carbon nitride p-n heterojunctions for enhanced photocatalytic nitrogen fixation. J Catal 395:273–281
87. Song M, Wang L, Li J, Sun D, Guan R, Zhai H, Gao X, Li X, Zhao Z, Sun Z (2021) Defect density modulation of La_2TiO_5: an effective method to suppress electron-hole recombination and improve photocatalytic nitrogen fixation. J Colloid Interf Sci 602:748–755
88. Donga S, Liua X, Tian G, Wang Y, Jin G, Zhao Y, Sun J, Fan M (2021) Surface oxygen vacancies modified Bi_2MoO_6 double-layer spheres: Enhanced visible LED light photocatalytic activity for ciprofloxacin degradation. J Alloys Comp 892:162217
89. Huang Y, Zhang N, Wu Z, Xie X (2020) Artificial nitrogen fixation over bismuth-based photocatalysts: fundamentals and future perspectives. J Mater Chem A 8:4978
90. Shi YB, Li H, Mao C, Zhan G, Yang Z, Ling C, Wei K, Liu X, Ai Z, Zhang L. Manipulating excitonic effects in layered bismuth oxyhalides for photocatalysis. https://doi.org/10.1021/acsestengg.1c00466
91. Gao X, Shang Y, Liu L, Fu F (2019) Chemisorption-enhanced photocatalytic nitrogen fixation via 2D ultrathin p–n heterojunction AgCl/d-Bi_2O_3 nanosheets. J Catal 371:71–80
92. Raizada P, Kumar A, Hasija V, Singh P, Thakur VK, Khan AAP (2021) An overview of converting reductive photocatalyst into all solid-state and direct Z-scheme system for water splitting and CO_2 reduction. J Indus Eng Chem 93:1–27
93. Xiao C, Zhang L, Wang K, Wang H, Y Zhou, Wang W (2018) A new approach to enhance photocatalytic nitrogen fixation performance via phosphate-bridge: a case study of SiW_{12}/K-C_3N_4. Appl Catal B: Environ 239:260–267

A Recent Review on Photocatalytic CO_2 Reduction in Generating Sustainable Carbon-Based Fuels

Tadele Negash Gemeda, Li-Hsiang Chang, Yu Tse Liang, Van Hoang Khang Phan, Gianna Fadhilah, Fery Prasetyo, and Mohamed Tarek Ahmed

Abstract Emissions of greenhouse gases from industrial activity, traffic, and solid waste landfills contribute directly to the air pollution crisis. To keep the quantities of carbon dioxide in the atmosphere at a safe level, air pollution must be reduced. Many studies have reported techniques for converting the main greenhouse gas, carbon dioxide, into viable fuels as a method of reducing air pollution. The trending topic of photocatalytic conversion of CO_2 to fuels has shown high potential. This process is environment-friendly due to its capability to be performed at ambient temperature and pressure and hence utilize less energy. This review focuses on the thermodynamic and reaction kinetics of photocatalytic CO_2 reduction. Various types of photocatalysts used for CO_2 reduction are discussed excessively by considering metallic, non-metallic, and composite systems. In addition, necessary experimental and product analysis parameters are covered and mentioned in detail.

Keywords CO_2 reduction · Carbon · Fuel · Photocatalysis · Greenhouse gas

T. N. Gemeda (✉) · L.-H. Chang · Y. T. Liang · V. H. K. Phan · M. T. Ahmed
Department of Materials Science and Engineering, National Taiwan University of Science and Technology, No.43, Sec. 4, Keelung Road, Taipei 10607, Taiwan
e-mail: tadelenegash@gmail.com

T. N. Gemeda
Department of Chemical Engineering, School of Mechanical, Chemical and Material Engineering, Adama Science and Technology University, P.O. Box 1888 Adama, Ethiopia

G. Fadhilah · F. Prasetyo
Department of Chemical Engineering, National Taiwan University of Science and Technology, No.43, Sec. 4, Keelung Road, Taipei 10607, Taiwan

H. Abdullah (ed.), *Photocatalytic Activities for Environmental Remediation and Energy Conversion*, Green Energy and Technology,
https://doi.org/10.1007/978-981-19-6748-1_4

1 Introduction

A significant amount of carbon dioxide (CO_2) has been produced and released into the atmosphere as a result of the development of industries. CO_2 levels are expected to rise roughly up to twice their current quantity [1–5]. CO_2 exhaust, being a highly disruptive greenhouse gas, can unbalance the atmosphere and the biosphere accordingly. As a result, encouraging CO_2 reduction and usage limitations is critical. Because CO_2 is a linear molecule with high ionization energy and a low electron affinity, it is easier to reduce than to oxidize [1–5]. To reduce CO_2 and enhance the reaction progress, various approaches have been used. Chemical reduction, photochemical reduction, electrochemical reduction, and biological transformations, for example, are investigated. The CO_2 photoreduction reaction is the most promising among these approaches.

Photosynthesis combines the creation of O_2 and carbohydrates with the fixation of CO_2 using solar light energy. This process creates a readily available carbon source as well as an aerobic environment that can support practically all life forms on the planet [2–5]. With an extra input of energy from photosystem I, photosystem II photoinduces water oxidation, which supplies a key supply of reducing equivalents (water-derived electrons and protons) to transform CO_2 into biomass, food, and fuel [1].

For more than 30 years, the idea of imitating the overall natural photosynthetic cycle of chemical conversion of CO_2 into hydrocarbon fuels has piqued interest. Including biological conversion [1], thermochemical conversion [2], electrochemical conversion [3], and photocatalytic reduction of CO_2 [4]. Due to CO_2's thermodynamic stability, a large amount of energy is required to break the $C = O$ bond during the conversion process [5]. Compared to other technologies, photocatalytic CO_2 reduction into hydrocarbon fuels is a difficult but promising path. It can be a source of a sustainable alternative to traditional fossil fuels, according to the following benefits: (i) It can be done under relatively mild settings, such as at ambient temperature and pressure; (ii) This process starts with a mass of abandoned CO_2 and is powered by unlimited and pure solar energy; (iii) CO_2 photoreduction may directly generate short-chain hydrocarbon fuels like CH_4, CH_3OH, C_2H_6, and so on, [3–5] alleviating the world's growing energy crisis; (iv) the commercialization of this technique will allow CO_2 to replace fossil fuels as a source of carbon in the chemical industry [5].

In this review, we focus on the reduction of CO_2 using irradiated photons and spotlighting different drawbacks as well. Section 2 depicts the fundamental mechanisms of photocatalytic CO_2 reduction by deeply exploiting the thermodynamics and kinetics of photocatalytic CO_2 reduction (PCR). In Sect. 3, the experimental parameters affecting the process are discussed in detail. Photocatalysts types, selection, and preparations are discussed in Sect. 4. Then, product characterization, analysis, and selectivity are explained in Sect. 5. Section 6 states the main challenges facing PCR and the promising opportunities of this topic ahead. Finally, the conclusion in Sect. 7 sums up the main points and declares the critical parameters to overcome the previously mentioned obstacles facing this research work.

2 Fundamentals, Mechanisms, and Kinetics of PCR

2.1 *Thermodynamics of PCR*

Table 1 illustrated the standard redox potential ΔE_0 and the Gibb free energy ΔG_0 of the multi-electron water splitting and CO_2 reduction, respectively [6]. The positive ΔG_0 substantiated that the CO_2 reduction process is the endothermic one and it is a high challenge to carry out at room temperature. It also proved that the CO_2 reduction reaction can archive much more energy than the water splitting reaction [6, 7].

The multi-electronic processes are more highly captivating than the mono-electronic process because the required energy for electron transfer is smaller. Besides that, the reaction by one electron possesses a bigger reduction potential of −1.9 V versus normal hydrogen electrode (NHE). Therefore, it calls for larger kinetics (overvoltage) [7, 9, 10].

A Latimer-Frost diagram shown in Fig. 1 depicted the multi-proton and multi-electron reduction of CO_2 in a water solution at pH 7 [11]. Table 1 also listed a summary of the standard reduction potentials of CO_2 for the half-cell reactions (at pH 7 in water solution versus NHE) [7, 9, 10, 12–14]. From Table 2 and Fig. 1, we can observe that the thermodynamic barrier was lowered significantly by a multi-electron and proton-assisted approach to the reduction of CO_2 [11]. On the other hand, the conversion from CO_2 to CO, then to H_2CO, and then to alcohols or hydrocarbons also has a lower kinetic barrier if it is collated into the mono-electron process [11, 15]. Thus, the proton-assisted multi-electron transfer is a promising candidate to reduce CO_2.

Via the multi-electron transfer process, several compounds like CH_3OH, HCOOH, and HCHO can be produced from the reduction of H_2CO_3 and the carbonate ions in the solution. The potentials of both H_2CO_3 and CO_3^{2-} mentioned in Table 2 are conclusive that the pathway of methanol formation from H_2CO_3 or CO_3^{2-} is more thermodynamically favorable than that from CO_2 [12, 13].

Table 1 ΔE_0 and ΔG_0 of the multi-electron water splitting and CO_2 reduction

Reaction	ΔG_o (kJ.mol^{-1})	ΔEo (V)
$H_2O_{(l)} \rightarrow H_{2(g)} + 0.5O_{2(g)}$	237	1.23
$CO_{2(g)} \rightarrow CO_{(g)} + 0.5O_2(g)$	257	1.33
$CO_{2(g)} + H_2O_{(l)} \rightarrow HCOOH_{(l)} + 0.5O_{2(g)}$	286	1.48
$CO_{2(g)} + H_2O_{(l)} \rightarrow HCHO_{(l)} + O_{2(g)}$	522	1.35
$CO_{2(g)} + 2H_2O_{(l)} \rightarrow CH_3OH_{(l)} + 1.5O_{2(g)}$	703	1.21
$CO_{2(g)} + 2H_2O_{(l)} \rightarrow CH_{4(g)} + 2O_{2(g)}$	818	1.06

Reproduced with permission [8]. (Copyright 2014, Springer. License Number 5311300338819)

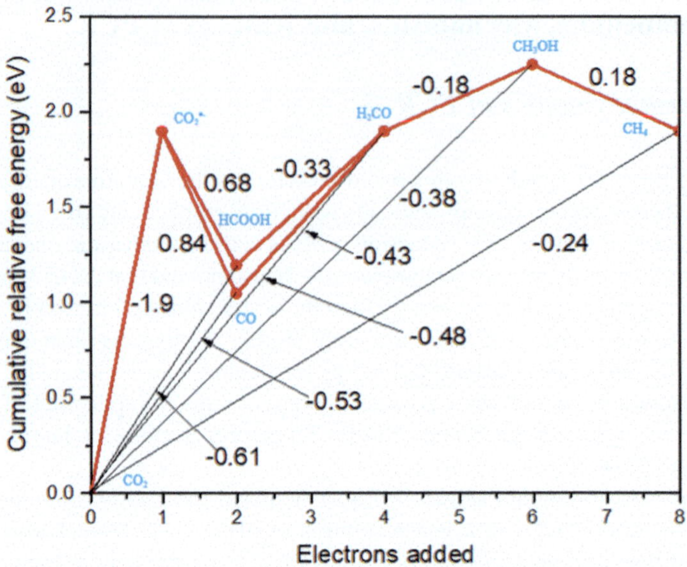

Fig. 1 Latimer–Frost photograph of the multi-proton and multi-electron processes of CO_2 in water solution at pH = 7. (Copyright 2014, Springer. License Number 5311300338819)

The thermodynamics of the CO_2 chemical change hindering increases due to the low energy class and high stability of CO_2. Consequently, the targeted conversions need much more energy. Furthermore, due to its inertness, the utilization of catalysts is a considerable requirement [10, 15]. Because the reduction of CO_2 is sophisticated and the transformation of CO_2 into hydrocarbon fuels utilizing assisted catalyst has been expansively reported with different aspects. Some of these approaches are thermochemical, electrochemical, photo-electrochemical (PEC), and photochemical reductions [15, 16]. The lack of releasing more greenhouse effect gas, eco-friendly and economical reduction of CO_2 into value chemicals will become a trending topic if the renewable energies are used widely. Meanwhile, the new routes of the artificial photosynthetic system (APS) e.g. PEC or photochemical reduction of CO_2 into solar fuel, are the urgent aims in the reduction of CO_2 [17]. For instance, Halmann was a pioneer to explore the p-type of GaP which was able to transform CO_2 to CH_3OH via the photo-electrochemical process in 1978 []. Then, Inoue and coworkers reported several chemical products such as HCHO, CH_3OH, and HCOOH by PCR utilizing TiO_2, ZnO, and CdS, GaP, and SiC aqueous sedimentations [14]. Thanks to APS systems, humans can use solar light or CO_2 as chemical energy. In the past decade, photoreduction of CO_2 was intensively studied because of the urgency to find out solutions for environmental pollution problems and their high applicability.

According to the semiconductor materials the production of the photogenerated (PG) charge carriers (electrons and holes), as shown in Fig. 2, are created by the absorption of photons with energy larger than or equal to its bandgap (Eg). The bandgap (energy difference between the valence band (VB) and the conduction band

Table 2 Reduction potentials of CO_2

Reaction	E0 (V) versus NHE at pH = 7
Reduction potentials of CO_2	
$2H^+ + 2e^- \rightarrow H_2$	−0.410
$CO_2 + e^- \rightarrow CO_2^-$	−1.900
$CO_2 + 2H^+ + 2e^- \rightarrow HCO_2H$	−0.610
$CO_2 + 2H^+ + 2e^- \rightarrow CO + H_2O$	−0.530
$CO_2 + 4H^+ + 4e^- \rightarrow C + 2H_2O$	−0.200
$CO_2 + 4H^+ + 4e^- \rightarrow HCHO + H_2O$	−0.480
$CO_2 + 6H^+ + 6e^- \rightarrow CH_3OH + H_2O$	−0.380
$CO_2 + 8H^+ + 8e^- \rightarrow CH_4 + 2H_2O$	−0.240
$2CO_2 + 8H_2O + 12e^- \rightarrow C_2H_4 + 12OH^-$	−0.340
$2CO_2 + 9H_2O + 12e^- \rightarrow C_2H_5OH + 12OH$	−0.330
$3CO_2 + 13H_2O + 18e^- \rightarrow C_3H_7OH + 18OH^-$	−0.320
Reduction potentials of H_2CO_3	
$2H^+ + 2e^- \rightarrow H_2$	−0.410
$2H_2CO_3 + 2H^+ + 2e^- \rightarrow H_2C_2O_4 + 2H_2O$	−0.800
$H_2CO_3 + 2H^+ + 2e^- \rightarrow HCOOH + H_2O$	−0.576
$H_2CO_3 + 4H^+ + 4e^- \rightarrow HCHO + 2H_2O$	−0.460
$H_2CO_3 + 6H^+ + 6e^- \rightarrow CH_3OH + 2H_2O$	−0.366
$H_2CO_3 + 4H^+ + 4e^- \rightarrow C + 3H_2O$	−0.182
Reduction potentials of CO_3^{2-}	
$2H^+ + 2e^- \rightarrow H_2$	−0.410
$2CO_3^{2-} + 4H^+ + 2e^- \rightarrow C_2O_4^{2-} + 2H_2O$	0.070
$CO_3^{2-} + 3H^+ + 2e^- \rightarrow HCOO^- + H_2O$	−0.099
$CO_3^{2-} + 6H^+ + 4e^- \rightarrow HCHO + 2H_2O$	−0.213
$CO_3^{2-} + 8H^+ + 6e^- \rightarrow CH_3OH + 2H_2O$	−0.201
$CO_3^{2-} + 6H^+ + 4e^- \rightarrow C + 3H_2O$	−0.065

Reproduced with permission [9, 12]. (Copyright 2014, Springer. License Number 5311300338819)

(CB)) plays a crucial key in forming photocatalytic behaviors of the semiconductors [18, 19]. On the surface of the semiconductors, the PG holes diffused from the VB react with water to induce either O_2 or hydroxyl radicals (•OH). Then, •OH radicals oxidize the surrounding organic contaminants on the semiconductors' surface [18, 19]. On the other hand, the electrons of the CB take part in the reduction process, which reacts either with water to yield H_2 or with CO_2 to fabricate hydrocarbon fuels [18–20]. To attain CO_2 photoreduction, there are many requirements for an elite photo-catalyst. The PCR have to be positioned more positively than the lowermost CB of the photocatalyst. The oxidation potentials of H_2O to O_2 ought to be located at a more negative potential than the uppermost VB. The redox reaction can be carried

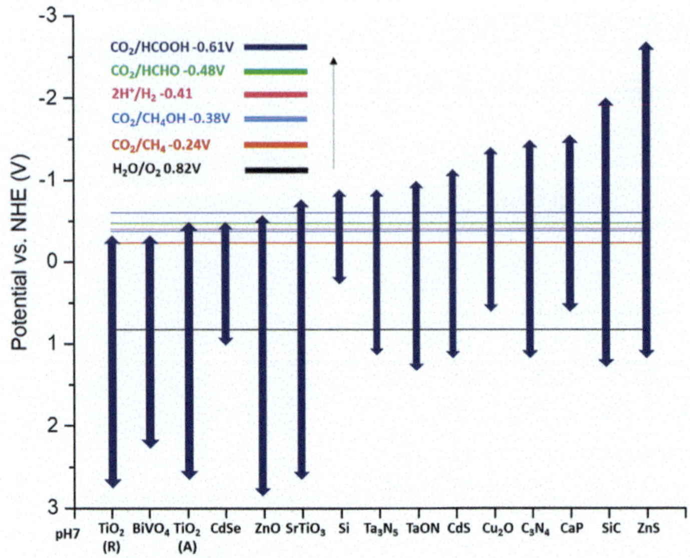

Fig. 2 Bandgap of some semiconductor photocatalysts and the redox potential of PCR at pH 7 in water solution. (Copyright 2014, Springer. License Number 5311300338819)

out under irradiation at a specific energy, which is equal to or greater than the Eg of the photocatalysts [21]. In 1979, the first published article in which the CB of the semiconductor approaches significantly more negative potential compared with the redox potential to perform a certain reaction of CO_2 reduction. Hence, the efficiency of obtaining products from CO_2 reduction escalated [14].

Many promising semiconductors shown in Fig. 3 such as TiO_2 [22, 23], ZnO [14, 24–27], ZnS [17, 28–31], $SrTiO_3$ [32–35], SiC [14, 36–38], Cu_2O [39–42], CdS [14, 43–48], GaP [14, 49], TaON [50–53], C3N4 [32, 54–56], $BiVO_4$ [57–61] and Ta_3N_5 [62–65] are appropriate for PCR. Among them, TiO_2 is a highly highlighted material, which was studied severally. TiO_2 attracts attention due to its many excellent properties such as being non-toxic and cheap, made up of abundant elements, and resistant to photo-corrosion. Even possessing a lot of good behaviors, TiO_2 is poor visible light absorption because of a wide bandgap, so it ought to be enhanced. Conversely, Cu_2O, CdS, GaP, TaON, C_3N_4, and Ta_3N_5 are promising materials for PCR under visible light irradiation. But, they are quite sensitive because of the weak photostability, so they also should be improved. From Fig. 3, on the right side, many photocatalysts are possessing more negative CB levels suitable for the PCR.

2.1.1 Processes of PCR

Not only do the appropriate Eg and CB potential play a salient role, but variously several other factors also significantly affect the yield of the PCR, for example, the

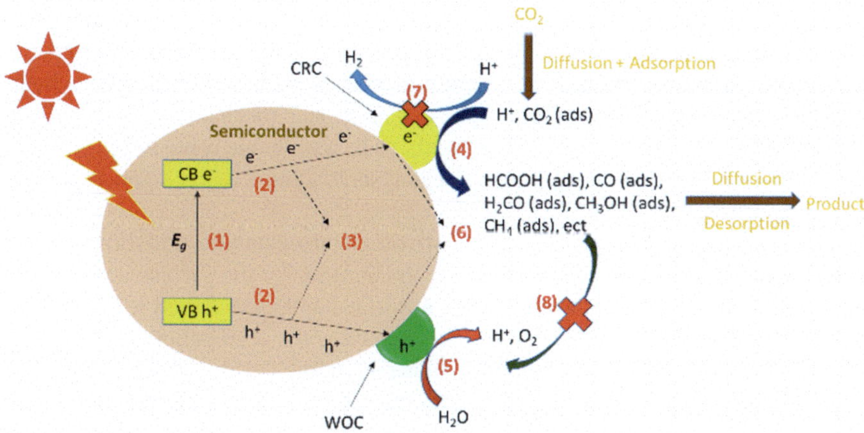

Fig. 3 Steps containing in PCR on a heterogeneous photo-catalyst

photocatalytic process and CO_2 reduction kinetics. According to a typical route of PCR over a semiconductor photo-catalyst, 4 main steps are composing the excitation, the transport, the segregation, and the electro-catalytic reduction of CO_2 and the water oxidation. In detail, the initial process (1) in Fig. 3, which indicates the formation of electron–hole pairs as the semiconductor absorbs the photon energy larger than or equal to the bandgap of the material, is the excitation. Hence, the materials with a narrow bandgap ($Eg < 3.0$ eV or $\lambda > 415$ nm) are high potential candidates for rising the excitation yields of electron–hole pairs by visible light. From these fundamentals, many new novels to synthesize the innovative photocatalysts utilizing the visible light region were intensively studied.

The second process (2) in Fig. 3 describes the segregation of the PG charge carriers and their relocation on the surface for specific chemical reactions. In the third process (3) in Fig. 3, the number of excited charge carriers is decreased dramatically because the holes and electrons recombine and form heat, as a consequence, the efficiency is decreased. This process can call by the deactivation one. To avoid these unwanted phenomena and improve the yield, the transfer of the PG charge carriers to the surface necessitates enhancing and prohibiting the recombination in the bulk plays a vital key. The segregation and the recombination process are dominated by structural and electronic behaviors or photocatalysts. Hence, several beneficial strategies are studied intensively to maximize the segregation and minimize the recombination such as the fabrication of semiconductor/nano-carbon heterojunctions or nanostructured semiconductors or semiconductor heterojunctions.

After the PG electrons come to the surface, the fourth process (4) in Fig. 3 or the electro-catalytic reduction of CO_2, which is frequently a multi-electron and multi-step process related to a cascade of reactions. Electron and photon transfer, C–O bond breaking, C–H/C–C bone creation, and numerous chemical compounds, will be taken place by the PG electrons trapped in the CO_2 reduction co-catalysts (CRC) or the surface active sites [66–71]. Typically, to result in a high yield, the surface of semiconductors ought to be fulfilled by the CRC. While the photocatalytic reduction of CO_2 occurs, does the formation of some kinds of stable products be to need a minimum of two electrons due to some intermediates being easily changed or hard to find out and enumerate. Due to the complicated multi-step mechanism, any approach which enhances CO_2 reduction kinetics ought to be a probable route to improve the yield. The synthesis of mesoporous photocatalysts loading CRC is a good example for such a purpose [9].

Simultaneously with the fourth process (4), the electro-catalytic oxidation of water via PG holes caught in the water oxidation co-catalysts (or the surface active sites) will happen once the PG holes approach the surface and are referred to as the fifth process (5). Enhancing water oxidation can encourage the segregation of PG charge carriers on the surface of semiconductors, hence inducing developed activity of CO_2 photoreduction. Additionally, the sixth step (6) in Fig. 3 shows the surface charge recombination process of the holes and electrons. The surface charge recombination and the efficiency of the photocatalytic CO_2 reduction process are inversely proportional. To reduce recombination, surface trapping ought to be enhanced by increasing the surface properties of the photocatalysts [9].

Besides that, the seventh process (7) and the eighth process (8) in Fig. 3 show the electro-catalytic H_2 evolution and reduction products of the electro-catalytic oxidation, respectively. In the seventh process (7), the consumption rate of PG electrons for CO_2 reduction will be decreased steadily by the H_2 evolution effect. In the eighth process (8), the oxidation of the CO_2 reduction products by the PG holes is detriment for both water oxidation and CO_2 reduction. The quantum yield will be unwantedly affected by these aforementioned two processes. Hence, the efficient approaches to enhance the PCR yield need to comprehensively consider such unfavorable factors [9].

2.2 *Possible Mechanisms of PCR*

In photocatalytic CO_2 reduction, there are many reduction products formed with the joining of several protons and electrons, thus, the photocatalysts need to meet extraordinarily specific requirements. Table 3 depicts several related reactions containing variable numbers of protons and electrons and their recorded reduction potentials [14, 18].

Furthermore, the adsorption of CO_2 on the photocatalysts as well as various reaction pathways also will affect the final products significantly. Nevertheless, existing of some ambiguous intermediate species has resulted in some challenges in the reaction mechanism study. Figure 4a and b show two possible CO_2 reduction routes through formaldehyde and carbene pathways [73].

In detail, the formaldehyde pathway is the fast hydrogenation pathway related to the simultaneous hydrogenation and deoxygenation reactions, namely displayed by the route: $CO_2^- \rightarrow HCOOH \rightarrow HCHO \rightarrow CH_3OH \rightarrow CH_4$. Whereas, HCOOH is generated by the combination of CO_2^-, protons, and electrons. Then, they are

Table 3 PCR reaction and corresponding redox potentials (versus NHE at pH 7)

Product	Reaction	E^0_{redox}
CO	$CO_2 + 2H^+ + 2e^- \rightarrow CO + H_2O$	−0.53
HCOOH	$CO_2 + 2H^+ + 2e^- \rightarrow HCOOH$	−0.61
HCHO	$CO_2 + 4H^+ + 4e^- \rightarrow HCHO + H_2O$	−0.48
CH_3OH	$CO_2 + 6H^+ + 6e^- \rightarrow CH_3OH + H_2O$	−0.38
CH_4	$CO_2 + 8H^+ + 8e \rightarrow CH_4 + 2H_2O$	0.24
CH_3CHO	$2CO_2 + 10H^+ + 10e^- \rightarrow CH_3CHO + 3H_2O$	−0.36
C_2H_4	$2CO_2 + 12\ H^+ +12e^- \rightarrow C_2H_4 + 4H_2O$	−0.34
C_2H_5OH	$2CO_2 + 12H^+ + 12e^- \rightarrow C_2H_5OH + 3H_2O$	−0.33
C_2H_6	$2CO_2 + 14H^+ + 14e^- \rightarrow C_2H_6 + 4H_2O$	-0.27
O_2	$H_2O \rightarrow 0.5O_2 + 2H^+ + 2e^-$	+ 0.81
H_2	$2H_2 + 2e^- \rightarrow 2H_2$	−0.42

(Reproduced with permission [72]. (Copyright 2021, Wiley–VCH GmbH. License Number 5311301363180)

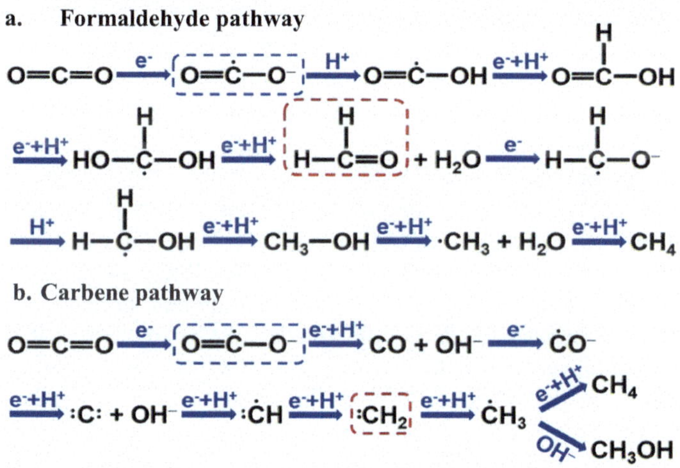

Fig. 4 Schematic demonstration for the **a** formaldehyde pathway and **b** carbene pathway for CO_2 reduction. (Reproduced with permission [73]. (Copyright 2020, Royal Society Chemistry License Number 1222250–1)

transformed into HCHO, dihydroxymethyl, two protons, and CH4 in the final step. This pathway is evidenced to be thermodynamically feasible [73]. On the other hand, the CO_2 reduction reaction, which is the fast deoxygenation, can also occur following the carbene pathway ($CO_2^- \rightarrow CO \rightarrow C \rightarrow CH_3 \rightarrow CH_3OH/CH_4$). Namely, there is the deoxygenation reaction in the initial step and then the hydrogenation reaction in the second one [8, 26].

In the carbene pathway, the CO intermediate usually is transformed easily into the final products because of the weak affinity of the CO products and the surface of the photocatalysts. The CO intermediate can interact with the protons or electrons to generate the CH_3OH or CH_4, the surface of the photo-catalyst possesses a CO strong adsorption capacity. According to the two possible mechanisms, C–O is attacked by protons in both routes. On the other hand, there are some highlighted feature characterizations belonging to each mechanism. For example, in the carbene pathway, the carbon or mixed coordination mode is probably to break the C–O linkage. Likewise, in the formaldehyde pathway, the CO_2^- will adapt to oxygen coordination mode [24, 27].

2.3 *Kinetics of PCR*

Based on the various mechanisms, many types of kinetic equations were generated. For modeling the photocatalytic reduction of carbonate by the TiO_2 under UV light (carried out in an aqueous solution), a Langmuir–Hinshelwood (L–H) equation was expressed. The results show that the photocatalytic reduction rate of carbonate is adsorption-controlled [74]. Whereas, a one-site L–H equation was used further to simulate the reduction rate of carbon dioxide to carbon monoxide (or hydrogen) and methane utilizing water over TiO_2 [74–77]. Additionally, the mechanism of Anpo, in which carbon monoxide is defined as the primary intermediate, was strongly supported by this model [74, 75, 77].

In conclusion, because of the complicated processes of PCR and their kinetics, the yield of PCR on the photo-catalyst surface is so low. The reduction potentials of carbon dioxide conversion to hydrogen are less feasible than to methane or/and methanol in order of thermodynamic aspects. Nevertheless, because of the multi-electron reduction process leading to the kinetics of CO_2 reduction, it is unfavorable. Thus, the enhancement reduction kinetics of CO_2 makes a significant contribution to improving the efficiency. Whereas there are many complicated processes in PCR. Water oxidation, CO_2 activation, and CO_2 reduction kinetics processes also contribute important roles in the overall efficiency. Hence, to improve the overall efficiency, the modification of the kinetics and photocatalytic processes of CO_2 reduction is critical. These can be achieved by the synthesis of high-efficient photocatalysts with significant surface areas [21, 78, 79]. Figure 5 demonstrates all of the critical factors which can influence the overall PCR efficiency.

3 Experimental Parameters Consideration

Photocatalytic conversion of CO_2 into value-added hydrocarbon fuels or useful chemical products has been the focus of active research. The photocatalytic conversion rate of CO_2 to methanol depends on the photocatalyst used, the photoreactor design, and

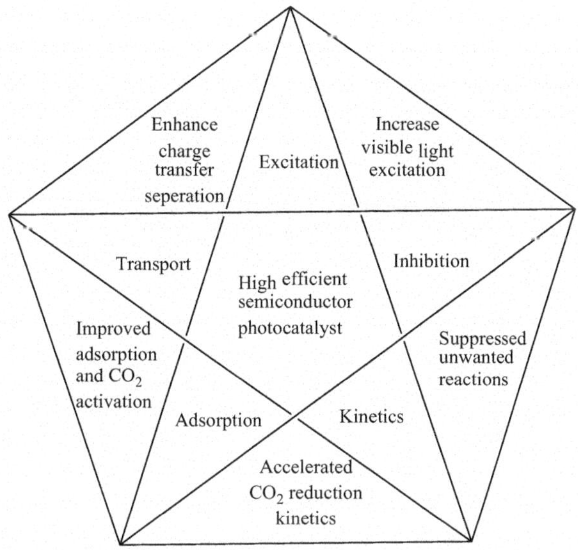

Fig. 5 Keys influencing photocatalytic yield and corresponding design strategies

experimental parameters. Here, we introduce several considerations for photocatalytic CO_2 reduction experiments. Preparation of catalysts, light source, and type of photoreactors as the PCR experiments preparation. Experimental parameters include reaction temperature, CO_2 pressure, and presence of impurities adsorbed on the photocatalyst.

3.1 Preparation of Catalysts

There are numerous articles have been reported, including pure, doped, metal–organic framework based and composite photocatalysts, synthesized by using various methods, and used for the photocatalytic conversion of CO_2 into fuels [80–87]. They are discussed excessively in section four.

3.2 Source of Light

Among the important criteria for photocatalytic CO_2 reduction are light intensity and irradiation nature [88]. The number of photons falling on a unit area in a unit of time determines the intensity of an illumination source [89]. According to the literature, most photocatalysts work better in the UV region, hence they can only capture a small portion of the solar spectrum irradiation [90]. To get over this limitation, semiconductors are being modified to harvest a wider range of sunlight spectrum.

Aside from that, light can be focused and diverted to increase irradiation photon flow [91]. By focusing light over TiO_2 and Pt/TiO_2, Han et al. were able to reduce CO_2. The authors tested CO_2 photoreduction with various concentrating ratios (CRs). CRs are defined as the ratio of concentrated light flux (amount of energy per unit time per unit area) on the photocatalyst surface to ambient flux (under non-concentrated conditions).

By varying the distance between the Fresnel lens (placed between the light source and the photocatalyst) and the photocatalyst surface, the light irradiation is concentrated and modulated, resulting in variable light intensities with different light concentrated focal regions. The optimum concentration ratio (CR) increased dramatically, according to their findings [91]. Based on these findings, better light interaction with the photocatalyst under ideally concentrated light will result in a significant improvement in the yield [92]. Employing a greater intensity radiation source, the geometrical design of the photoreactor, the lamp to photoreactor distance, and the use of fiber optics are all appropriate techniques for maximizing the radiation intensity on the reaction medium [89].

3.3 Geometry and Design of Photoreactors

Photo-reactor geometry and design also encourage maximum photon flux distribution, allowing for a large active surface area with a high mass transfer rate and minimal light diffusion effects. However, in most reactor geometries when light impinges on the photocatalyst's surface from the center or side, a shadow is created on the other side, preventing a significant amount of the photocatalyst from being activated. Fabricating suitable reactor designs to achieve a consistent distribution of light and greater photocatalyst dispersion could improve light-photocatalyst contact [93]. A variety of ways have been documented in the literature to achieve this, including the use of various reactor geometries and catalyst supports [93–95].

Monoliths have been used as a photocatalyst support material in innovative photoreactors, attempting to overcome mass transfer restrictions and limited light distribution efficiency seen in immobilized photocatalysts. Monoliths have consistent structures, supported with parallel channels that come in a variety of shapes and sizes depending on how they are extruded. These materials have a large surface area per unit volume and other appealing characteristics such as minimal pressure drop, high mechanical strength, and thermal stability, making them superior to traditional catalyst arrangements (powders and pellets). By comparing the performance of TiO_2 coated micro channel monolith and cell type support, Tahir et al. investigated the effect of photocatalyst dispersion (dispersed as a single layer over stainless steel cell). The TiO_2-coated monolith showed a considerable increase in CO output in their research. This increase was mostly due to a larger exposed photocatalyst surface that was available for photocatalytic activity. This improvement was attributed to the monolith's increased illuminated surface allowing for more efficient photon usage [97].

Fiber optic reactors have an advantage over packed bed reactors in terms of photocatalyst dispersion and light spreading across a vast surface area. Nguyen et al. investigated the yield of photocatalytic CO_2 reduction using photocatalyst coated on optical fiber against photocatalyst coated on a glass plate. For the same amount of photocatalysts, their research showed 15.2 times CH_4 yield and increased C_2H_4 yield 11.6 times more. The synergistic effects of catalyst dispersion and effective light utilization could explain this [97].

Wang et al. used a fiber optic reactor to perform CO_2 photoreduction and ascribed the increased output to the progressive and uniform distribution of light during irradiation. Optical fibers have the advantages of catalyst support and effective light distribution, but they also have the disadvantages of limited reactor capacity utilization and shorter light transit distance from the point of incidence. They occupy 20–30% of the reactor capacity, but the effective use of incident light is limited because of the limited catalyst-coated area [99]. Ola et al. fabricated an internally illuminated monolith reactor by combining the mutual effects of greater monolith surface area and effective light distribution of fiber optics and compared the CO_2 reduction performance of this system to that of a slurry reactor. Due to the higher surface area of the monolith and the equal dispersion of light by optical fibers, it was discovered that internal illumination of the monolith reactor by optical fibers increased quantum efficiency by 23 times [97]. Optical fibers constructed of carved polymethylmethacrylate were placed into a $NiO/InTaO_4$ coated monolith (honeycomb structure) by Liou et al. When used for photocatalytic CO_2 reduction, this reactor increased product yield (methanol and acetaldehyde). Increased surface area, better photocatalyst loading, and effective light use are all factors that contribute to a higher yield [99].

3.4 Effect of Temperature Variation

Due to long-wavelength irradiation, concentrating solar light raises the temperature depending on the CR [12, 28, 29]. Photocatalytic CO_2 reduction at high temperatures is promising because it bypasses the thermal barriers that cause slow reaction rates and low yields [96]. The effectiveness of temperature rise in photoreaction can be demonstrated by increased effective collisions between photogenerated charges and reactants, which are directly proportional to the reaction rate [97]. Furthermore, increasing temperature increases the desorption of products, allowing CO_2 to adsorb on unoccupied sites, resulting in a faster reaction rate [98].

The reaction temperature influences photocatalytic CO_2 to methanol conversion. Thus, determining the ideal temperature is difficult since temperature affects the methanol production rate in four different ways, as shown in Table 4. First, the amounts of CO_2 that can dissolve in water is affected by temperature, therefore decreasing the temperature increases CO_2 solubility in water. The increasing samount of dissolved CO_2 in water can speed up the synthesis of methanol. When water is

Table 4 Effect of temperature on various factors indirectly affecting the photocatalytic CO_2 reduction rate [89]

Temperature	CO_2 solubility	Reactant adsorption	Catalyst availability	Reaction rate constant
Increase	Decrease	Decrease	Increase	Increase
Decrease	Increase	Increase	Decrease	Decrease

chilled from 25 to 0 °C, the solubility of CO_2 in water increases by around 2.5 times, resulting in a 2.5-fold increase in dissolved CO_2 [89, 97, 98]. Second, the temperature affects the ease with which reactants adhere to the surface of the catalyst.

Because there is less thermal agitation at low temperatures, reactants soak more quickly onto the catalytic surface. The photocatalytic reaction rate naturally increases with increasing the amounts of reactants adsorbed on the catalyst surface. Third, temperature influences the rate of product desorption, affecting catalyst poisoning and, as a result, catalyst availability. Catalyst poisoning occurs when reactive intermediates and products are more likely to remain adsorbed on the catalyst surface at lower temperatures, preventing further catalytic activity. As a result of the scarcity of vacant catalytic adsorption sites on the surface, fresh reactants are unable to adsorb on the surface, slowing the photocatalytic CO_2 reduction [89, 97, 98]. Fourth, the decreased temperature has an undesirable influence on the photocatalytic CO_2 reduction process because it reduces the diffusion rates and collision frequencies of the reactants, resulting in a lower reaction rate constant for methanol generation. These four temperature impacts of the photocatalytic reaction show that the negative effects of high and low temperatures can be avoided without sacrificing the favorable effects [89]. When the temperature was raised from 25 to 75 °C, Wang et al. discovered that the production rate nearly doubled [99]. However, the reaction temperature should not rise too high, as this may cause the CO_2 to desorb, slowing down the photoreduction process.

3.5 Flowing CO_2 Gas Pressure

The CO_2 gas pressure in the reaction chamber, like reaction temperature, plays an essential role in controlling the rate of CO_2 reduction in product production. Increased CO_2 solubility in water with pressure causes an increase in reaction rate and, as a result, an increase in reaction product generation rate. Because cooling down the solvent can lower the product desorption rate, clog the catalyst surface, and slow down the reduction process, boosting CO_2 solubility by raising pressure is preferred over reducing the temperature for the goal of enhancing photocatalytic CO_2 reduction. Aside from enhanced reaction products, increasing CO_2 concentration caused by increased CO_2 pressure in aqueous media has also been shown to improve product selectivity. The increase in methanol formation rate, on the other hand, does not increase endlessly with pressure. It increases to a maximum value at optimum pressure and then begins to drop as pressure is increased further. Mizuno et al. investigated the effect of CO_2 gas pressure on the rate of reaction product formation and found that methanol was produced at ambient pressure [100]. However, as CO_2 pressure increased, the rate of methanol formation increased sharply, peaking at 1.0 MPa, and then declining significantly with further CO_2 pressure increases. When choosing CO_2 reduction reactor designs [101], the expense of producing advanced high-pressure systems must be taken into consideration.

3.6 *Effect of Contaminant*

Organic contaminants in the catalyst or on the catalyst surface might cause a falsely positive result since they are often reduced more efficiently than CO_2 to create diverse reduction products. This is especially true when the concentration of products obtained from photocatalytic CO_2 reduction is low, as it has been in almost all photocatalysis experiments.

Small particle size and wide surface area of the TiO_2 photocatalyst (and this also applies to other photocatalysts) are crucial elements in adsorbing airborne organic pollutants, which are more reactive than even CO_2, according to Neatu et al. [102]. As a result, they are more easily reduced than CO_2 and become the source of photoproducts, which are sometimes mistaken for CO_2 reduction reaction products [103]. When the photocatalyst amount is large but the produced yield is low this can be attributed to the adsorbed contaminants. This condition can be avoided by calcining the photocatalyst before the photocatalytic experiment to remove any carbon-containing impurities that may be present on its surface. Because CO_2 is more stable than other carbon-containing organic contaminants, organic impurities linked to photocatalyst surfaces degrade more quickly, resulting in an overestimation of photocatalytic activity and yield. In summary, photocatalyst contamination can lead to false results, hence rigorous surface cleaning is required to obtain accurate and precise results.

4 Photocatalyst Types, Selection, and Preparation

4.1 *Metal System*

CO_2 photocatalysts are divided into five types: metal, mixed oxides, metal sulfides, polymeric materials, metal–organic framework, and others. Metal and mixed oxides, metal sulfides, and the metal–organic framework are included in the metallic system, which reduces CO_2 into several forms, such as carbon monoxide, methane, methanol, and formic acid for fuel. Some examples included in the metal and mixed oxides are titanium oxide, iron oxide, tantalum oxide, copper oxide, niobium oxide, and strontium oxide. Furthermore, some examples of metal sulfides included are cadmium sulfide, zinc sulfide, MoS_2, SnS_2, Bi_2S_3, In_2S_3, and $Znln_2S_4$. Meanwhile, MIL 101, PMOF-AI, Co-ZIF9, and MOF 525-Co are examples of metals included in the metal–organic framework (MOF) group [104]. Activity and selectivity are essential properties in the selection of photocatalysts. To obtain a photocatalyst with high activity and selectivity, several things must be considered, including band structure, surface state, and photoreaction conditions. Increasing the activity and selectivity of photocatalysts can be done by optimizing carbon dioxide absorption, optimizing light harvesting, charge separation effectiveness, and synergistic effects [104].

4.1.1 Metal and Mixed Oxides

As previously described, TiO_2 is the most widely used CO_2 photocatalyst of metal oxides and mixed oxides. This is due to its non-toxicity, stability, slightly corrosive nature, and low cost. This metal is widely used to convert CO_2 to methane and carbon monoxide. However, TiO_2 is known to have a large bandgap (3.2 eV); due to this large bandgap, TiO_2 only shows photocatalytic activity under UV radiation. In contrast, sunlight as a source of radiation is still not qualified. Therefore, the development of this photocatalyst to have a narrower bandgap is still being carried out [104]. Several strategies to increase the ability of TiO_2 are doping using Ag and Cu or using co-doping [104, 105]. This doping can reduce bandgap energy, improve interfacial charge transfer, trap electrons, and allow the use of visible light [104, 105]. Another strategy is to use heterostructured crystal growth. The next modification is to insert the defect chemistry into the forbidden gap on the catalyst surface by thermal treatment [104, 105]. Modification through the formation of nanomaterials on TiO_2 also shows advantages in the catalytic behavior. Modifying the nanostructure can increase the diffusion rate and surface area to increase the catalytic activity [105]. Furthermore, to increase the efficiency of the excitation process, modifications using a dye sensitizer can be made by increasing the wavelength so that the catalytic activity increases [104, 105]. There are several methods for TiO_2 synthetizing including sol–gel, hydrothermal, impregnation, one-pot, and co-precipitation. Sol–gel process, hydrothermal synthesis, and one-pot synthesis were implemented using TiO_2 and multi-walled carbon nanotube composites to improve the photocatalytic performance [104, 105]. A co-precipitation method can narrow the bandgap to absorb large amounts of energy from visible light. Synthesis using the impregnation method was performed by increasing irradiation and doping with metal oxides such as CuO, CoO, and Fe_2O_3 [105] (Fig. 6).

Silver, rhodium, gold, palladium, and platinum are metals that are often used as co-catalysts with TiO_2 to increase TiO_2 occupation. The rate of methane formation is reported to be increased due to the use of noble metal co-catalysts. One of the noble metals, Palladium, acts as a co-catalyst on the TiO_2 surface and provides active sites for CO_2 adsorption and activity [106]. In addition, the size of metal nanoparticles is a critical factor to determine the activity and the rate of methane formation [106].

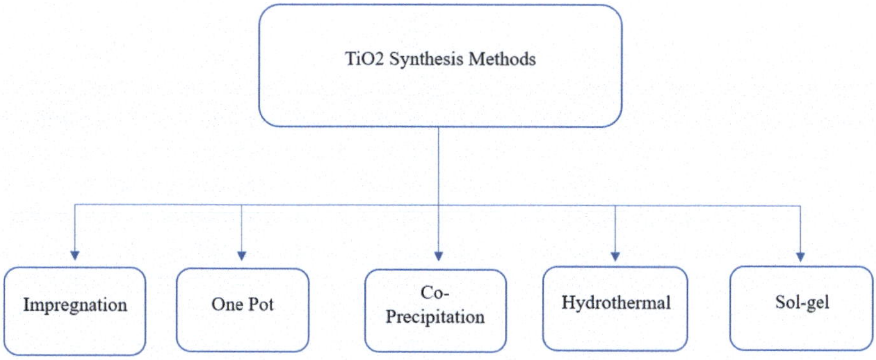

Fig. 6 TiO_2 synthesis methods [105]

Copper is a transition metal that is often used as a TiO_2 co-catalyst. This is because they are abundant, non-toxic, low in cost, and potential alternatives to the use of noble metals. The use of this co-catalyst showed an increase in the formation of methane. In addition, the use of Cu doping on TiO_2 under UV-rich illumination can convert CO_2 into formic acid [106]. The use of Cu-TiO_2 was also reported to increase the efficiency of light harvesting under UV conditions to produce carbon monoxide and methane from CO_2. Copper has also recently been reported to depress hydrogen development in CO_2 photoreduction, thereby selectively producing CO [106].

Subsequently, the use of binary co-catalysts (Cu_2O-Pt/TiO_2 and MgO-Pt/ TiO_2 systems) was investigated. Platinum in TiO_2 is intended to promote electron capture and inhibit charge pair recombination. However, the use of Platinum increased H_2, so it takes Cu_2O or MgO to suppress the formation of H_2. As a result, CH_4 will be obtained with high selectivity. Recently, multi-heterojunctions were fabricated on TiO_2-MnOx-Pt films. This multi-heterojunction can efficiently separate charged pairs to produce three times higher CH_4 and methanol than pure TiO_2 [106].

Au shows the effect of surface plasmon resonance on TiO_2. Synergistic mixing of the plasmonic effects of Au and Pt nanoparticles as electron absorbers with TiO_2 nanofibers was reported to increase visible light harvesting and inhibit the recombination of photoexcited TiO_2. The reduction of CO_2 under UV light and visible light for the Au-Cu alloy on TiO_2 as a photocatalyst showed excellent performance. The electron selectivity for CH_4 evolution can reach 97% under visible light irradiation. This high light harvesting ability comes from the plasmonic effect of Au [106].

Cu is one of the most commonly used metals for CO_2 reduction photocatalyst. This is because Cu has low bandgap energy and a high conductive band. CuO and Cu_2O are reported to have small band gaps of 1.7 and 2.2 eV. CuO nanomaterials are known to absorb visible light effectively and produce photogenerated electrons and holes. Cu-based photocatalysts have three pathways for CO_2 reduction. There is the formaldehyde, carbinol, and glyoxal pathway. The Formaldehyde pathway can produce formic acid, methane, and methanol from Cu-based photocatalysts for CO_2 reduction. The carbene pathway is usually used to produce methane and methanol. While the glyoxal pathway is usually used to produce formic acid [107].

The reduction activity and selectivity of CO_2 can be affected by the morphology, particle size, and dispersibility of Cu when different methods are used. CuO which has a high density with poor dispersion will affect light absorption due to masking between particles. To address this problem, it is possible to utilize CuO thin film to enhance the catalytic effect. In addition, the use of glass fiber mesh coating CuO can also be done to increase the production of CH_4. The photocatalytic activity and selectivity of CuO products can also be increased through size modification or create quantum dots of CuO. Modification of the size of CuO to be smaller will increase light harvesting and charge transfer separation due to the risen uniform distribution [106, 107]. The modification with CuO quantum dots plays a significant role in the CO_2 adsorption and activation [107]. There are several methods to synthesize CuO. Among them, the solvothermal method and impregnation are quite famous [107].

The use of pure Cu is also carried out to minimize semiconductor energy bandgap. This reduced band gap energy can make full use of visible light to improve light utilization and encourage the practical application of photocatalysis. Cu can be prepared by microwave hydrothermal method and secondary calcination. Cu_2O was studied as a photocatalyst to reduce CO_2. This metal oxide is known to have high photocatalytic activity. Doping using Cu_2O is known to improve its energy band structure and increase photocatalytic activity. Cu_2O can be prepared by hydrothermal deposition-reduction method and microwave-assisted in situ reduction chemistry [107] (Fig. 7).

Perovskite is a metal oxide that is also used as a photocatalyst for CO_2 reduction [108]. Perovskite is used as a photocatalyst because it provides a broad spectrum for CO_2 conversion. In addition, perovskite also has good stability, flexible composition, efficient catalytic activity, long charge diffusion, low cost, and easy preparation [108]. Perovskite converts CO_2 through light harvesting, which then separates electrons and transfers them from VB to CB. Then charge photogeneration is carried out before the redox reaction occurs on the catalyst surface until finally the product is formed

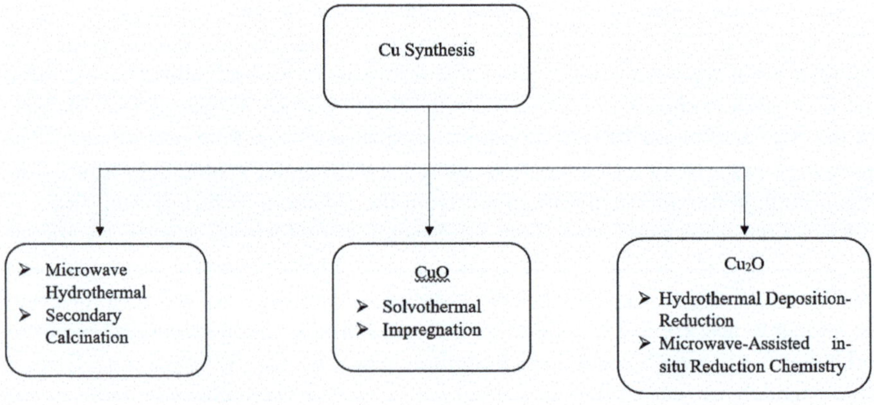

Fig. 7 Cu synthesis methods

[108]. There are several classifications of perovskite oxides, including ABO_3, halide, layered, ruddlesden-popper phase, aurivillius phase, (110) layered, (111) layered, Dion-Jacobson phase, and oxynitrides [108].

ABO_3 perovskite consists of alkali metal cations at site A and transition metal cations at site B. These transition metal cations regulate the perovskite catalytic activity and electron mobility. One way to narrow the perovskite band gap can be done by forming a solid solution. This is because the solid solution has a lower CB with a higher VB. In addition, the formation of this solid solution can also reduce the CO_2 induced due to some perovskites having a band gap not suitable for CO_2 reduction under visible light. This solid solution formation also offers advantages such as band gap control, charge transfer, and chemical stability [108].

Perovskite halides (ABX_3) are another type of perovskite group where the cation at site A is usually Cs^+ or Rb^+, while the cation at site B is Pb, Ge, or Sn and the halide is located at site X. Perovskite halides are usually used to produce CH_4 and CO at the surface of the catalyst. There is also layered perovskite which has flexibility in its structure, effectiveness in charge transport, and attractive optoelectronic characteristics [108].

The Aurivillius phase is a perovskite with cations such as Na, K, Ca, Sr, Ba, and Bi located at site A while, cations such as Fe, Cr, Ti, Ga, Nb, V, Mo, and W are located at site B. On the other hand, the Dion-Jacobson phase ($A_{n1}B_nO_{3n+1}$) is a perovskite with cations such as Rb, K, Ag, and Cs at site A while Pr, Sm, Nd, and La at site B. This perovskite has a band gap of about 3.8–4.3 eV which allows it to work under UV light. Ruddlesden-Popper phase ($A_2A_{n-1}B_nO_{3n+1}$) is a perovskite consisting of alkali metals at site A and transition metals at site B. Finally, perovskite oxynitrides ($ABO_{2-x}N_{1+x}$) is a perovskite consisting of alkali metals at site A with transition metal at site B [108] (Fig. 8).

Besides being used as the main catalyst, several metal oxides can also be used as co-catalysts including Cu_2O, CuO, NiO, MgO, CO_3O_4, and Fe_2O_3. The use of copper oxide as a co-catalyst can intensify CO_2 adsorption on the surface and suppress charge recombination. Cu is also able to increase the electron density at the active site when it is excited and facilitates CO_2RR which requires multi-electron transfer. Photocatalyst activity can be increased through the formation of oxygen vacancies at the active sites of Cu_2O. Not only that, but this formation is also able to reduce the rate of charge recombination. CuO is usually used as a co-catalyst together with SiC, TiO_2, $K_2Ti_6O_{13}$, and $NaTaO_3$ as the main catalyst. Magnesium oxide itself is usually used as a co-catalyst to increase the activation and adsorption of CO_2 because of its ability to interact with CO_2 strongly. This metal oxide is usually used with TiO_2 to catalyze the reduction of CO_2. Meanwhile, nickel oxide is usually used as a co-catalyst to inhibit the reverse reaction and increase charge carrier separation. This co-catalyst can increase the production of methanol, methane, and carbon monoxide. Nickel oxide is often used as a co-catalyst along with $InTaO_4$, $InNbO_4$, $KTaO_3$, $K_2Ti_6O_{13}$, and $InVO_4$ as the main catalyst. Last but not least, is the use of cobalt oxide as a co-catalyst for the promotion of the oxygen evolution reaction and suppressing corrosiveness due to the buildup of photogenerated holes [109].

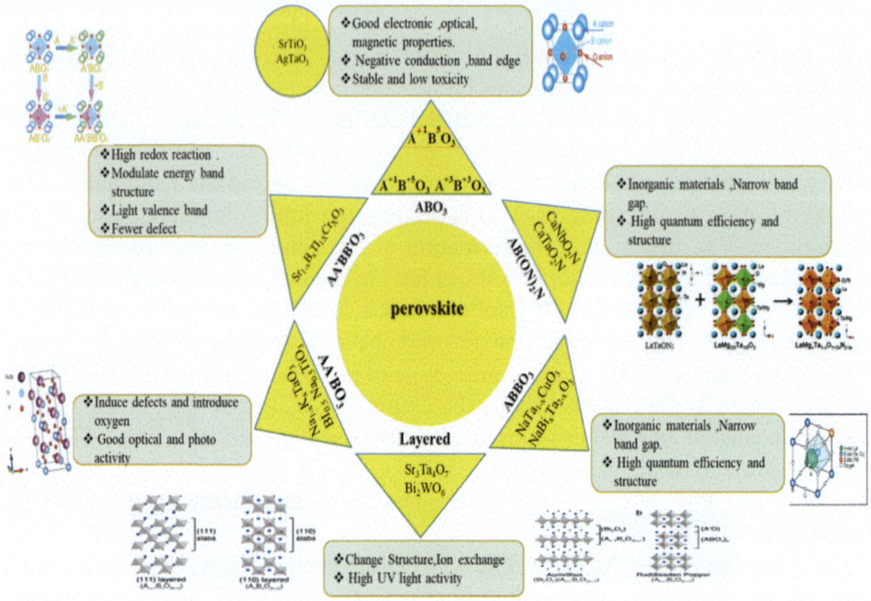

Fig. 8 Perovskite classification [108]. (Copyright Elsevier License Number 5326340974814)

4.1.2 Metal Sulfides

On the other hand, metal sulfides such as ZnS and CdS have also been reported. Similar to TiO_2, ZnS only absorbs visible light due to its large bandgap (3.66 eV). In contrast to CdS which has a smaller bandgap (2.4 eV), this metal is preferred for CO_2 reduction. The catalytic performance of metal sulfide can be improved by combining metal sulfide photocatalysts with other photocatalysts. Incorporating metal sulfide photocatalysts with different photocatalysts will provide more advantages in bandgap regulation. The use of metal sulfides as photocatalysts is known to have great research potential. Unfortunately, metal sulfides have less stability during photocatalytic processes leading to structural damage [104].

Molybdenum disulfide (MoS_2) is a layered binary sulfide widely applied in photocatalytic due to its excellent optical/electrical properties and flexible electronic band structure. The layer on MoS_2 has a high d electron density, so it has potential in gas–solid CO_2 photoreduction systems for methanol production. Bismuth sulfide (Bi_2S_3) is known to have a narrow bandgap (≈1.3 eV), so it is also getting attention in its role as a photocatalyst. In addition, this photocatalyst is also known to have low toxicity, high biocompatibility, good performance in reducing CO_2, and an absorption coefficient in the visible light region (>105 cm^{-1}). Like Bi_2S_3, In_2S_3 is also widely used for CO_2 reduction. This metal has an expansive light response range due to its narrow bandgap (2.0–2.3 eV), and low toxicity [72] (Fig. 9).

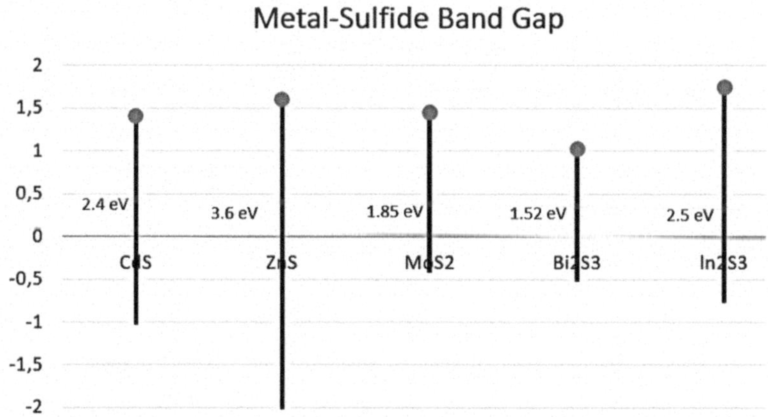

Fig. 9 Metal-sulfide band gap

One method that can be used to synthesize ZnS is heat-free synthesis with Cu+ and Cd^{2+} co-catalysts doping to maximize light utilization [72]. There is also the synthesis of ZnS with zinc hydroxide and thiourea using the surface of Ru nanoparticles to form formic acid, which is known to provide high product selectivity [72]. Metal sulfide type MoS_2 can be synthesized by hydrothermal method and Chemical vapor deposition (CVD) synthesis. Likewise, with the synthesis of In_2S_3 through the hydrothermal cationic exchange method [72]. SnS_2/SnO_2 type metal sulfide photocatalyst can be synthesized by hydrothermal method and sequential template to maximize CO formation [72]. The ion-exchange method between Bi and Cd_2S in ethylene glycol solution can be carried out to synthesize Bi_2S_3/CdS/FeTCPP photocatalysts [72] (Fig. 10).

Besides being used as the main catalyst, metal sulfide can also be used as a co-catalyst because it can increase charge separation and add more active sites. The co-catalysts commonly used in this group are MoS_2 and NiS_2 [9, 109]. MoS_2 is usually used together with Bi_2WO_6 under visible light to produce both ethanol and methanol. One method that can be used for the preparation of this co-catalyst is impregnation-calcination method. In addition, MoS_2 is also used with TiO_2 to produce methanol using the in situ grown method [109]. On the other hand, NiS_2 is used as a co-catalyst with graphite carbon nitride (g-C_3N_4) to produce CO due to its ability to accelerate photogenerated electron–hole pair separation. In addition, NiS_2 can also be used with ZnO to produce CO and CH_4. However, the use of metal sulfide as a co-catalyst is also reported to have drawbacks due to its poor stability [109].

4.1.3 Metal–Organic Framework

Another metal system that is widely used is MOF. MOF can be used as co-catalysts as well as stand-alone photocatalysts. The MOF structure is potentially a photocatalyst

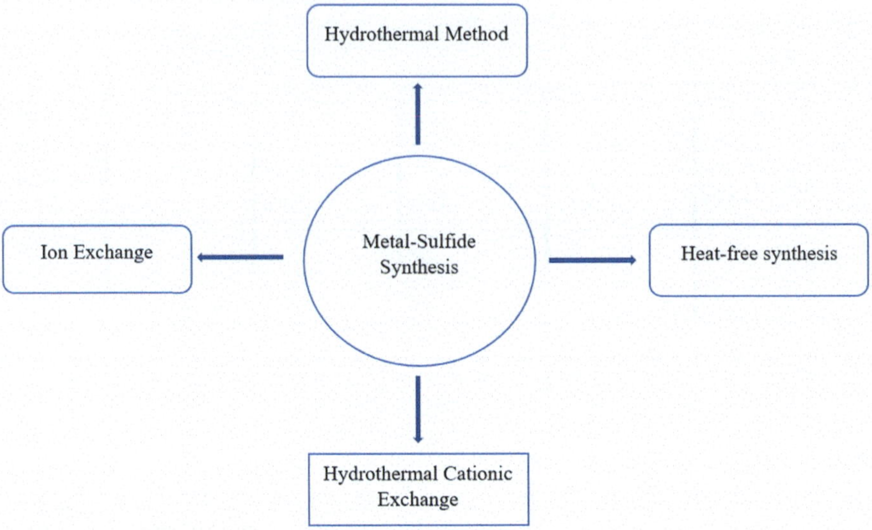

Fig. 10 Metal-sulfide synthesis

because the network is porous, regular, and heterogeneous so that ions and CO_2 can move freely into the matrix. MOF is also known to increase the reactivity of other catalysts. However, as with metal sulfides, MOFs have poor stability and problems in economic viability on an industrial scale. The use of MOFs as photocatalysts can still be improved by modifying the structure and adjusting the reactive functional groups. The metal-added MOF photocatalyst has a bandgap from 1.52–2.4 eV. Meanwhile, if MOF is used as a composite, the bandgap variation is 1.6–3.1 eV [104].

MOF can be synthesized through solvothermal, slow evaporation, microwave assistance, electrochemistry, sonochemistry, and mechanochemistry. These five methods generally combine three main components: metal salts, ligands, and solvents. The solvothermal method is the most frequently used method for MOF synthesis by involving high boiling point heating between organic linking ligands and metal salts in a solvent. Slow evaporation synthesis is preferred because it does not require external energy and uses only room temperature, even though it takes a long time. Microwave-assisted synthesis involves heating a solution with microwaves to produce nano-sized crystals. The electrochemical method is carried out by adjusting the pH of the solvent at room temperature without metal salts and consisting only of a mixture of organic linkages and electrolytes to provide metal ions. Synthesis using the sonochemical method is based on molecular changes due to ultrasonic wave radiation to produce fine crystalline materials. Finally, the mechanochemical synthesis method is based on applying a mechanical force without a solvent to form a porous MOF [110].

Metal systems have several drawbacks, including some metals toxicity, being not environmentally friendly, and have low selectivity. However, metal can still be used, such as doping carbon material as a photocatalyst. In addition, the use of bimetallic MOF is also known to be more efficient than the monometals (Fig. 11).

Besides being used as the main catalyst, MOF can also be used as a co-catalyst for CO_2 reduction in nanocomposites. MOF acts as a kinetic process driver in the catalytic reaction and increases CO_2 adsorption while other components act as light harvesters. For example the use of Co-Zif-9 as a co-catalyst together with $[Ru(bpy)_3]Cl_2.6H_2O$ as a photosensitizer, and TEOA as an electron donor under 380–700 nm irradiation. Although the use of this co-catalyst is known to have high catalytic efficiency, the selectivity of each product is quite low. Not only that, but photo-bleaching also causes a decrease in catalytic activity. So, it is necessary for coupling with semiconductors such as $g\text{-}C_3N_4$ and CdS as light harvesters. In addition, MOF is also used as a co-catalyst with TiO_2. This system shows that the photocatalytic activity for methane production increases up to five times [111].

Homogeneous metal complexes have been widely used for CO_2 reduction photocatalysts. However, this metal complex does not have long-term stability, is difficult to separate from the reaction mixture, the product will be contaminated, and it is difficult to recycle. Therefore, the use of MOF as a host to support homogeneous metal complexes is considered the right choice. This is based on the nature of the MOF which has a high surface area and uniform pores, so it can be adapted for the diffusion of reactants. MOFs also have well-defined and isolated sites for anchoring catalytic species, so they can be used to construct single-site catalysts [111].

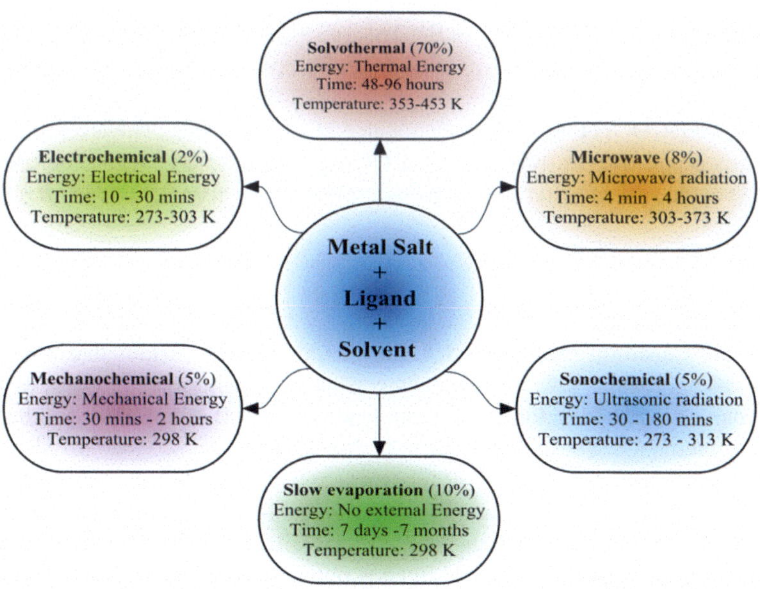

Fig. 11 MOF synthesis [110]. (Copyright Elsevier License Number 5326341140976)

4.1.4 Metal Complex

Transition metal complex ions have been widely used as photocatalysts. Some examples of metal complexes used as photocatalysts include complexes of ruthenium, osmium, rhenium, cyclometallation iridium, metalloporphyrins, and organic dyes. This photocatalyst has several properties, including selective light absorption, relatively inefficient reverse electron transfer rate, fast decomposition, sufficient reducing power for electron transfer, stability in ground and excited states, and reacts selectively to CO_2. The metal complex-based homogeneous photocatalytic system consists of a photosensitizer and a CO_2 reduction catalyst. The Ru(II)-Re(I) photocatalyst is reported to reduce CO_2 efficiently when installed on solid materials. Photocatalysts made with photosensitizers and metal complex catalysts showed efficient, selective, and long-lasting results. Building a photocatalyst with a hybrid system can be done by connecting a photosensitizer and a catalyst to a solid material so that electron transfer will take place faster [111].

The Re dynamic complex has CO_2 reduction activity and high product selectivity. The Re(bpy)$(CO)_3$Cl complex has a stable and adjustable structure so that it has the potential as a CO_2RR photocatalyst. When photocatalytic CO_2RR is carried out in a mixed solution of DMF/TEOA, the addition of TEOA can capture CO_2 even at the atmospheric level. Re(bpy-)$(CO)_3$ has also been reported to react with CO_2 in the dark. In addition, the use of a photosensitizer compatible with the Re complex can also increase the efficiency and durability of the catalyst [112]. Unlike the Re complex, which only reduces CO_2 to CO, the Ru photocatalyst can catalyze the conversion of CO_2 to CO and HCOOH. This reaction occurs through the capture of one CO_2 molecule by $Ru(bpy^-)_2(CO)$ to form $Ru(bpy^-)_2(CO)(CO_2)$ so that one H^+ can be added to create the complex $[Ru(bpy^-)_2(CO)(COOH)]^+$. This complex will then produce HCOOH through the acceptance of two electrons. However, it doesn't stop at this point, the $[Ru(bpy^-)_2(CO)(COOH)]^+$ complex will then be protonated in the CO cycle so that water molecules are released and form CO molecules [112].

4.1.5 MXenes

MXenes are transition metal carbides, nitrides, or carbon nitrides that have great potential as photocatalysts. These metals consist of transition metals such as Scandium, Titanium, and Vanadium, elements from groups III A and IV A, and carbon or nitride elements. The general formula for MXenes is $M_{n+1}X_n$ or $M_{n+1}X_nT_x$ where T is a surface functional group such as oxygen, fluorine, and hydroxyl. Despite their potential as photocatalysts, MXenes cannot be used directly due to their non-semiconducting nature. Therefore, most of the MXenes are used as co-catalysts. MXene is commonly used in CO_2 reduction to improve photogeneration of charge carriers, photogenerated species separation, photo-corrosion inhibition, enhance CO_2 adsorption and activation, enhance light absorption, and photothermal effects. MXene can be applied in reactions of gas and liquid phase systems [114].

The photocatalytic activity and photostability of the semiconductor can be enhanced through the combination of the metallic conductivity of MXene with a suitable band structure. This combination allows for efficient electron migration from the semiconductor to the MXene. In addition, the regulation of the surface functional group of MXene can maximize CO_2 adsorption and activation. This is because the surface functional groups act as active sites for the photocatalytic reactions. MXenes can also improve light and photothermal harvesting for metal nanoparticles. MXenes as co-catalysts have a significant role in regulating product selectivity so diesel fuel production can be enhanced [114].

MXene as a co-catalyst is usually used with nitrides, metal oxides, metal salts, perovskite, and MOF as the main catalyst. One example of the use of an MXene/Nitride catalyst is the use of g-C_3N_4/Ti_3C_2. This heterojunction is reported to improve CO_2 adsorption, photogenerated charge carrier separation, and hybrid stability. There is also the use of MXene with metal oxides such as TiO_2 which is known to accelerate the efficiency of electron–hole separation. This combination will enhance the photocatalytic reduction reaction with 3.7 times higher methane production than commercial TiO_2. Other combinations such as perovskite $CsPbBr_3$ with MXene Ti_3C_2 were reported to produce higher CO and CH_4 than $CsPbBr_3$ NCs. This is due to MXene's ability to increase the photocatalytic activity [115] (Fig. 12).

Generally, the basic synthesis of MXenes goes through the etching process. There are three etching approaches, namely fluorination etching, fluorine-free etching, and electrochemical etching. In the fluorination etching process, materials derived from hydrofluoric acid (HF) or salt solutions containing fluorine are used. Meanwhile, in the fluorine-free etchant method, Ti_3C_2Tx is fluorine-free in NaOH or KOH solutions. This method is considered to have a higher capacity for electrochemical properties. The electrochemical etching method is carried out through a redox reaction between the anode and cathode in an electrolytic cell through the application of voltage. Wet

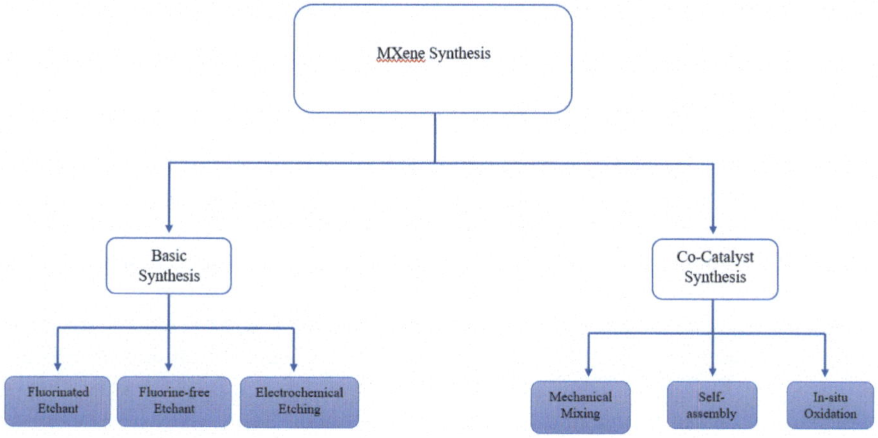

Fig. 12 MXene synthesis

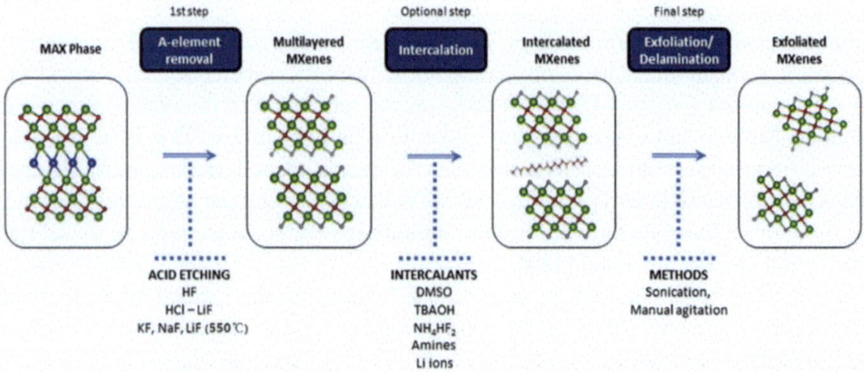

Fig. 13 MXenes synthesis schematic [116]. (Copyright Elsevier License Number 5326340653842)

etching using HF is a method that is often used to synthesize MXenes. This is because HF-based etching has high selectivity and can break M-A metal bonds [115].

The MXenes as a photocatalytic co-catalyst synthesis is divided into three methods, namely mechanical mixing, self-assembly, and in situ oxidation. Mechanical mixing is a simple method for catalyst synthesis. This method includes advantages such as saving energy and low cost. Mechanical mixing is carried out through the process of grinding solid powder and mixing the substances in solution to deposit MXene on the surface of the photocatalyst. The self-assembly method is considered to have advantages over the mechanical mixing method, due to its ability to have closer contact and more uniform dispersion. Lastly, there is the direct in situ oxidation method of MXene. This method produces MXene in the form of MO/MXene or MO/MXene/C, where C is an amorphous carbon that can be used as a co-catalyst. Unfortunately, this method can only be used if the MO is a semiconductor [116] (Fig. 13).

4.1.6 Metal Phosphide

Metal phosphides are widely used because of their abundance, low prices, and high electrical conductivity. Some examples of catalysts belonging to this group include Co_xP, Ni_2P, Fe_2P, Cu_3P, WP, and InP. The presence of this P group moderates the strength of the metal phosphide bond for the catalytic product conversion process, in which the active site facilitates product desorption. One example of a metal phosphide catalyst for CO_2 conversion is InP which has a band gap of 1.35 eV. Due to this large band gap, some modifications need to be made. One of them is to minimize the InP particle size to the quantum dot (QD) scale. The advantage of this QD modification is an increase in the redox potential so that the photocatalytic activity will also increase. This modification is also able to provide a more active site for the photocatalytic reaction [117].

In addition, metal phosphides are also commonly used as co-catalysts. This material can create sufficient and strong contact with the photocatalyst to effectively separate the electron–hole pairs. Metal phosphides are also considered to have an electronic structure like Pt so that they can be used as a substitute for precious metal co-catalysts. The bond between metal and phosphide can induce a small ligand effect so that the product will separate from the catalyst surface. Not only that, but this bond is also able to capture reaction intermediates so that the catalytic activity will increase [109].

Some examples of metal phosphide co-catalysts commonly used are FeP with polymeric carbon nitride as the main catalyst and WP with g-C_3N_4. The use of FeP co-catalyst with carbon nitride polymer was reported to be able to produce five times higher CO than the use of pure carbon nitride. The use of metal phosphide-based co-catalysts also has good stability. In addition, there was a narrowing in the band gap from 2.77 to 2.40 eV due to a change in the position of the VB resulting in increased light absorption under UV light. WP which was used as a co-catalyst with g-C_3N_4 also showed satisfactory results. CO production is reported to have increased considerably due to the use of this co-catalyst. Life cycle stability has also been reported to improve. This is due to the promotion of electron transfer due to the presence of P-N bonds at the WP and g-C_3N_4 interfaces [109].

4.1.7 Bismuth-Based p-Block Semiconductors

Most photocatalysts such as TiO_2 and ZnO use a d-block semiconductor. In this discussion, p-block elements will be used to construct a CO_2 reduction photocatalyst. This is because p-blocks have the potential to increase their photocatalytic activity in visible light. The p electrons in the hybridized state can decrease the CB state and increase the VB state so that the band gap becomes narrower. Not only that, another advantage that can also be obtained is the mobility of the photoexcited charge carriers and the charge separation will be increased [118].

Research conducted by Cui et al. [118] discusses the use of bismuth as a semiconductor by utilizing p electrons as a CO_2 reduction photocatalyst. According to them, there are three strategies to increase photoreduction activity, constituent adjustment, vacancy engineering, and heterostructure construction. The adjustment of the constituents is carried out through the adjustment of the halogen component and the ratio of Bi:O:X. Meanwhile, vacancy engineering is carried out through the engineering of oxygen, bismuth, and halogen vacancies. Finally, the heterostructure construction was carried out through the $Bi_xO_yX_z$ construction of nanomaterial heterostructures and 2D semiconductors. Through this strategy, it is hoped that the band structure can be controlled so that the catalytic reduction conditions are met along with the introduction of high-level active sites on the photocatalyst surface. Improvements in energy conversion efficiency are also expected through the promotion of charge separation and transportation [118].

4.2 Non-metallic Photocatalyst

Photocatalysts in application to reduce CO_2 by using non-metallic materials exhibit rapid growth for photoconversion of CO_2. In addition, this material also shows tremendous potential as an alternative to photocatalysts owing to its abundance on earth, cost-effectiveness, high electrical conductivity, as well as environmental friendliness. Moreover, they also exhibit outstanding catalytic activity, durability, and exceptional selectivity in transforming CO_2 into solar fuel. In this section, we will explain and elaborate the non-metallic photocatalysts for CO_2 reduction [119].

4.2.1 Graphene

Graphene has been studied in-depth and has shown considerable interest in its application in photocatalytic CO_2 reduction. In addition, many experiments and studies have tried to design, develop, and apply graphene as a photocatalyst to reduce CO_2 [119]. The structure of graphene-based catalyst is unique. Aside from its large specific surface area, graphene also has many other desirable properties, including excellent electron transfer, transparency, high resistance to temperature changes, flexibility, and good CO_2 adsorption, which have made it ideal to be used as CO_2 photocatalysts [120]. The separation of charge carriers in photocatalytic CO_2 reduction can also be facilitated by graphene. In addition, graphene has a large contact area and interacts strongly with other catalysts, so the photocatalytic reaction and performance for reducing CO_2 can be promoted and improved by graphene [121].

Some common graphene includes pure graphene, graphene oxide (GO), and reduced graphene oxide. The three materials possess different functional groups, which gives them different properties as well [122]. A modified Hummers' method is the common method to synthesize these materials using graphite as their raw material. The synthesis process with this technique begins with the chemical oxidation of graphite. The next step for preparing GO is exfoliation. GO typically has abundant oxygen-carrying functional groups, such as carboxyl, epoxy, hydroxyl, and carbonyl groups. The bandgap value of GO is about 2.2 eV while graphene pristine has a bandgap of 0 eV and does not have surface functional groups. The reduction of graphene oxide will produce rGO [123]. The reduction techniques are thermal treatment and chemical methods. The bandgap of rGO is variable, depending on the degree of reduction, from ~1.00 to 1.69 eV. The surface functional groups of rGO are the same as GO. The performance of graphene oxide can be improved by surface modification, doping, morphology control, and defect introduction. Tan et al. [124] investigated TiO_2 doped GO. The wet chemical impregnation technique is the technique used to combine GO with TiO_2 in this experiment. A modified Hummers' technique was used to prepare graphite oxide powder. Followed by the exfoliation process through ultrasonication for 1.5 h and separation of the graphite layer. Within 8 h of reaction time, GO-TiO_2 successfully yielded 3.45 mol g^{-1} h^{-1} of CH_4. This performance represents a 14-fold performance increase over commercially available TiO_2 P25 [124].

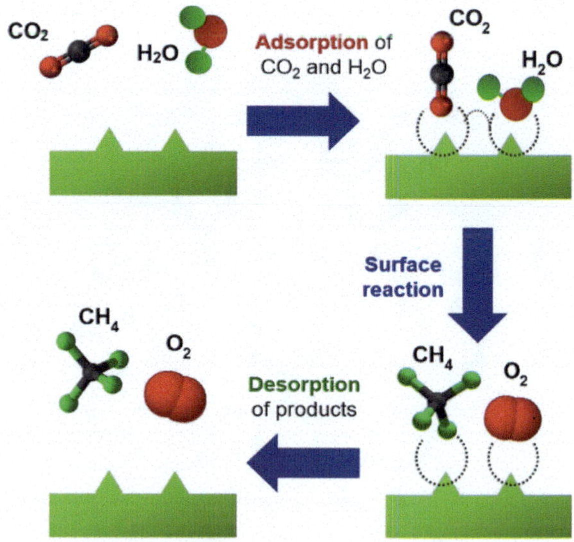

Fig. 14 The Langmuir–Hinshelwood mechanism [124]. (Copyright Elsevier License Number 5321211075597)

The reaction scheme of the CO_2 reduction through GO-TiO_2 is shown in Fig. 14. The photocatalytic CO_2 reduction mechanism is suggested followed Langmuir–Hinshelwood (L–H) model. First, reactant molecules, in the form of carbon dioxide and water molecules, adsorb onto the photocatalyst surface. The next process is the surface reaction. At this stage where $\bullet CO_2^-$ radicals are generated from photogenerated electrons. Then, these $\bullet CO_2^-$ radicals will be transferred to the adsorbed CO_2. Meanwhile, on the other side, the hole reacts with the adsorbed H_2O molecules. The product of this reaction is $\bullet OH$ radicals and H^+ ions. Then, it will form O_2. To produce intermediate radicals and hydrocarbon products, carbon radicals will react with H radicals formed by proton reduction on the photocatalyst surface [124].

4.2.2 g-C_3N_4

Graphite carbon nitride is a layered semiconductor material. g-C_3N_4 is a non-metallic material in the form of a polymer material and is composed of a polymer combination of tri-s-triazine/heptazine (C_6N_7) or triazine (C_3N_3) [120]. According to its wavelength of 460–430 nm, the g-C_3N_4 band gap value is narrow, ~2.7–2.9 eV. g-C_3N_4 also shows a CBat a negative position of about 1.1 V versus NHE. This property meets the thermodynamic requirements of photocatalytic CO_2 reduction reactions [125]. Therefore, under visible light and sunlight CO_2 will be activated and reduced successfully. The easy fabrication process, metal-free composition, durability, low cost, environmental friendliness, and good chemical and thermal stability are the superiority as a semiconductor material for the photocatalyst of g-C_3N_4. Furthermore, the conduction band of 1.14 eV and the valence band of 1.57 eV, make it

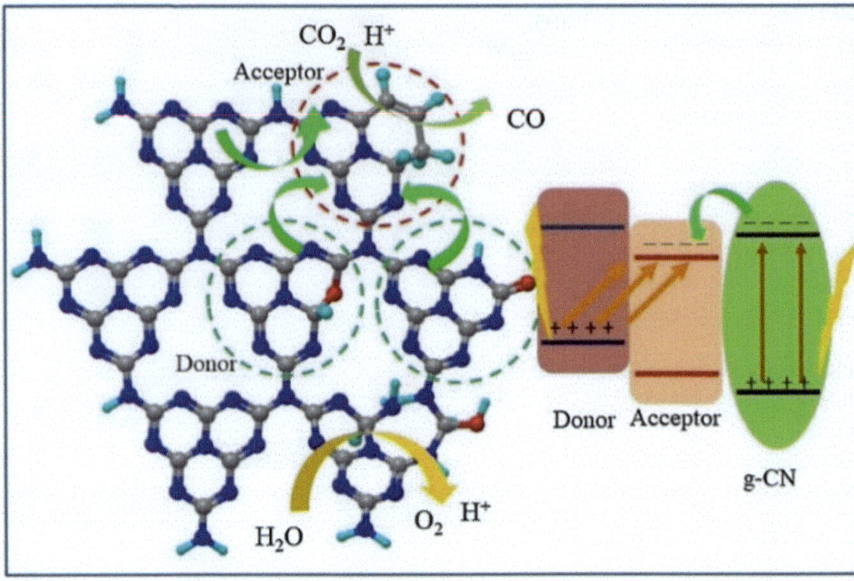

Fig. 15 Photocatalytic reduction of CO_2 mechanism through g-C3N4 [128]. (Copyright Elsevier License Number 5321220110079)

a polymeric semiconductor suitable for visible light. However, g-C_3N_4 has drawbacks namely low exposed surface area, insufficient use of visible light, and fast photo-induced electron–hole pair recombination [126, 127].

Song et al. [128] developed the use of g-C_3N_4 as a photocatalyst in CO_2 reduction applications. They obtained g-CN as a catalyst. g-CN was incorporated with C and O simultaneously. The donor–acceptor structure was developed using hexa-methyl melamine (HMM) as a doping agent that provides C and O. Simple thermal condensation method of melamine was used to fabricate g-CN-coupled Samples C and O with HMM at elevated temperatures. The catalyst exhibits efficient photocatalytic CO_2 reduction activity and durability. Their experiment resulted in a production rate of CO at 34.97 $\mu mol \cdot g_{cat}^{-1}$ in a 4-h-reaction-time. This result is 4.3-fold greater than that obtained by pristine g-CN [128] (Fig. 15).

4.2.3 Boron Nitride (BN)-Based Photocatalyst

Boron nitride has similarities to graphene, in particular the boron nitride with a hexagonal structure or *h*-BN. *h*-BN is a 2D material whose structure imitates a honeycomb, consisting of boron (B) and nitrogen (N) atoms with the same amounts and has excellent thermal conductivity and a large surface area. This material shows great potential in catalysis applications, especially in photocatalytic CO_2 reduction. *h*-BN has a broad bandgap of 5.2–5.5 eV. Similar to *h*-BN, porous BN also has a wide bandgap (~5.5 eV) [131].

Cao et al. [130] successfully fabricated the h-BN (O/BN) nanosheets. This nanosheet is modified with ultra-thin oxygen-containing a single bond between the B atom and O atom. When O is associated with the B atom, it has a greater electronegativity, so the introduction of the O atom into the *h*-BN aims to form a single bond between the B atom and the O atom. These nanosheets exhibit a great performance for CO_2 conversion to H_2 at a production rate of 3.3 $\mu mol \cdot g_{cat}^{-1} \cdot h^{-1}$ and CO at 12.5 $\mu mol \cdot g^{-1} \cdot h^{-1}$ [130].

4.2.4 Black Phosphorus (BP)

Black phosphorus, a 2D layered material, is an emerging remarkable photocatalytic semiconductor, especially for CO_2 reduction. BP nanosheets have advantages such as a tailorable band gap value of about 0.3–2.0 eV as the layer decreases from bulk to monolayer, suitable electronic structure, excellent optical properties, and high mobility of charge carriers, which is about 1000 $cm^2 \cdot V^{-1} \cdot s^{-1}$ [131].

For example, a previous study succeeded in making a CO_2 reduction photocatalyst from the fabrication of a monolayer of black phosphorus with a modified hydroxyl functional group, also known as M-BP-OH. In situ preparation of the photocatalyst can be achieved through the green exfoliation technique. In the synthesis using this technique, nitrogen is needed. From this research, the M-BP-OH exhibits an excellent photocatalytic activity performance. 112.6 $mol \cdot g^{-1} \cdot h^{-1}$ of carbon monoxide (CO) can be produced by this photocatalyst by reducing CO_2. Furthermore, the M-BP-OH exhibit excellent stability under humid condition (90% humidity) for 24 h, ambient conditions for a month, and cycling test using a Xe lamp at 300 W for 60 h [132].

Recently, a hybrid material constructed of BP QDs and g-C_3N_4 was prepared to convert CO_2 to CO. As a result of the research, the material can catalyze the CO_2 photochemical conversion reaction to produce CO. The CO product produced at high production levels was 6.54 $mol \cdot g^{-1} \cdot h^{-1}$ [125]. Compared to whole g-C_3N_4 which was only able to produce 2.65 μmol per gram of catalyst used for a 1-h basis, the yield achieved was much higher. Chen et al. (2021) fabricated a heterojunction with an S-scheme in which a stable ambient-based 2D/2D material (Pt/BP-Bi_2WO_6) for photocatalytic CO_2 reduction to syngas. The Pt/BP-OvMBWO heterojunction is electrostatically assembled. Pt/BP nanosheets are negatively charged and the OvMBWO nanosheets are positively charged. These two nanosheets were electrostatically assembled to obtain the heterojunction. As a result, this material can generate CO and H_2 as high as 20.5 $\mu mol \cdot g^{-1} \cdot h^{-1}$ and 16.8 $\mu mol \cdot g^{-1} \cdot h^{-1}$, respectively [131].

4.2.5 Silicon Carbide (SiC)

Silicon carbide is a kind of environmentally friendly semiconductor. Great chemical and thermal stability are other advantages of SiC. The band gap of SiC is wide, about 2.4–3.2 eV, and has a conduction band potential of −1.40 eV. Due to these properties, SiC has become a promising semiconductor material as a photocatalyst for

CO_2 reduction [119]. Compared to the CO_2 reduction potential of some products, the potential of SiC is more negative. Owing to its properties, SiC is a suitable material for the reduction of CO_2 that requires a high potential. SiC has about 250 polytypes. However, in general, there are 2 kinds of SiC polytype, β-SiC, and α-SiC. Cubic 3C-SiC, which belongs to the β-SiC polytype is SiC. Meanwhile, the hexagonal SiC, such as 2H-, 4H-, 6H-SiC, and others belong to the α-SiC polytype [133]. Among these polytypes, the polytype that has the highest bandgap value of 2H-SiC is 3.3 eV while the lowest band gap of 3C-SiC is 2.4 eV [134].

Wang et al. [135] developed a β-SiC polytype of SiC. The developed SiC is a hollow-Sphere with an open mouth and is synthesized using environmentally friendly techniques. Glucose is used as a carbon source in this fabrication. In the application of photocatalytic CO_2 reduction with purified water, the prepared SiC showed a very active performance for converting CO_2 to methane due to the uniqueness of its electronic structure. Besides that, it is also supported by its hollow structure and its large BET surface area (28 m^2/g). The performance of β-SiC Hollow-Sphere with Pt charge is greatly improved. Under sunlight, 2.0 wt% of Pt showed that the conversion of CO_2 to CH_4 reached 67.2 mol·g^{-1}. While the results with bare SiC only reached 28.1 mol/g, which means the results using β-SiC Hollow-Sphere with Pt loading were 2 times higher. Moreover, for the evolution of CH_4 from CO_2, the performance shown by this material is also more than that of many metal oxides. Figure 16 illustrates the charge transfer process for reducing CO_2 with water to produce methane under simulated sunlight in a Pt/SiC hollow sphere [135].

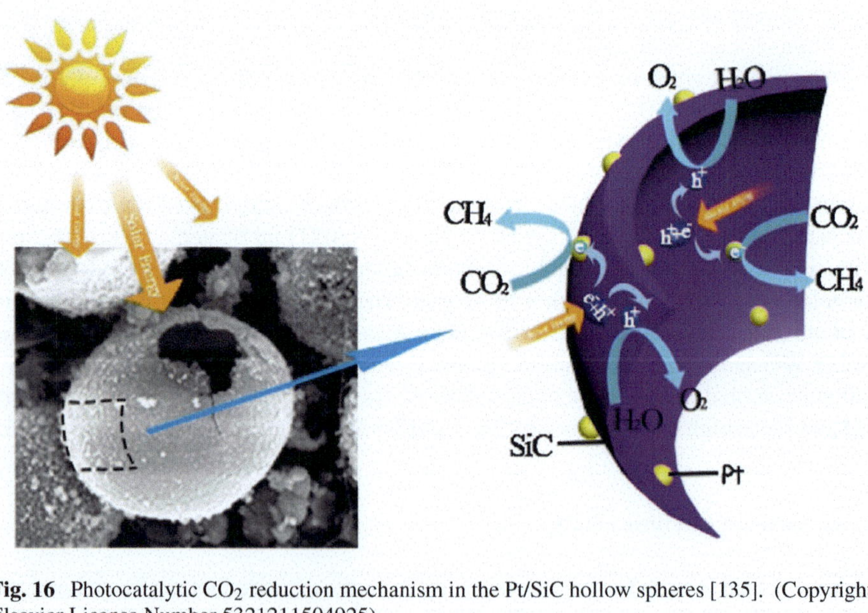

Fig. 16 Photocatalytic CO_2 reduction mechanism in the Pt/SiC hollow spheres [135]. (Copyright Elsevier License Number 5321211504925)

The development of polytype β-SiC, in the form of 3C-SiC, was also carried out by Li and Sun [136] on a gram scale. 3C-SiC is produced in the form of nanoparticles by using the ball milling method with a top-down approach. A cheap 3C-SiC crystalline powder is used for the fabrication of 3C-SiC nanoparticles. Ball-milling machine with high energy is used to grow nanoparticles by chemical vapor deposition. A 420 nm wavelength filter light from a Xe lamp with a power of 300 W is used in photocatalytic CO_2 reduction. 3C-SiC nanoparticles were dispersed into a glass reactor for the reduction of CO_2. 3C-SiC is capable of harvesting visible light. The band gap value is relatively small, about 2.36 eV, so 3C-SiC has an ideal band position straddling CO_2 reduction. In addition, 3C-SiC has high stability, thermal conductivity, and relatively low price. The achieved selectivity for reducing CO_2 to CH_4 is as high as 90% combined with efficient water oxidation. Therefore, 3C-SiC has great potential and deserves consideration in photocatalytic CO_2 reduction applications [136].

Liu et al. [133] fabricated a nanocage with a heterophase junction of 2H/3C-SiC. They prepared this nanocage by a simple magnesiothermic reduction technique mediated by low-temperature molten salt. SiO_2 was used as a template. Uniform heterophase junctions in the SiC nanocage will be formed by the 2H and 3C crystalline phases. This accelerates photogenerated electron transfer effectively and is very important in improving the performance to generate CO through the photocatalytic reaction of CO_2 reduction. CO produced by SiC nanocage was recorded to be 4.68 $mol{\cdot}g^{-1}{\cdot}h^{-1}$. This result is 3.25-fold more than commercially available SiC [133].

4.2.6 Covalent Organic Frameworks (COFs)

Known as COFs, covalent organic frameworks contain crystalline organic pores with orderly crystal structures connected with organic compounds by covalent bonds. COFs are polymeric materials. COF is built by elements that tend to be lightweight, such as carbon, hydrogen, oxygen, nitrogen, boron, and silica. As a new material class, COFs have some advantages. The advantages of COF include tailorable pore size, large specific area, tailored functionalities, and good thermal stability [119].

For example, Lu and coworkers designed a sequence of COFs in the form of crystalline 2D porphyrin-tetrathiafulvalente, TTCOF-M. The notation M indicates other elements used in the material, such as 2H, zinc (Zn), nickel (Ni), and copper (Cu). This material was synthesized as a photocatalyst for artificial photosynthesis including reducing CO_2. In this study, no sacrificial agents, precious metal co-catalysts, or supplementary photosensitizers were added to the photocatalyst system. Schiff base condensation technique between 5,10,15,20-tetrakis (4-aminophenyl)-porphinato] (TAPP-M) which is metalized is the technique used in this study for the fabrication of TTCOF- M. Meanwhile, the solvothermal method was used for the preparation of 2,3,6,7-tetra (4-formylphenyl)-tetrathiafulvalene [137].

The mechanism of the CO_2 reduction reaction process with H_2O oxidation is shown in Fig. 17b. Visible light irradiation encourages the CO_2 reduction process with H_2O oxidation. Before the photo-induced electron transfer process from the tetrathiafulvalene (TTF) part to the TAPP part, the photons will be absorbed first. After the electrons are excited, they travel to the active catalytic site in TAPP, which is Zn/Cu. The excited electrons that have been transferred are used to reduce CO_2. Meanwhile, TTF can generate O_2 through H_2O oxidation by the act of its photogenerated holes, in which electrons are obtained by the photocatalytic system from H_2O to maintain the charge balance. This research was able to reduce CO_2 with TTCOF-Zn/Cu COF using water as an electron donor. The highest performance is shown by TTCOF-Zn with CO production as high as 12.33 mmol and with high selectivity which is also combined with excellent durability [137].

Fu et al. [138] successfully fabricated a crystalline bipyridine-containing sp^{2}c-COF as a photocatalyst to reduce CO_2. Rhenium complex [Re(CO)$_5$Cl] was used to modify the COF and form a homogeneous catalyst Re-Bpy-sp^{2}c-COF

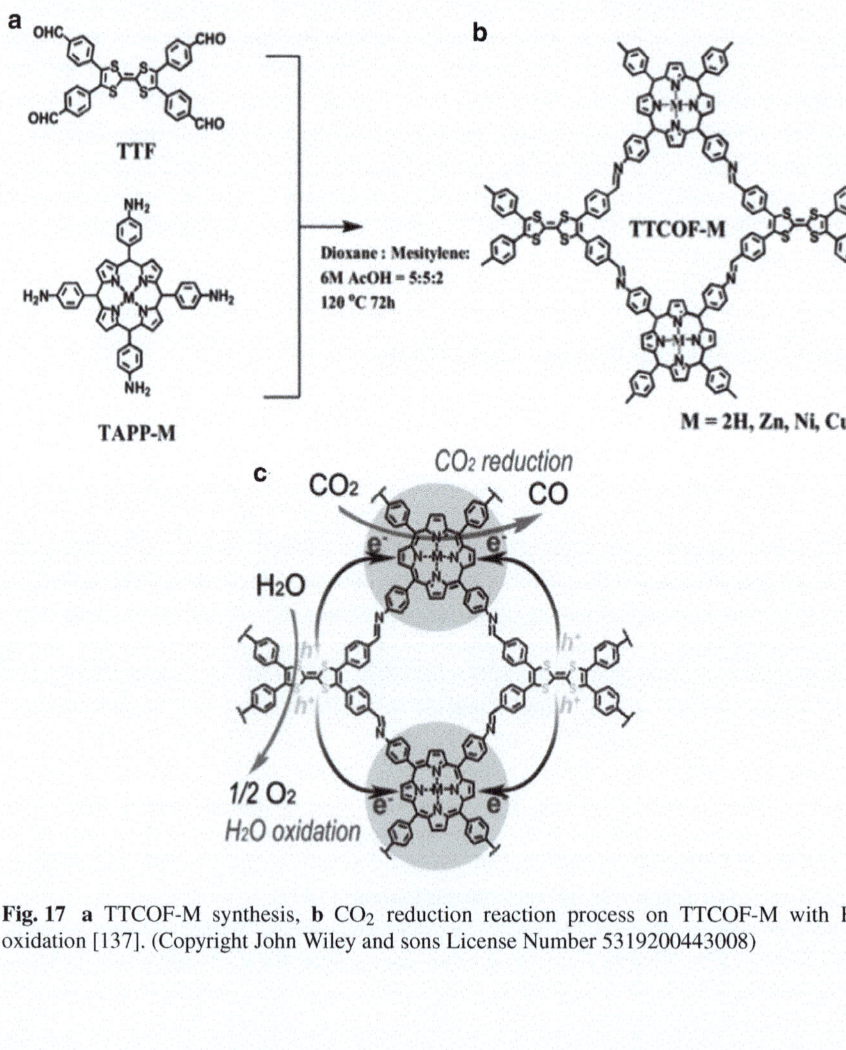

Fig. 17 **a** TTCOF-M synthesis, **b** CO_2 reduction reaction process on TTCOF-M with H_2O oxidation [137]. (Copyright John Wiley and sons License Number 5319200443008)

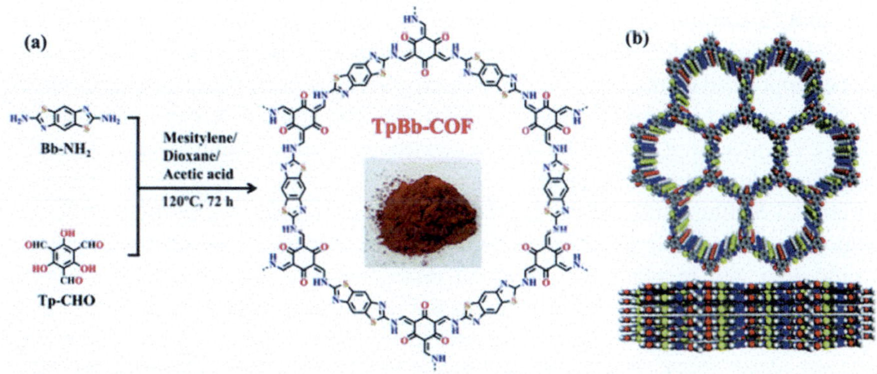

Fig. 18 **a** TpBb-COF Synthesis **b** Stacking mode of TpBb-COF from the top and side view [139]. (Copyright Royal society of chemistry License Number 1226810–1)

or [Re(bpy)$(CO)_3$Cl]. This COF photocatalyst showed outstanding photocatalytic performance to convert CO_2 into CO with a production rate of CO 1040 mol·g^{-1}·h^{-1} and selectivity of 81% (H_2) during 17.5 h of illumination. With dye sensitization, the photocatalytic performance was increased to reach 84% selectivity for more than 5 h of illumination and produce CO as much as 1400 mol·g^{-1}·h^{-1} and 86% selectivity (CO/H_2) [138].

Recently, Cui et al. [139] developed a ketoenamine-linked 2D COF, also known as TpBb-COF. TpBb-COF was prepared under solvothermal conditions by an acid-catalyzed Schiff-base condensation (Fig. 18a). There are two reactants used in this condensation. The first reactant is 2,6-diaminobenzo[1,2-d:4,5-d']bisthiazole or Bb-NH2, the second reactant is 1,3,5-triformylphloroglucinol or Tp-CHO. The reactant is in the binary solvent mesitylene and dioxane. TpBb-COF was used for photocatalytic reaction to convert CO_2 to produce CO. This material has a bandgap value of 1.72 eV. As a result, TpBb-COF showed excellent photocatalytic performance where as much as 52.8 mol·g^{-1}·h^{-1} CO was able to be generated in pure CO_2. Meanwhile, in conditions of 30% CO_2 at 80 °C, the amount of CO that can be produced is 89.9 mol·g^{-1}·h^{-1} [139].

4.3 Composites System

Low et al. [140] designed nanoparticles of TiO_2 on the MXene Ti_3C_2 for photocatalytic CO_2 reduction. The nanoparticles grew in situ on MXene Ti_3C_2 through a simple calcination technique. This composite has a unique structure like rice crust. Due to the unique multilayer structure of TiO_2 to Ti_3C_2, there is a large specific surface area which in turn provides a large active site for photocatalytic reduction. By using the optimized TiO_2/Ti_3C_2 composite, the performance to reduce CO_2 in photocatalytic reaction to produce methane was 3.7-fold more than commercial TiO_2

or about 0.22 μmol·h^{-1}. Ti3C2 has superior conductivity to facilitate the photogenerated electron transfer. Another consequence is that the photogenerated holes will also be suppressed by the effect of this property. TiO_2 has a conduction band of about −0.25 eV against a normal hydrogen electrode. The composite band gap value is −0.04 eV concerning normal hydrogen electrodes, a more negative value than the Fermi Ti_3C_2 level (Fig. 19a) [140].

Figure 19b illustrates the photocatalytic process to reduce CO_2 in TiO_2/Ti_3C_2 composites. First, with an interesting morphology in which a surface with a large specific area will increase the ability of this composite to adsorb CO_2. Second, every component involved contributes to the outstanding performance. The black color properties of Ti_3C_2 increase the light absorption of the composite and create a photothermal effect. Third, on conductive Ti_3C_2 2D, a heterogeneous interface will be formed from in situ growth, where this interface can change the properties of the original material. The tight and strong contact facilitates the migration of charge carriers so that electron holes separate efficiently. In addition, photogenerated electrons also accumulate in Ti_3C_2 so it is very supportive of the multi-electron CO_2 reduction reaction [140].

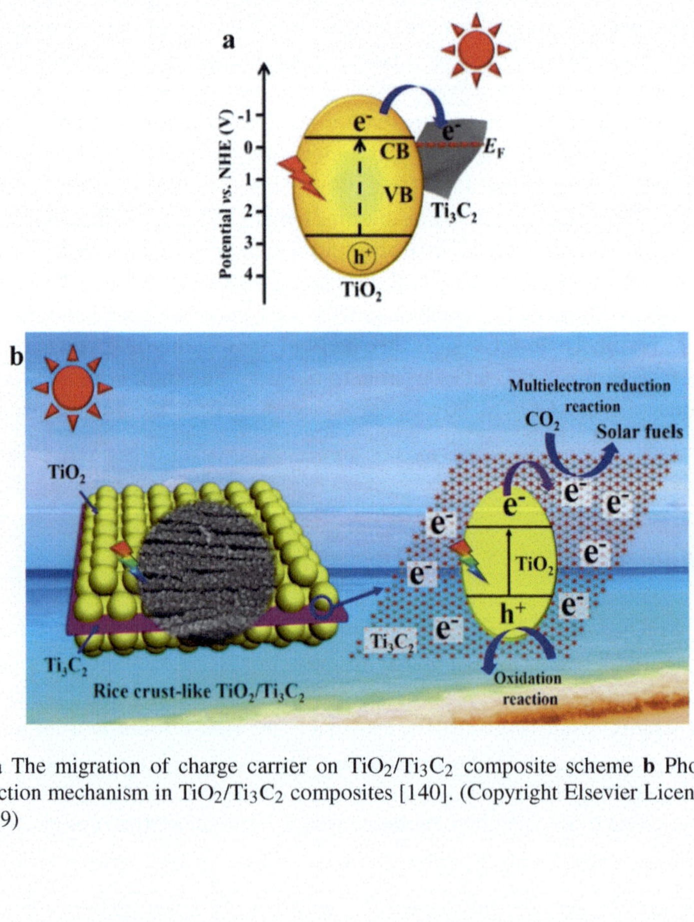

Fig. 19 **a** The migration of charge carrier on TiO_2/Ti_3C_2 composite scheme **b** Photocatalytic CO_2 reduction mechanism in TiO_2/Ti_3C_2 composites [140]. (Copyright Elsevier License Number 501735419)

Wang et al. [126] developed a photocatalyst to reduce CO_2 in the form of a nanocomposite z scheme between g-C_3N_4 and α-Fe_2O_3. The two materials are linked by Al-O. The first step to make this photocatalyst is to make a g-C_3N_4/α-Fe_2O_3 nanocomposite with an optimal ratio. The hydrolysis-solvothermal method with phase separation was used in the synthesis of nanocrystalline α-Fe_2O_3. To prepare g-C_3N_4, urea was heated in an alumina combustion vessel. The mass ratio of the two components was adjusted to obtain the optimal nanocomposite. Then, the photocatalyst composite of g-C_3N_4 and α-Fe_2O_3 which is linked by the Al-O bridge was fabricated by a two-step wet chemical technique. As the source of the Al-O bridge in the nanocomposite, an aqueous solution of $AlCl_3$ was used. Compared to using pristine g-C_3N_4 and pristine Fe_2O_3, nanocomposite photocatalysts exhibit higher photocatalytic activity for enhanced CO_2 conversion. After irradiation for 1 h, α-Fe_2O_3 showed low photocatalytic activity in reducing CO_2 to several products, such as CO, CH_4, and O_2 with production rates of ~ 5.7 mol·g^{-1}·h^{-1}, ~ 0.3 mol·g^{-1}·h^{-1}, and ~ 6.2 mol·g^{-1}·h^{-1} respectively. However, the addition of a large amount of g-C_3N_4, especially 15CN-F, give a positive contribution to the photocatalytic activity where the production rate for each product increased to ~ 15.8 mol CO g^{-1}·h^{-1}, ~ 3.1 mol CH4 g^{-1}·h^{-1}, and ~ 18.5 mol O2 g^{-1}·h^{-1}. This photocatalytic activity performance can be further enhanced by building a link or bridge between g-C_3N_4 and α-Fe_2O_3 (15CN-6Al-F) on 15CN-F nanocomposite in the form of Al-O. As a result, the amount of CO produced is 4 times higher than α-Fe_2O_3, or about 24 mol·g^{-1}·h^{-1}. The charge transfers and schematic of the photo-induced charge separation process for a g-C_3N_4/α-Fe_2O_3 nanocomposite linked by an Al-O bridge are illustrated in Fig. 20. Initially, the photoinduced electrons of α-Fe_2O_3 will join the g-C3N4 photoinduction holes in the g-C_3N_4/α-Fe_2O_3 nanocomposite when these two materials are excited simultaneously under a wavelength of 460 nm. In this case, a redox reaction will be induced because the holes in α-Fe_2O_3 and electrons in g-C_3N_4 are spatially separated and have thermodynamically sufficient energy. This reaction will result in a promoted charge separation which will increase the photocatalytic activity. Al-O bridges in nanocomposites will promote charge transfer from the α-Fe_2O_3 side to the g-C_3N_4 side and lead to increased photocatalytic performance for CO_2 reduction. Photo-induction of electrons in g-C_3N_4 stimulates a reduction reaction with the H_2O component to produce hydrogen atoms. The resulting hydrogen atoms attack the CO_2 molecules to produce products in the form of CO and CH_4. Meanwhile, another product, namely the radical hydroxyl group, could evolve O_2 through the photoinduced hole reaction of α-Fe_2O_3 with H_2O [126].

Photocatalyst made from g-C_3N_4 was also recently fabricated by Xu and coworkers. In their research, they developed a photocatalyst nanocomposite with a ternary system between g-C_3N_4, Ag_2CrO_4, and graphene oxide having a Z-scheme mechanism. This nanocomposite photocatalyst system with g-C_3N_4 was developed because its ability to absorb visible light is low and its charge recombination is so fast. In this photocatalyst system, silver chromate (Ag_2CrO_4) nanoparticles are used which act as photosensitizers to drive charge transfer and harvest light. Silver chromate (Ag_2CrO_4) was used as a photosensitizer due to its narrow bandgap (-1.8 V). This is favorable for the g-C_3N_4 light absorption so that the amount of absorbed light by g-C_3N_4 can be increased. In addition, Ag_2CrO_4 may also pair with

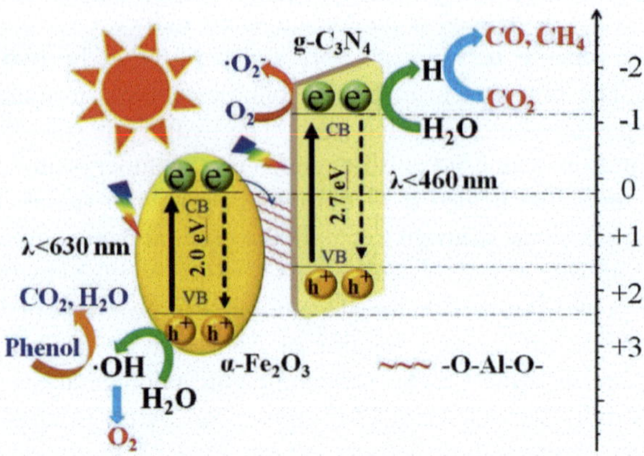

Fig. 20 Mechanism of photoinduction charge transfer and separation in $g\text{-}C_3N_4/\alpha\text{-}Fe_2O_3$ nanocomposites linked by Al-O bridges [126]. (Copyright Elsevier License Number 501735385)

$g\text{-}C_3N_4$ to form binary heterojunctions to promote photogeneration charge separation. Graphene oxide (GO) was used as a co-catalyst. Apart from being a co-catalyst, GO is also an electron collector and provides a reactive site for CO_2 conversion. A composite photocatalyst with a ternary system was recently prepared to reduce CO_2 in photocatalytic reactions under simulated light irradiation to produce methane and methanol. The precipitation self-assembly technique was used for the fabrication of $Ag_2CrO_4/g\text{-}C_3N_4/GO$ composites (Fig. 21) [141].

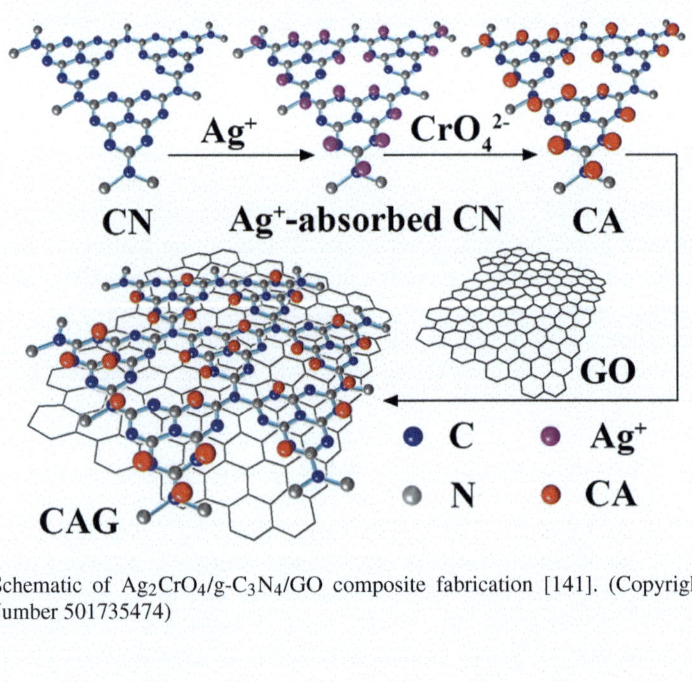

Fig. 21 Schematic of $Ag_2CrO_4/g\text{-}C_3N_4/GO$ composite fabrication [141]. (Copyright Elsevier License Number 501735474)

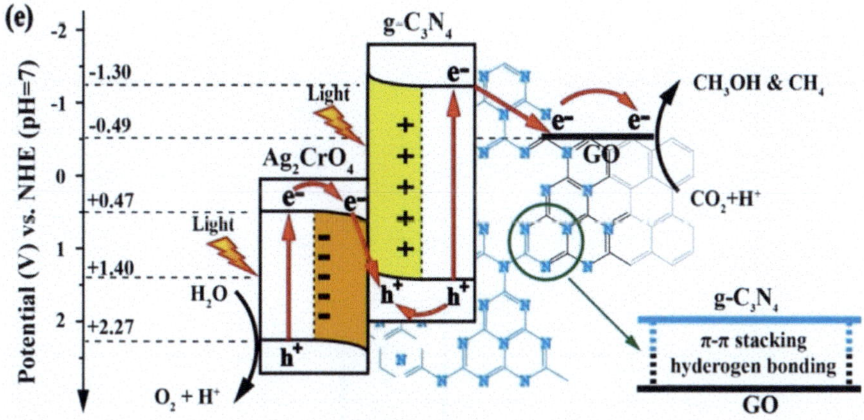

Fig. 22 Photocatalytic process in Ag_2CrO_4/g-C_3N_4/GO ternary composite through Z-scheme [141]. (Copyright Elsevier License Number 501735474)

Under full spectrum sunlight simulation, the photocatalytic activity of the ternary composite Ag_2CrO_4/g-C_3N_4/GO in reducing CO_2 was very good, as high as 1.03 mol·g^{-1}. This result is 2.3-fold more than using the whole g-C_3N_4. The direct Z-scheme mechanism is a charge transfer process that occurs between Ag_2CrO_4 and g-C_3N_4 (Fig. 22). Apart from promoting charge separation, this charge transfer process enhances the reduction–oxidation capability of the photocatalytic system as well [141].

Crake et al. [142] developed a composite photocatalyst of inorganic and organic materials for CO_2 reduction applications. In this study, they used titanium oxide and carbon nitride. Carbon nitride is used in the form of nanosheets (CNNS). The two materials, namely TiO_2 and CNNS were prepared by hydrothermal in situ growth technique (Fig. 5a). This photocatalyst system does not require an expensive co-catalyst and CO_2 reduction takes place in a single process. First, the preformed carbon nitride was exfoliated by sonication in water to form carbon nitride (CNNS) nanosheets. This CNNS was used to obtain TiO_2 and CNNS composites. In the PTFE autoclave, titanium butoxide, hydrofluoric acid, and CNNS were added and mixed. The process was continued by heating at 180 °C for 24 h. The addition of hydrofluoric acid aims to control the formation of crystal facets. This photocatalyst exhibits an excellent photocatalytic CO_2 reduction performance compared to its constituent materials, approximately an increase of more than tenfold. In addition, this composite also increases the capacity to adsorb CO_2 [142] (Fig. 23).

4.3.1 TiO_2-MOF Composite

The combination of TiO_2 metal with MOF covers the shortcomings of each material as a CO_2 reduction photocatalyst. These metals have their respective roles; TiO_2 acts

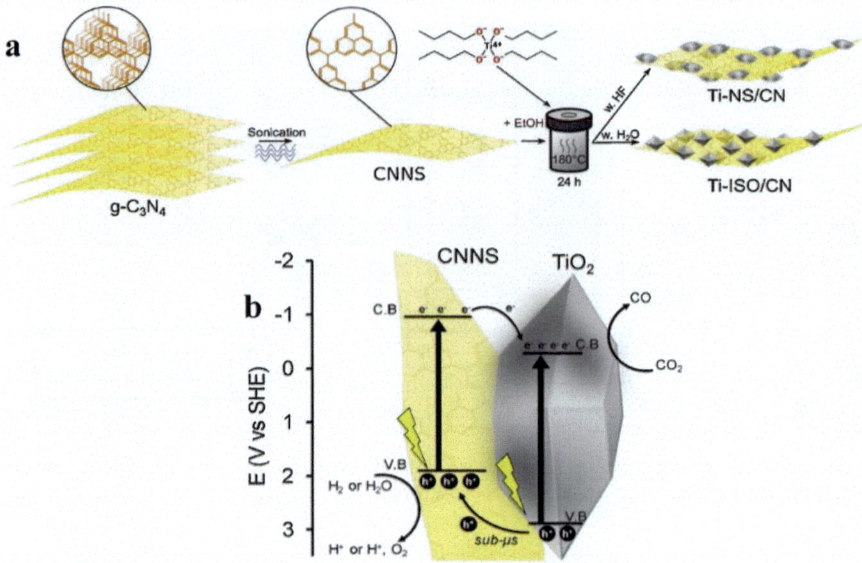

Fig. 23 TiO_2/Ti_3C_2 composite **a** Schematic of the fabrication process of TiO_2/CNNS composites **b** Photocatalytic CO_2 reduction reaction pathway [142]. (Copyright Elsevier License Number 5321210450514)

as a primary photocatalyst. Meanwhile, MOF acts as a photosensitizer for efficient light harvesting and electron transfer. Combining the two improves light absorption, electron–hole pair separation, and photocatalytic efficiency. The results obtained can be 5 times higher than pure metal TiO_2 even though CO_2 adsorption is reduced [143].

TiO_2-MOF composites with TiO_2 particles graphed onto ($Cu_3(BTC)_2$) microcrystals (TiO_2 in HKUST-1) were found to have a bandgap of 1.6–3.1 eV under UV light. The use of this composite reduces the adsorption capacity of CO_2, but the production of CH_4 is five times higher than the use of pure TiO_2 due to electron–hole separation [144].

Crake et al. [145] performed the synthesis of TiO_2 through the solvothermal method. Then the in situ method was used to produce TiO_2-MOF composites under UV–Vis irradiation. This composite prevents TiO_2 aggregation by maintaining a high surface area. Furthermore, the porosity of the MOF results in a sizeable CO_2 adsorption capacity so that the amount of adsorbed CO_2 on the surface increases. Wang et al. [143] explain four types to produce TiO_2-MOF composites. The first type uses organic ligands and metal templates of different morphology to form MOFs. The second type uses a precursor to produce TiO_2 into a prepared MOF. The third type is the in situ method, as done by Crake et al. [145], and the last one uses a partial organic linker from MOF.

4.3.2 Graphene-MOF Composite

Another MOF combination, graphene, makes up for the lack of electrical conductivity in MOF and improves the photothermal stability. This composite can also reduce aggregation at high-temperature calcination and increase the catalytic activity of the active sites. The MOF in the composite acts as a substrate for the diffusion of the substrate because of its porous structure so that there is more room for additional components. In addition, MOF also has high crystallinity and stability during the photocatalytic process. Graphene is a conductor for light harvesting and electron transfer [110].

Briefly, there are two methods for making MOF-Graphene composites. The first way is physical mixing with pH adjustment to increase electrostatic adsorption. The second method is in situ growth to increase the interaction between graphene and MOF. The performance of these two methods can be improved in several ways, namely MOF aminase, polydopamine, heteroatom doping, use of graphene aerogel, and constructing heterogeneous structures [146].

4.3.3 Titanium-Based Perovskites

Modification of TiO_2 catalyst with perovskite as host material was used to adjust optical properties and induce visible light absorption. This composite is also intended to form a catalyst that has good photo-corrosion resistance and thermal stability. In addition, this composite is applied as a candidate for producing hydrogen and hydrocarbons as well as water separation processes through photocatalytic reduction of CO_2 [108].

Strontium titanate oxide ($SrTiO_3$) and calcium titanate oxide ($CaTiO_3$) are the most widely used perovskite oxides. $SrTiO_3$ perovskite composites with TiO_2 are known to have the ability to selectively separate CO_2 reduction reactions and enhance charge transfer synergism. In addition, this perovskite can escalate the selectivity of CH_4 and CO production. The addition of Pt and Pd nanoparticles to the composite material increased the photoconversion efficiency of CO_2 to CH_4. As a result, the production of CH_4 and CO increased by 4.7 and 1.6 times, respectively, compared to ordinary composites. This is because Pt contributes to increase the efficiency of charge separation [108] (Fig. 24).

4.3.4 Graphene-Based Perovskites

Graphene is one of the most widely used photocatalysts for CO_2 reduction. However, this material has a cypher bandgap which makes it incapable to perform charge separation. However, graphene offers high charge mobility and electron–hole pair separation through its interface. The use of composites with graphene enables increasing its light absorption, charge mobility, and catalytic activity in visible light through their synergistic effect [108].

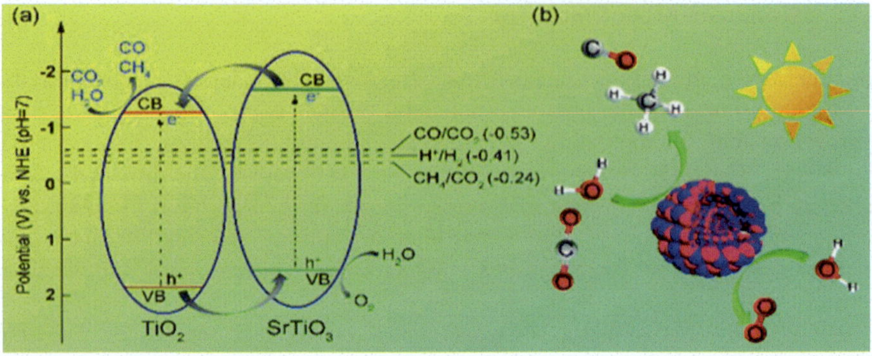

Fig. 24 Photocatalytic Mechanism over $SrTiO_3/TiO_2$ Composite. Reproduced with permission [108]. (Copyright Elsevier License Number 5326340974814)

The use of perovskite Bi_2MoO_6 composite with graphene for photocatalytic CO_2 reduction was carried out to produce CH_3OH and C_2H_5OH. Bi_2MoO_6 electrons will migrate from VB to CB, then the Mo atoms in the perovskite will trap photogenerated electrons to attract CO_2. These electrons will then move towards graphene which has an electron-rich interface. Finally, CO_2 absorption occurs due to the large surface area of graphene [147].

4.3.5 Graphitic Carbon Nitride-Based Perovskites

$KNbO_3$ is a perovskite used as a composite with g-C_3N_4 and Pt. g-C_3N_4 functions as a sensitizer while Pt acts as a co-catalyst. The electrons will be migrated to the perovskite which is then switched to the Pt co-catalyst to reduce CO_2 into CH_4. The use of this composite system produces four times higher methane product than the use of pure g-C_3N_4, under visible light radiation. In addition, the use of perovskite as a composite material is proficient to increase the catalytic activity along with the increase in the concentration of perovskite. In another study, the use of perovskite $NaNbO_3$ composites with g-C_3N_4 was able to increase methane production eightfold compared to the use of pure g-C_3N_4-Pt or $NaNbO_3$. This is because there is a faster separation and charge transfer so that the photocatalytic activity also increases. In addition, g-C_3N_4 CB (-1.13 eV) was located at more negative potential than that of $NaNbO_3$ (-0.77 eV) [108] (Fig. 25).

4.3.6 MXenes-Based Perovskites

MXenes are transition metal carbides/nitrides and carbonitrides that have the general formula $M_{n+1}X_n$, where M is a transition metal such as scandium, titanium, zirconium, hafnium, vanadium, niobium, tantalum, and molybdenum while X is carbon

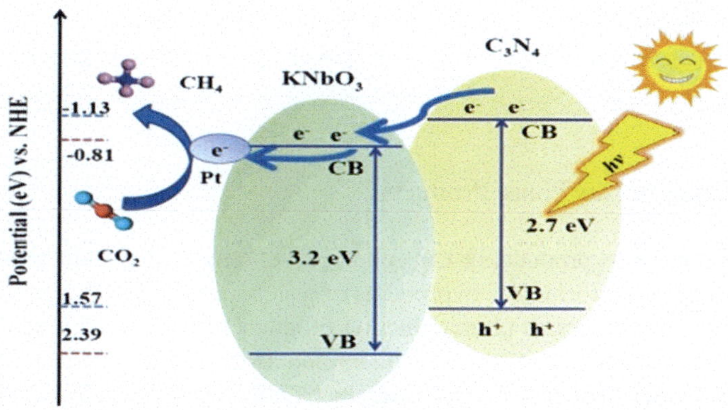

Fig. 25 Photocatalytic Mechanism over g-C_3N_4/ $KNbO_3$ Composite. Reproduced with permission [108]. (Copyright Elsevier License Number 5326340974814)

or nitrogen. Perovskite Bi_2WO_6 composite with MXene Ti_3C_2 yielded 4.6 times higher CH_4 and CH_3OH than the use of pure Bi_2WO_6 perovskite. This is because CB perovskite is more negative than Ti_3C_2 so there is a sudden electron migration as a result of photo induction to the heterojunction interface. In another study, $CsPbBr_3$ NC/MXene perovskite nanocomposites were able to selectively produce CO and CH_4 [148].

5 Product Characterization, Analysis, and Selectivity

5.1 Product Characterization and Analysis

CO_2 photoreduction involves multi-electron reactions that can result in a wide range of products in the gas phase, including CO, CH_4, and higher hydrocarbons, as well as numerous, oxygenates in the liquid phase, including alcohols, aldehydes, and carboxylic acids. Table 3 depicts a few reaction examples that result in gas and liquid products. According to a recent analysis by Izumi, certain reduction products with reasonably high yields that are not thoroughly verified may not be entirely made from CO_2. While our study is focused on producing effective photocatalyst systems, there are a few critical challenges that need to be addressed in terms of CO_2 reduction product analysis: (1) Is it possible to reliably detect the most likely reduction products? After accurate quantification of the products and elimination of the contribution from carbon contamination, a reliable conclusion on photocatalytic activity can be reached. (2) Will organic additions have an impact on the product analysis? Organic additives are frequently utilized in homogenous molecular systems as sacrificial reagents, solvents, photocatalysts, or photosensitizers. It is necessary to evaluate their impact on the analysis of CO_2 reduction products. (3) Is it true that the measured products are the result of CO_2 reduction? Verification techniques include 13C nuclear magnetic resonance (NMR) and gas chromatography-mass spectrometry (GC–MS).

5.1.1 Analysis of Gas Phase Products

The principal gaseous products of CO_2 photoreduction are CH_4 and CO, while H_2 and/or O_2 can also be formed as byproducts of water splitting [149]. Varghese et al. also found higher alkanes as photoreduction products of CO_2 in their study [150]. Aside from the products, CO_2 must be quantified as the primary species in the gas phase due to limited conversion to estimate the reaction progress and/or the extent of CO_2 dissolution in the liquid phase [151]. The use of infrared spectroscopy (IR) or diffuse reflectance infrared Fourier transform spectroscopy (DRIFT) to check CO_2 consumption and generation has been done on occasion [152]. The most widely used method for quantifying these gas species is gas chromatography (GC) [152]. GC–MS was also employed for routine CH_4 and CO analysis in addition to GC [153] or to confirm the source of carbon [154].

5.1.2 Analysis of Liquid Phase Products

Although GC with TCD/FID is still the most commonly used technology for analyzing various forms of liquid oxygenates (mainly alcohols) [155], Other approaches have been used, such as MS, high-performance liquid chromatography (HPLC) [156], ion-exchange chromatography (IEC) [157], ultraviolet–visible (UV–Vis) spectroscopy (colorimetric assay) after reacting with chromotropic acid [158] or Nash reagent [159] and NMR [160]. Table5 summarizes the substances that can be evaluated and their detection limits utilizing various techniques, as well as the accompanying constraints.

5.2 *Product Selectivity*

Photocatalytic CO_2 reduction can directly convert CO_2 and H_2O into hydrocarbon solar fuels utilizing endless solar energy as the only source of energy [161]. CO_2, as the highest oxidation state of carbon, can be reduced into a variety of products by obtaining different numbers of electrons and protons, such as CO, HCOOH, HCHO, CH_3OH, and CH_4 [162]. Besides these C1 products, some C2 products (C_2H_4, C_2H_6, CH_3CH_2OH) derived from the C–C coupling reaction also can be detected in the photocatalytic CO_2 reduction reactions [163].

In thermodynamics and kinetics, Fig. 26 summarizes some key influencing factors. To begin, light must be absorbed by a semiconductor photocatalyst to produce effective photogenerated electrons and holes. Photon energy and light intensity are the

Table 5 Comparison of techniques used for liquid phase analysis in CO_2 photoreduction

Technique	Compounds	Detection limit	Limitation
GC	Alcohols Aldehydes	3 $\mu mol \cdot L^{-1}$ 100 $\mu mol \cdot L^{-1}$	Much higher detection limits for aldehydes
HPLC	Carboxylic acids Aldehydes	5 $\mu mol \cdot L^{-1}$ 0.07 $\mu mol \cdot L^{-1}$	Aldehydes need to be derivatized before analysis
IEC	Carboxylic acids	0.1 $\mu mol \cdot L^{-1}$ 10 $\mu mol \cdot L^{-1}$	For acids only
UV–vis	HCHO after reaction with Nash's reagent or HCOOH	0.17 $\mu mol \cdot L^{-1}$ 0.08 $\mu mol \cdot L^{-1}$	Not applicable for other aldehydes and acids, HCOOH can only be analyzed when no other organics are present
1H & 13C NMR	All oxygenates	Not available	High cost, difficult for quantification although some work demonstrated quantification mainly for product quantification, carbon source verification
GC/LC–MS	All oxygenates	Not available	Mainly for product qualification, carbon source verification

most common light-excitation attributes. The former determines whether a semiconductor can absorb excited photons and thus influence the product selectivity of thermodynamic reactions. The latter impacts the number of photogenerated electrons and holes generated by excitation, and subsequently affects the reaction rate and product selectivity of multiple electron processes kinetically.

5.2.1 Product Pathways and Selectivity

Figure 27 depicts the various routes of complex photocatalytic CO_2 reduction processes, which include electron and proton transfer, hydrogenation and deoxygenation, C–O bond breaking, and C–H bond formation [165]. CO_2 adsorption on the surface of the photocatalyst initiates photocatalytic CO_2 reduction processes. CO_2 adsorption mechanisms in general include carbon and oxygen coordination with photocatalyst surface atoms [166]. The adsorption/desorption of reactants/intermediates, as well as the sequencing of hydrogenation and deoxygenation processes, are all affected by the C, H, and O affinities of photocatalysts, resulting in diverse product paths and selectivity [164].

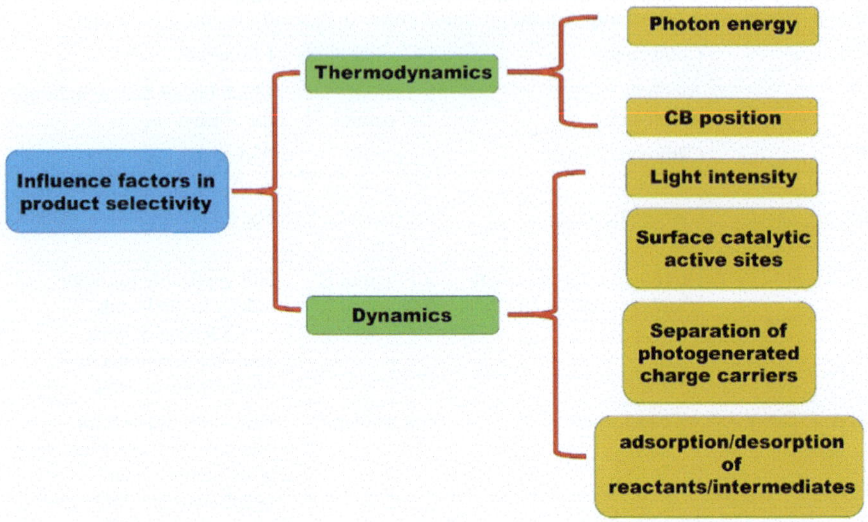

Fig. 26 Important thermodynamic and kinetic influence factors in product selectivity of photocatalytic CO_2 reduction reactions [163]. (Copyright Elsevier License Number 5327490662508)

5.2.2 Main Strategies for Controlling Product Selectivity

Complex multi-step mechanisms are involved in photocatalytic CO_2 reduction reactions. In terms of reaction activity and product selectivity, each procedure is critical. The photocatalytic CO_2 reduction process can be broken down into the following phases in general: (1) Photocatalysts are activated by the right amount of sunlight and produce a large number of electrons and holes. (2) Charge carriers generated by photocatalysts separate and transfer from the bulk to the surface. (3) On the surface of photocatalysts, reactants (CO_2 and H_2O) are simultaneously captured and activated. (4) Photogenerated holes are consumed by H_2O or additional sacrificial reagents to form oxidation products, whereas effective photogenerated electrons reach the surface active sites and catalytic reduction processes occur. (5) The resulting products desorb from the photocatalyst surface, and the re-exposed surface active sites engage in the subsequent catalytic events.

By adjusting the key phases of the photocatalytic CO_2 reduction process, the product selectivity of photocatalytic CO_2 reduction reactions can be controlled. The primary engineering strategies for enhancing photocatalytic CO_2 reduction product activity and selectivity are given in Fig. 28.

5.2.3 Light-Excitation Attributes

In general, the wavelength requirement of incident light that can excite semiconductors is determined by the band gaps of classic semiconductor photocatalysts. By

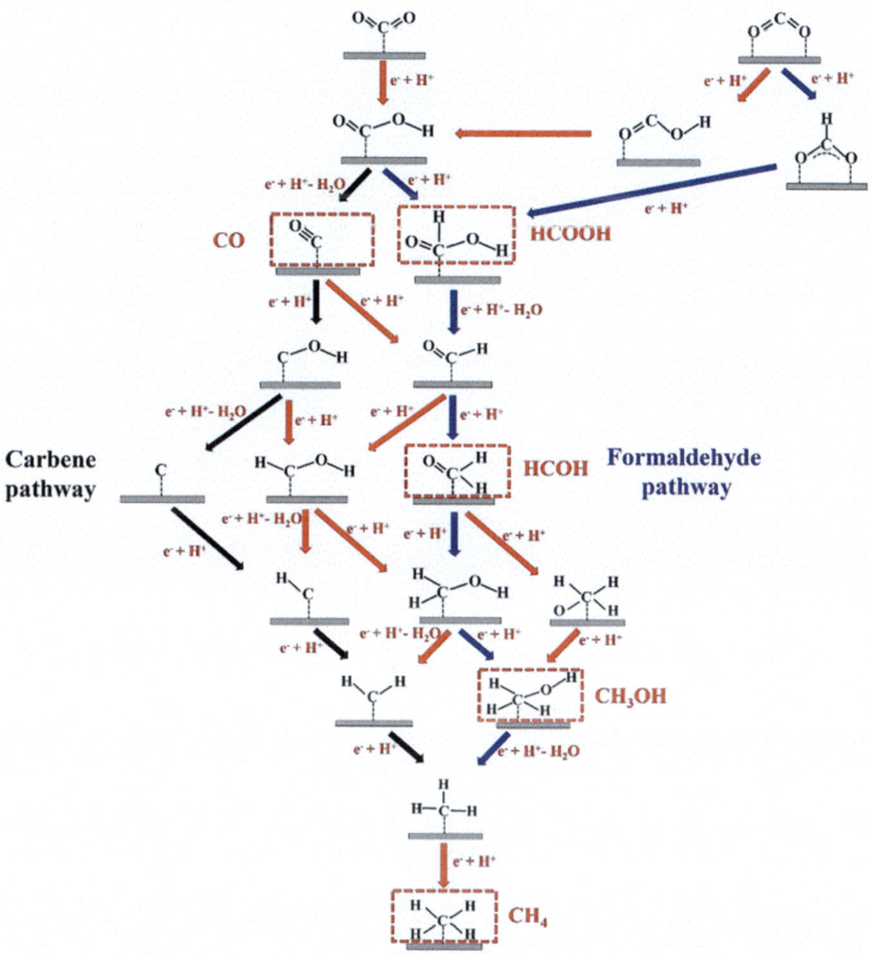

Fig. 27 Possible pathways of photocatalytic CO_2 reduction to CH_4 on the surface of photocatalysts [164]. (Copyright Elsevier License Number 5327490662508)

altering the light-excitation properties, it is difficult to manage the product selectivity of CO_2 reduction processes. Plasmonic photocatalysts have different modes of light absorption and excitation than standard semiconductor-based photocatalysts. The collective free electron oscillations induced by light irradiation, known as localized surface plasmon resonances (LSPRs), can produce strong light absorption in plasmonic nanoparticles (NPs) [167].

Adsorbed reactants on the surface of plasmonic NPs can be activated by the energetic (hot) electrons obtained. Excess hot electrons stored on the surface can participate in CO_2 reduction processes with the help of an effective hole scavenger, and oxidation products result from the oxidation of the hole scavenger. The distribution of photoexcited hot electrons is highly influenced by the energy and density of

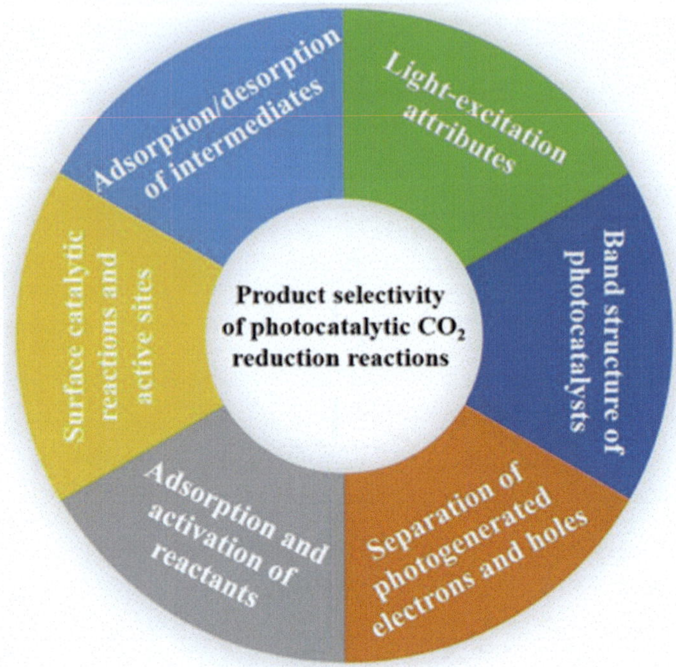

Fig. 28 Main factors influencing the product selectivity of photocatalytic CO_2 reduction reactions. (Copyright Elsevier License Number 5327490662508)

photons, resulting in distinct reaction paths and rates. It should be possible to tune light-excitation properties to alter the product selectivity of plasmonic photocatalysts.

5.2.4 Band Structure of Photocatalysts

The redox abilities of photogenerated electrons and holes are highly dependent on the locations of photocatalysts' CB and VB during photocatalytic processes. CO_2 conversion to various reduction products necessitates certain reduction potentials of photogenerated electrons, as shown in Fig. 29. Regulating the band structure of photocatalysts to regulate the CB positions allows efficient management of product selectivity with respect to thermodynamics by obtaining designated reduction potentials of electrons.

5.2.5 Separation of Photogenerated Electrons and Holes

CO_2 reductions by photocatalysis are multi-electron processes. The density of photogenerated electrons on photocatalyst surfaces can alter the rate of surface catalytic

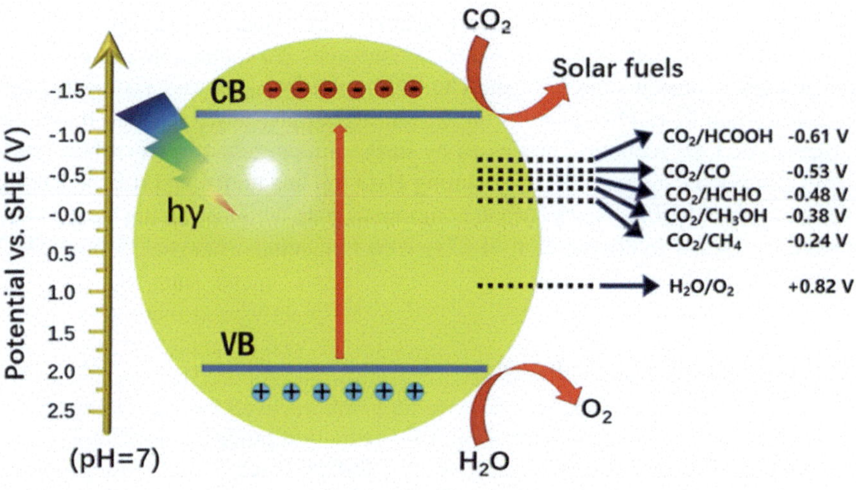

Fig. 29 Reduction potentials of photocatalytic CO_2 reduction to different products (versus SHE, pH = 7) [168]. (Copyright Elsevier License Number 5327490662508)

reactions, which in turn affects the overall product selectivity of the reaction. The separation efficiency of photogenerated electrons and holes in photocatalysts determines the density of surface photogenerated electrons [169]. As a result, increasing the effectiveness of charge separation of photocatalysts should be a key technique for controlling product selectivity [170].

5.2.6 Adsorption and Activation of Reactants

In a catalytic reaction, the adsorption and activation of reactant molecules is a crucial step. The only two reactant molecules in photocatalytic CO_2 reduction processes are H_2O and CO_2. Their adsorption and activation capabilities on the photocatalyst surface have a significant impact on activity and product selectivity. CO_2 is a nonpolar linear molecule that has a bond energy of 799 $kJ{\cdot}mol^{-1}$. CO_2 is difficult to activate and diminish due to these properties [171]. CO_2 adsorption on photocatalyst surfaces can change their linear molecular conformation, allowing the CO_2 molecule to be activated for photocatalytic reduction processes [172]. CO_2 adsorption and activation are usually synchronized. CO_2 adsorption on the surface of photocatalysts is indicated by typical carbon coordination, oxygen coordination, and mixed coordination modes.

A partially charged species CO_2 radical dot will arise as a result of the interaction between CO_2 and the surface atoms of photocatalysts, which is the typical activation intermediate. The process of further reduction by acquiring electrons can occur smoothly after activation. The surface atomic states and characteristics of photocatalysts influence the coordination modes of CO_2 radical dots. Better CO_2 adsorption and activation increase CO_2 usage and lower the reaction barrier, affecting the product selectivity of photocatalytic CO_2 reduction processes [173].

H_2O is a key reactant in the photocatalytic CO_2 reduction process [174]. To complete the oxidation half-reaction, H_2O consumes photogenerated holes. H_2O also serves as a proton source for the CO_2 hydrogenation process. The adsorption and dissociation of H_2O to protons on the surface of photocatalysts can boost the photocatalytic CO_2 reduction processes by further reducing the hydrocarbon products. Pt is a good candidate for dissociating H_2O to H adatoms, and it is beneficial in the treatment of HER. Hydrocarbon compounds with a high reducing state can be obtained by further hydrogenation of CO_2 with H adatoms obtained through H_2O dissociation.

5.2.7 Control Reaction Active Sites

The product selectivity of photocatalytic CO_2 reduction processes is influenced by the quantity and species of surface reactive sites. Different active sites are appropriate for various surface catalytic processes, yielding various products. High-index facets can bring more surface low-coordinated atoms than low-index facets, which can serve as good active sites for CO_2 activation and catalytic processes.

The product selectivity of CO_2 photocatalytic reduction was heavily impacted by the size of Pt NPs. The smaller Pt NPs resulted in better H_2O reduction selectivity for H_2 production. The bigger Pt NP-modified HTSO can achieve a better CH_4 selectivity. Surface terrace sites were more common in bigger Pt NPs, while low-coordinated locations (corner and edge) were more common in smaller Pt NPs (Fig. 30a). The DFT simulations also revealed that terrace sites can operate as active sites for CH_4 production, while low-coordinated areas are better for converting H_2O to H_2.

The partial CO-modified 1.8PHTSO (1.8 nm Pt modified HTSO) may cover the corner and edge sites of Pt particles, as shown in Fig. 30b, c. The photocatalytic performance of 1.8PHTSO and CO-1.8PHTSO is compared in Fig. 30d, demonstrating that the terrace sites were favored for CH_4 production. The size of Pt modified the geometric feature's surface site percentage, which varied the product selectivity of photocatalytic CO_2 reduction processes [164].

5.2.8 Intermediates/Side Products of Adsorption and Desorption

Good reactant adsorption and simple product desorption are required to yield targeted products with high selectivity. Many different intermediates or side products can be generated during the photocatalytic CO_2 reduction process. Intermediate and side products differ in their strong interactions with photocatalysts. The species are difficult to desorb from the surface of the photocatalyst if the contact was too strong. It is more likely that a further hydrogenation reduction reaction will be carried out [9, 164]. Intermediaries is the name given to these species. The weak interaction, on the other hand, will make desorption and release from photocatalyst surfaces easier, and

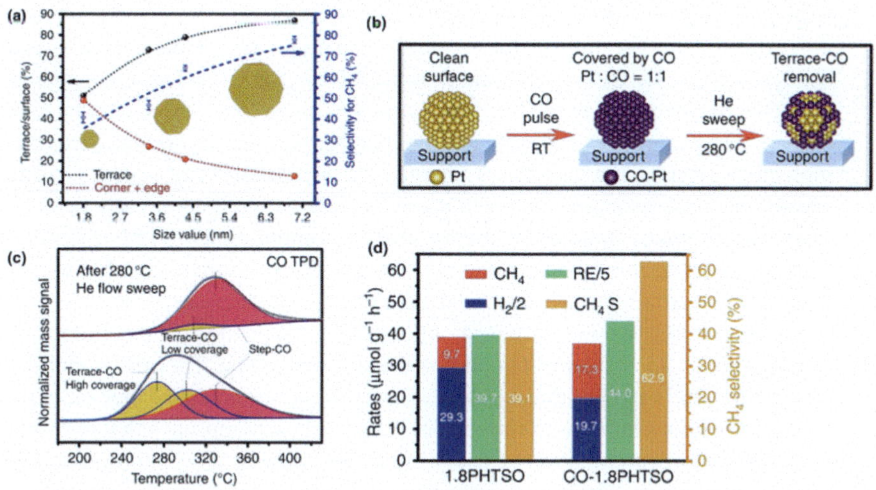

Fig. 30 **a** Relationship between selectivity of CH4, the ratio of terrace/surface, and size of Pt particles. **b** Process of partial CO-modified 1.8PHTSO by CO stepwise adsorption/desorption. **c** CO-TPD results of 1.8PHTSO after CO pulse adsorption and partial CO-modified 1.8PHTSO by He flow desorption at 280 °C. **d** Photocatalytic CO_2 reduction performance of 1.8PHTSO and CO-1.8PHTSO [164]. (Copyright Elsevier License Number 5327490662508)

these species act as key side products. In this regard, the adsorption and desorption of intermediates/side products can have a significant impact on photocatalytic CO_2 reduction reaction product selectivity [9, 164, 173].

6 Challenges and Opportunities

Energy scarcity and environmental issues are becoming increasingly serious global issues and major challenges for humanity in the twenty-first century. Solar-driven CO_2 reduction to value-added chemical fuels is a viable option for influencing the global carbon balance, with tremendous potential rewards. The current goal is to create photoactive materials that can chemically tie these light-driven redox processes together, achieving conversion efficiencies and selectivity that are superior to natural photosynthesis. Thermodynamically upward, multi-electron, multi-hole, and multi-proton reactions occur on a multicomponent photocatalyst in the photoconversion of CO_2. In the fields of catalysis, energy science, semiconductor physics and engineering, and green chemistry, long-standing difficulties are posed. Several key considerations must be balanced to develop effective CO_2 fixation and conversion, including:

(i) A thorough understanding of the processes that take place on the surface of photocatalysts during CO_2 reduction, such as CO_2 and intermediate product adsorption and desorption, as well as the role of adsorbed water. In this regard, the research would combine experimental and computational methods to assess catalyst activity for CO_2 photoreduction, advancing fundamental knowledge in the fields of material science, surface chemistry, photocatalysis, and nanotechnology. STM would be used to determine CO_2 adsorption locations and explore molecular orbitals and transfer of charge. The local density of states, lowest-energy molecule structure, and vibrational modes would theoretically be clarified using density functional theory (DFT) computations. STM and DFT working together would give a natural framework for future research targeted at gaining an atomic-scale knowledge of CO_2 photoreduction.

(ii) A significant increase in the lifespan of the charge-separated state could result in a significant increase in CO_2 photoreduction efficiency. This electron–hole recombination is two to three orders of magnitude faster than other electron transfer processes in terms of time scale. As a result, any approach that prevents electron–hole recombination will considerably improve CO_2 photoreduction efficiency and rates. To decrease electron–hole pair recombination, the photocatalyst's electrical conductivity and diffusion length should be as high as possible. Charge-carrier transfer to surface reaction sites for reductive chemistry may be aided by ultrathin nanostructures. The electron–hole recombination can be decreased by spatial separation of photoexcited electrons and holes, which can be done via co-catalysts, Z-schemes, or heterostructures linking two semiconductors with properly aligned band structures.

(iii) Light absorption by the photocatalyst affects the kinetics of CO_2 photoreduction. Many optical approaches, such as structuring for multiple light scattering to increase the effective optical path length and up-conversion to convert nonabsorbed infrared light to absorbed visible light, could be useful for harvesting light to improve efficiency.

(iv) To maximize the adsorption, transport, and desorption of reactants, intermediates, and products, a large surface area and porosity are necessary.

(v) Because defects like oxygen vacancies regulate the majority of the chemistry at many metal oxide surfaces, oxygen vacancies are thought to play a critical role in electron trapping and CO_2 activation. These defects can be discovered by in situ electron paramagnetic resonance spectroscopy, ultraviolet photoemission spectroscopy, and metastable impact electron spectroscopy, among other techniques.

(vi) A high-efficiency process must be based on photoactive materials that are abundant on the planet, non-toxic, light-stable, scalable, and affordable in cost. Furthermore, with prolonged irradiation, the efficiency of photocatalytic CO_2 reduction may be deactivated.

7 Conclusion

It may be argued that environmental issues are no longer localized issues, but have evolved into global issues concerning climate change. We are still far from having a superb as well as a cost-effective photocatalyst for photocatalytic CO_2 reduction with H_2O to overcome these challenges. As previously stated, photocatalytic processes are quite intricate in terms of both physical mechanism and product distribution, which limits their usefulness. The current state of research in this field is perplexing, and evaluating the efficiency of various photocatalysts is challenging due to the wide range of influencing factors and reaction circumstances. Many of the studies neglected some details like mass balance (moles of CO_2 transformed into the specified product), product distribution, and the amount of reducing agent, as well as time requirements, solution pH, temperature, CO_2 pressure, light power, and activity decay with time. Comprehensive research into this method is required before it can be put into practice on a large scale. A single, uniform and widely agreed standard format should be created.

The major challenges responsible for the current significantly low rate of average productivity in photocatalytic reduction of CO_2 with H_2O are low photocatalytic efficiency, low response to sunlight, inefficient electron transport between reduction and oxidation catalysts, and a high recombination rate of photogenerated species. The short lifetimes of one electron-reduced species and the photoexcited state in the presence of O_2 produced by H_2O oxidation are considered also more drawbacks. Even though UV light has more energy than visible light, visible light harvesting photocatalysts are the most desirable for this process due to the plentiful visible light from natural sunlight. Plasmonic photocatalysts, which are still in their inception, have already shown promise in overcoming the first two flaws listed above. Because the boundaries of these extremely effective noble metal photocatalysts are quickly growing, it is reasonable to expect plasmonic photocatalysts to play a substantial role in future environmental solutions.

Fabrication of ideal structured visible light-responsive photocatalysts with a wide bandgap, high rate of photogenerated electron–hole transport, high rate of photogenerated electron–hole transport, and low rate of recombination, to increase the possibility for practical application of photocatalytic CO_2 reduction with H_2O should be the focus of future research.

References

1. Bard AJ, Fox MA (1995) Acc Chem Res 28:141
2. Aresta M, Dibenedetto A (2007) Dalton Trans 2975
3. Halmann M, Fox MA (1978) Nature 275:115
4. Inoue T, Fujishima A et al (1979) Nature 277:637
5. Meduri A, Fuoco T et al (2014) Macromolecules 47:534–543
6. Bolton JR (1978) Solar Fuels Sci 202:705–711
7. Lehn JM, Ziessel R (1982) Proc Natl Acad Sci USA 79:701–704

8. Hori Y, Vayenas CG, White RE, Gamboa-Aldeco ME (2008) Modern aspects of electrochemistry. Modern aspects of electrochemistry. Springer, New York, 89–189
9. Li X, Wen J, Low J, Fang Y, Yu J (2014) Sci China Mater 57:70–100
10. Fujita E (1999) Coordin Chem Rev 185:373–384
11. Morris AJ, Meyer GJ, Fujita E (2009) Acc Chem Res 42:1983–1994
12. Bard AJ, Parsons R, Jordan J (1985) Standard potentials in aqueous solution. Routledge, New York, pp 195–197
13. Yahaya AH, Gondal MA, Hameed A (2004) Chem Phys Lett 400:206–212
14. Benson EE, Kubiak CP, Sathrum AJ, Smieja JM (2009) Chem Soc Rev 38:89–99
15. Song C (2006) Catal Today 115:2–32
16. Kumar B, Llorente M, Froehlich J et al (2012) Annu Rev Phys Chem 63:541–569
17. Kanemoto M, Shiragami T, Pac C, Yanagida S (1992) J Phys Chem 96:3521–3526
18. Linsebigler A, Lu G, Yates JJ (1995) Chem Rev 95:735–758
19. Nakata K, Fujishima A (2012) J Photochem Photobiol C 13:169–189
20. Chen X, Shen S, Guo L, Mao SS (2012) Chem Rev 110:6503–6570
21. Maeda K, Domen K (2007) J Phys Chem C 111:7851–7861
22. A. Dhakshinamoorthy, S. Navalon, A. Corma A, H. Garcia, Energy Environ Sci 5 (2012) 9217–9233.
23. Indrakanti VP, Kubicki JD, Schobert HH (2009) Energy Environ Sci 2:745–758
24. Guan GQ, Kida T, Harada T, Isayama M, Yoshida A (2003) Appl Catal A-Gen 249(1):1–18
25. Mahmodi G, Sharifnia S, Rahimpour F, Hosseini SN (2013) Sol Energ Mat Sol C 111:31–40
26. Nunez J, O'Shea VAD, Jana P, Coronado JM, Serrano DP (2013) Catal Today 209:21–27
27. Xi GX, Ouyang SX, Ye JH (2011) Chem Eur J 17:9057–9061
28. Fujiwara H, Hosokawa H, Murakoshi K, Wada Y, Yanagida S (1998) Langmuir 14:5154–5159
29. Inoue H, Moriwaki H, Maeda K, Yoneyama H (1995) J Photoch Photobio A 86:191–196
30. Koci KK, Reli M, Kozak O et al (2011) Catal Today 176:212–214
31. Kuwabata S, Nishida K, Tsuda R, Inoue H, Yoneyama H (1994) J Electrochem Soc 141:1498–1503
32. Aurian-Blajeni B, Halmann M, Manassen J (1980) Sol Energy 25:165–170
33. Lee WH, Liao CH, Tsai MF, Huang CW et al (2013) J Appl Catal B-Environ 132:445–451
34. Sui DD, Yin XH, Dong HZ et al (2012) Catal Lett 142:1202–1210
35. Zhou H, Guo JJ, Li P et al (2013) Sci Rep 3:1667
36. Li H, Lei Y, Huang Y et al (2011) J Nat Gas Chem 20:145–150
37. Yamamura S, Kojima H, Iyoda J, Kawai W (1987) Electroanal Chem 225:287–290
38. Jamshidi K, Hyon S-H, Ikada Y (1988) Polymer 29(12)
39. Bessekhouad Y, Robert D, Weber JV (2005) Catal Today 101:315–321
40. Li T, Wang WN, Zhan ZL et al (2010) Appl Catal B-Environ 100:386–392
41. Robert D (2007) Catal Today 122:20–26
42. Tseng IH, Chang WC, Wu JCS (2002) Appl Catal B-Environ 37:37–48
43. Chaudhary YS, Woolerton TW, Allen CS et al (2012) 48:58–60
44. Fujiwara H, Hosokawa H, Murakoshi K et al (1997) J Phys Chem B 101:8270–8278
45. Li X, Chen J, Li H et al (2011) J Nat Gas Chem 20:413–417
46. Li X, Liu HL, Luo DL et al (2012) Chem Eng J 180:151–215
47. Liu BJ, Torimoto T, Yoneyama H (1998) J Photoch Photobio A 113:93–97
48. Praus P, Kozak O, Koci K, Panacek A, Dvorsky R (2011) J Colloid Interf Sci 360:574–579
49. Barton EE, Rampulla DM, Bocarsly AB (2008) J Am Chem Soc 130:6342–6344
50. Hara M, Nunoshige J, Takata T, Kondo JN, Domen K (2003), Chem Commun 3000–3001
51. Kim ES, Nishimura N, Magesh G et al (2013) J Am Chem Soc 135:5375–5383
52. Maeda K, Higashi M, Lu DL, Abe R, Domen K (2010) J Am Chem Soc 132:5858–5868
53. Sekizawa K, Maeda K, Domen K, Koike K, Ishitani O (2013) J Am Chem Soc 135:4596–4599
54. Mao J, Peng T, Zhang X et al (2013) Catal Sci Technol 3:1253–1260
55. Wang X, Maeda K, Thomas A et al (2009) Nat Mater 8:76–80
56. Zhang JS, Chen XF, Takanabe K et al (2010) Angew Chem Int Ed 49:441–444
57. Jiang H, Dai H, Meng X et al (2011) Appl Catal B-Environ 105:326–334

58. Jiang H, Meng X, Dai H et al (2012) J Hazard Mater 217–218:92–99
59. Kudo A, Omori K, Kato J (1999) J Am Chem Soc 121:11459–11467
60. Kudo A, Ueda K, Kato H, Mikami I (1998) Catal Lett 53:229–230
61. Tokunaga S, Kato H, Kudo A (2001) Chem Mater 13:4624–4628
62. Chun WJ, Ishikawa A, Fujisawa H et al (2003) J Phys Chem B 107:1798–1803
63. Higashi M, Domen K, Abe R (2011) Energy Environ Sci 4:4138–4147
64. Li Y, Takata T, Cha D et al (2012) Adv Mater 25:125–131
65. Ma SSK, Hisatomi T, Maeda K, Moriya Y, Domen K (2012) J Am Chem Soc 134:9993–19996
66. Anpo M, Yamashita H, Ichihashi Y, Ehara S (1995) Electroanal Chem 396:21–26
67. He H, Zapol P, Curtiss LA (2012) Energy Environ Sci 5:6196–6205
68. Liu G, Hoivik N, Wang K, Jakobsen H (2012) Sol Energ Mat Sol C 105:53–68
69. Pipornpong W, Wanbayor R, Ruangpornvisuti V (2011) Appl Surf Sci 257:10322–10328
70. Sasirekha N, Basha SJS, Shanthi K (2006) Appl Catal B-Environ 62:69–180
71. Subrahmanyam M, Kaneco S, Alonso-Vante N (1999) Appl Catal B-Environ 23:169–174
72. Wang J, Lin S, Tian N, Ma T, Zhang Y, Huang H (2021) Adv Funct Mater 31:2008008
73. Kong T, Jiang Y, Xiong Y (2020) Chem Soc Rev 49:6579–6591
74. Ku Y, Lee W, Wang W (2004) J Mol Catal A-Chem 212:191–196
75. Lo CC, Hung CH, Yuan CS, Wu JF (2007) Sol Energ Mat Sol C 91:1765–1774
76. Tahir M, Amin NS (2013) Appl Catal A-Gen 467:483–496
77. Tan SS, Zou L, Hu E (2008) Catal Today 131:125–129
78. Wasielewski MR (2009) Acc Chem Res 42:1910–1921
79. Zhou H, Qu Y, Zeid T, Duan X (2012) Energy Environ Sci 5:6732–6743
80. Mahmoud Idris A et al (2022) J Colloid Interface Sci 607(2):1180–1188
81. Li N et al (2022) ACS Nano 16(2):3332–3340
82. Wang F et al (2022) ACS Nano 16(3):4517–4527
83. Usman M et al (2022) J Envir Chem Engin 10(3)
84. Tahir M, Tahir B (2022) J Mat Sci Techno 106:195–210
85. Pachaiappan R et al (2022) Chem Engin Res Des 177:304–320
86. Guo K et al (2023) J Envir Sci 125:290–308
87. Chen P et al (2022) ACS Catal 12(8):4560–4570
88. Camera-Roda GF, Santarelli CA (2005) Martin. Sol Ener 79(4):343–352
89. Lais A et al (2018) Int J Ener Res 42(6):2031–2049
90. Paulino et al (2016) Appl Catal B: Envir 185:362–370
91. Han S et al (2017) J Ener Chem 26(4):743–749
92. Ali S et al (2019) Catalysts 9(9)
93. Kočí K et al (2011) Catal Today 176(1):212–214
94. Ola O et al (2012) Appl Catal B: Envir 126:172–179
95. Liou et al (2011) Ener Envir Sci 4(4)
96. Saladin et al (1997) Chem Soc Faraday Trans 93:4159–4163
97. Tan et al (2017) Chem Engin J 308:248–255
98. Zhang et al (2009) Catal Today 148(3–4):335–340
99. Wang W, Ku Y (2003) J Photochem Photobio A Chem 159(1):47–59
100. Mizuno T et al (1996) J Photochem Photobiol A Chem 98(1–2):87–90
101. Ola et al (2015) J Photochem Photobio C Photochem Rev 24:16–42
102. Neatu et al (2014) Int J Mol Sci 15(4):5246–5262
103. Yui T et al (2011) ACS Appl Mater Interfac 3(7):2594–2600
104. Ochedi et al (2021) Envir Chem Lett 19:941–967
105. Rehman ZU et al (2022) Molecules 27(2069):1–30
106. Nguyen TP et al (2020) Nanomaterials 10(337):1–24
107. Wang W, Wang L, Su W, Xing Y (2022) J CO_2 Util 61:1–21
108. Madi M, Tahir M, Tasleem S (2021) Advances in structural modification of perovskite semiconductors for visible light assisted photocatalytic CO_2 reduction to renewable solar fuels: a review. J Envir Chem Engi 9:1–39
109. You J, Xiao M, Wang Z, Wang L (2022) J CO_2 Util 55:1–22

110. Ikreedeegh RR, Tahir M (2021) A critical review in recent development of metal-organic frameworks (MOFs) with band engineering alteration for photocatalytic CO_2 reduction to solar fuels. J CO_2 Util 43:1–41
111. Chen Y, Wang D, Deng X, Li Z (2013) J Name 00:1–3
112. Kumagai H, Tamaki Y, Ishitani O (2021) Acc Chem Res 55:978–990
113. Luo et al (2019) Coord Chem Rev 390:86–126
114. Shen J, Wu Z et al (2021) Flat Chem 28:1–11
115. Zhao Y, Que M, Chen J, Yang C (2020) J Mat Chem C 1–53
116. Li K, Zhang S, Li Y, Fan J, Lv K (2021) MXenes as noble-metal-alternative co-catalysts in photocatalysis. Chi J Catal 42:3–14
117. Do KH, Kumar et al (2020) Chem Cat Chem 12(18):1–9
118. Cui D, Wang et al (2018) ASC 6(12):15936–15953
119. Shen H, Peppel T, Strunk J, Sun Z (2020) Solar RRL 4(8):1900546
120. Ochedi FO, Liu D, Yu J, Hussain A, Liu Y (2021) Envir Chem Lett 19(2):941–967
121. Hasani A, Teklagne MA et al (2020) Carbon Energy 2(2):158–175
122. Vu NN et al (2019) Adv Funct Mat 29(31):1901825
123. Mondal A, Prabhakaran A, Gupta S, Subramanian VR (2021) ACS Omega 6(13):8734–8743
124. Tan LL, Ong WJ, Chai SP, Mohamed AR (2017) Chem Engin J 308:248–255
125. Han C, Li J, Ma Z et al (2018) Sci Chi Mat 61(9):1159–1166
126. Wang J, Qin C, Wang H et al (2018) Appl Catal B Envir 221:459–466
127. Li G, Lian Z, Wang W, Zhang D, Li H (2016) Nano Energy 19:446–454
128. Song X, Li X, Zhang X, Wu Y, Ma C, Huo P, Yan Y (2020) Fabricating C and O co-doped carbon nitride with intramolecular donor-acceptor systems for efficient photoreduction of CO_2 to CO. Appl Catal B Envir 268:118736
129. Shankar R, Sachs M, Francàs L et al (2019) J Mat Chem A 7(41):23931–23940
130. Cao Y, Zhang R, Zhou T, Jin S, Huang J, Ye L, Huang Z, Wang F, Zhou Y (2020) ACS Appl Mat Interf 12(8):9935–9943
131. Chen C, Hu J, Yang X, Yang T, Qu J, Guo C, Li CM (2021) ACS Appl Mat Interf 13(17):20162–20173
132. Zhu X, Huang S, Yu Q, She Y, Yang J, Zhou G, Li Q, She X, Deng J, Li H, Xu H (2020) Appl Catal B Envir 269:118760
133. Liu Y, Wang B, Li D, Shen J, Zhang Z, Wang X (2022) J Coll Interf Sci 622:31–39
134. Wu R, Zhou K, Yue CY, Wei J, Pan Y (2015) Progr Mat Sci 72:1–60
135. Wang Y, Zhang L, Zhang X et al (2017) Appl Catal B Envir 206:158–167
136. Li H, Sun J (2021) ACS Appl Mat Interf 13(4):5073–5078
137. Lu M, Liu J, Li Q, Zhang M, Liu M et al (2019) Angew Chem Int Ed 58(36):12392–12397
138. Fu Z, Wang X, Gardner AM, Wang X et al (2020) Chem Sci 11(2):543–550
139. Cui J et al (2021) J Mat Chem A 9(44):24895–24902
140. Low J, Zhang L, Tong T, Shen B, Yu J (2018) J Catal 361:255–266
141. Xu D, Cheng B, Wang W, Jiang C, Yu J (2018) Appl Catal B Envir 231:368–380
142. Crake A, Christoforidis KC, Godin R, Moss B, Kafizas A, Zafeiratos S, Durrant JR, Petit C (2019) Titanium dioxide/carbon nitride nanosheet nanocomposites for gas phase CO_2 photoreduction under UV-visible irradiation. Appl Catal B Envir 242:369–378
143. Wang CC, Wang X, Liu W (2019) Chem Engin J
144. Li R et al (2014) Adv Mater 26(28):4783–4788
145. Crake A, Christoforidis KC, Kafizas A, Zafeiratos S, Petit C (2017) Appl Catal B Envir 210:131–140
146. Wang Z, Huang J, Mao J, Guo Q, Chen Z, Lai Y (2020) J Mat Chem A 1–30
147. Zhang W, Mohamed AR, Ong WJ (2020) Angew Chem Int Ed 59(51):22894–22915
148. Cao S, Shen B, Tong T, Fu J, Yu J (2018) Bioorg Med Chem Lett 28(21):2109–2115
149. Lizuka K, Wato T, Miseki Y, Saito K, Kudo A (2011) J Am Chem Soc 133:20863–20868
150. Varghese O, Paulose M, LaTempa T, Grimes C (2009) Nano Lett 9:731–737
151. Peng YP, Yeh YT, Shah SI, Huang C (2012) Appl Catal B 414–423
152. Lin W, Han H, Frei H (2004) J Phys Chem B 108:18269–18273

153. Asi MA, He C, Su M, Xia D, Lin L, Deng H, Xiong Y, Qiu R, Li X (2011) Catal Today 175:256–263
154. Teramura K, Iguchi S, Mizuno Y, Shishido T, Tanaka T (2012) Angew Chem Int Ed 51:8008–8011
155. Nguyen TV, Wu JCS (2008) Sol Energy Mater Sol Cells 92:864–872
156. Matsuoka S, Yamamoto K, Ogata T, Kusaba M, Nakashima N, Fujita E, Yanagida S (1993) J Am Chem Soc 115:601–609
157. Yanagida S, Kanemoto M, Ishihara K, Wada Y, Sakata T, Mori H (1997) Bull Chem Soc Jpn 70:2063–2070
158. Koci K, Matejka V, Kovar P, Lacny Z, Obalova L (2011) Catal Today 161:105–109
159. Rakibuddin M, Kim HY (2019) Beilstein J Nanotechnol 10:448–458
160. Irvine J, Eggins B, Grimshaw J (1990) Sol Energy 45:27–33
161. Li K et al (2016) ACS Catal 6:7485–7527
162. Fu J et al (2017) Small 13:1603938
163. Fu J, Jiang K, Qiu X, Yu J, Liu M (2020) Product selectivity of photocatalytic CO_2 reduction reactions. Mater Today 32:223
164. Vasileff A et al (2018) Chem 4:1809–1831
165. Low J et al (2018) J Catal 361:255–266
166. Chang X et al (2016) Energy Environ Sci 9:2177–2196
167. Hattori Y et al (2018) Mater Today 21:590–593
168. Ran J et al (2018) Adv Mater 30:1704649
169. Xu F et al (2018) Appl Catal B 230:194–202
170. Shi G et al (2018) Appl Surf Sci 427:1165–1173
171. Zheng Y et al (2017) Nano Energy 40:512–539
172. D'Alessandro DM et al (2010) Angew Chem Int Ed 49:6058–6082
173. Han B et al (2018) Angew Chem Int Ed 57:16811–16815
174. Karamian E, Sharifnia S (2016) J CO_2 Util 16:194–203

Semiconductor-Based Photocatalytic Oxygen Evolution Performance for Water Splitting: Light-Driven Energy Conversion and Storage

Habib Gemechu, Kebede Bekele, Woldesenbet Bafe, Prawesti Ambar, Rapita Astriani, Farghani Fariz, Farah Meutia, and Riski Titian Ginting

Abstract Global energy demand continues to rise due to population expansion and economic success. As a result, annual population increases require high energy demand to provide essential needs and perform daily processes. To achieve sustainable development, energy sources ought to be reliable and have slight environmental problem. Photocatalytic water splitting by producing H_2 is an alternative approach to replace the traditional fossil fuels and providing environment friendly energy storage. Design in photocatalysts is essential to achieve economically applicable photocatalytic water splitting with higher efficiency. Sustainable hydrogen production would be a crucial first step toward powering fuel-cells. In addition, implementation on a larger industrial scale for attaining mature technological operations to lower N_2 and CO_2 emissions while producing desired products is achieved. The main challenge is harnessing solar radiation to produce hydrogen from water. In the coming decades, it is anticipated that wind-powered electrolysis and photovoltaics will be prioritized for producing renewable hydrogen. However, integrating catalysis with light harvesting for photoelectrolysis has various benefits. Among those benefits, thermal management and employment of lower current densities are preferable. Nowadays, semiconductor photocatalysis using solar energy is considered as a prospective way for disbanding the environment pollution and the global energy crises. It has been thoroughly studied in a variety of photocatalysis domains, including the reduction of pollution, the production of H_2 and O_2, etc. The recent rising attention toward photocatalytic water splitting can be attributed to four-electron reaction mechanism.

Keywords OER · Water splitting · Photocatalysis · Energy · Storage

---

H. Gemechu · K. Bekele · W. Bafe · P. Ambar · R. Astriani · F. Fariz · F. Meutia
Department of Materials Science and Engineering, National Taiwan University of Science and Technology, Taipei 10672, Taiwan, ROC

R. T. Ginting (✉)
Department of Electrical Engineering, Universitas Prima Indonesia, Medan, Indonesia
e-mail: titiangt@unprimdn.ac.id

H. Abdullah (ed.), *Photocatalytic Activities for Environmental Remediation and Energy Conversion*, Green Energy and Technology,
https://doi.org/10.1007/978-981-19-6748-1_5

1 Introduction

Energy is required worldwide to provide basic needs like domestic activities and support high energy for industrial demands. The quality of energy in the development process is a critical matter. For sustainable development, energy sources are required to be safe and have less negative impact on the environment. Economic and social growth on the long term depends on the reliability and affordable access to the energy resources [1]. Global energy demand is still increasing as a result of population increase and economic success. The increasing energy crisis and environmental pollution due to excessive utilization of fossil fuels is detrimental to climate changes. Dependency on fossil fuels is urgently sought to be minimized by conducting research on an extended scale to explore alternative renewable and clean alternative energy resources. In addition to the renewable energy source, solar energy is widely abundant for tackling environmental concerns, because it is free and environmentally friendly. Solar fuel production from sunlight captured by man-made photocatalysts is both practical and highly desirable [2].

Researchers have studied solar energy conversion and storage techniques in depth in order to make the most of the massive quantity of energy of 4.3×10^{20} J h^{-1} that hits the surface of the Earth. Solar energy can be utilized for chemical conversion for energy storage or converted into electricity. Photocatalytic water splitting is a potential and environmentally friendly method of storing solar energy in the form of H_2. When used in fuel cells to produce electricity, H_2 is a green source of energy because the resultant byproduct is water. Additionally, hydrogen has a high energy density, ease of storage and transport with current technologies. Photocatalytic water splitting is conducted in an aqueous solution containing a photocatalyst to capture photonic energy and induces the water splitting process. Researchers have investigated and reported various materials that are capable of overcoming the thermodynamic and kinetic obstacles needed for photoconversion of solar energy via water splitting. Suitable energy band gap with appropriate CB and VB values, high resistivity of photocorrosion, and an adequate charge transfer system are all criterions for an effective photocatalyst [3].

Solar energy can be converted and stored into various beneficial forms of energy. It can be used in chemical fuel conversion via H_2 through solar water splitting, CH_4 production via solar-driven CO_2 reduction, and ammonia synthesis from N_2. Besides, it also generates electricity (photovoltaic effect) and heat production (thermal collectors or phase changing materials). H_2 fuel is an excellent approach as a green chemical energy carrier due to its high specific energy with byproducts containing only heat and water. The most promising method for generating H_2 is photocatalytic water splitting via solar energy. However, production of highly effective photocatalysts still faces significant technical challenges. Solar light-driven photocatalysts stand out as a solution to the recent energy and environment crisis. TiO_2 photoelectrode was first found to be capable of decomposing H_2O into H_2 and O_2 under UV irradiation in 1972. Since then photocatalytic semiconductors have gained extensive attention toward environmental purging and energy conversion [4]. The photocatalytic process consists typically of 3 steps:

(i) Electron–hole pairs formation in the photocatalyst in response to light irradiation ($h\nu = E_g$);
(ii) The migration of the photogenerated species from the bulk to the surface of the photocatalyst; during this step some of the generated species are recombined; and
(iii) Non-recombination electrons and holes will interact with adsorbed reactants by either reduction or oxidation processes. Overall, photocatalytic water splitting ($H = 285.5$ kJ mol^{-1}) is thermodynamically hindered. Hydrogen evolution reaction (HER) and oxygen evolution reaction (OER), both two half reactions, sum up the Gibbs free energy via photon energy. Photocatalysts are irradiated with light-induced reduction and oxidation reactions in which the equations are as follows:

$$1.\ 2H^+ + 2e^- \rightarrow H_2 \quad \text{(reduction reaction)} \tag{1}$$

$$2.\ 4OH^- + 4h^+ \rightarrow O_2 + 2H_2O \quad \text{(oxidation reaction)} \tag{2}$$

OER in which $4e^-$ are transferred in the water splitting process is the rate determining step with a higher reaction barrier than HER. Bandgap values of the photocatalysts should be higher than 1.6 eV for enabling multi proton/electron processes of water oxidation. The bandgap value is essential to overcome the sluggish kinetics, which is referred to as the multistep reaction of the existing intermediates during the O_2 production process. The pH of the medium has an influence on the OER, and the equations are as follows:

$$2H_2O \rightarrow O_2 + 4e^- + 4H^+ \quad \text{(acidic condition)} \tag{3}$$

$$4OH^- \rightarrow O2 + 4e^- + 2H_2O^+ \quad \text{(alkaline condition)} \tag{4}$$

OER efficiency is influenced by various parameters such as illumination absorption, photogenerated charge separation, and the catalytic surface reaction, which are the three fundamentals of a photocatalytic process.

1.1 Water Splitting for Energy Conversion and Storage

Fossil fuels have been indiscriminately used to convert the chemical energy contained in these fuels into heat and power, thus dominating the global energy landscape, despite rapid depletion and the impact of carbon dioxide emissions on the Earth's climate. Since the 1970s, interest in renewable energy sources like wind, solar, and biomass has increased, leading to the implementation of several strategies, many of which are ideal carbon neutral, meaning that only few CO_2 is released into the environment [5]. Limitations in energy storage technology and the degree of integration with the power grid system prevent these intermittent energy sources from being used to its full potential. An appealing approach to ensure a continuous, robust energy supply is the conversion of surplus electrical energy to chemical bonds (hydrogen gas, methane, and liquid ethanol) via electrochemical or photo-electrochemical conversion by renewable electricity or direct sunlight through water splitting as shown in Fig. 2. In this scenario, much of research interests focus on the well-known water splitting reaction [6]. This reaction can be described as two half-cell reactions as shown in Fig. 1; the oxygen evolution reaction (OER), as follows:

$$2H_2O \rightarrow 4H^+ + 4e + O_2 E^\circ = 1.23\,\text{V vs. NHE} \quad (5)$$

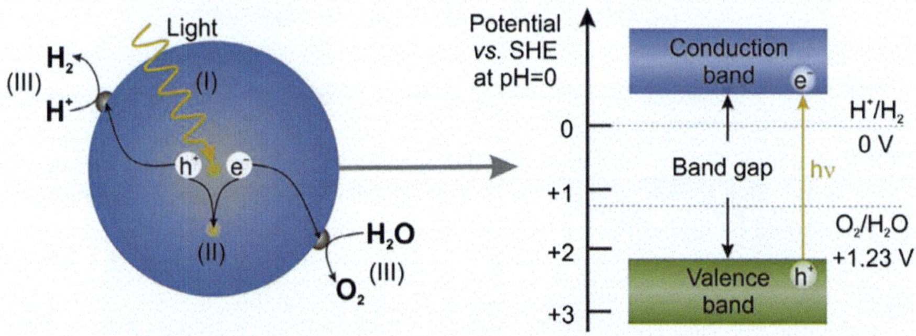

Fig. 1 The main processes and the principle of photocatalytic water splitting in semiconductor photocatalysts [2]

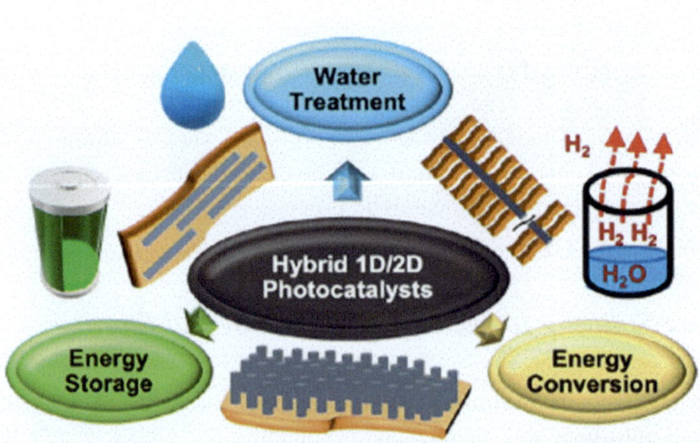

Fig. 2 Hybrid Photocatalyst for light harvesting water splitting for energy conversion and storage application

As well as the hydrogen evolution reaction:

$$\text{HER}, 2\text{H}^+ + 2\text{e} \rightarrow \text{H}_2\text{E}^\circ = 0.00\,\text{V vs. NHE} \quad (6)$$

$$\text{Overall}: \;\; 2\text{H}_2\text{O} \rightarrow 2\text{H}_2 + \text{O}_2 \;\; \text{E}^\circ\text{cell} = -1.23\,\text{V};\; \text{DG} = 475\,\text{kJ/mol}$$

In either case, a suitable catalyst is required to lower the huge overpotential barrier, hence speeding up the electrochemical kinetics. A four proton-coupled electron transfer process and the formation of oxygen–oxygen bonds, both of which are necessary for the reaction, are required for the catalysis of OER, which is more complex and kinetically challenging, in contrast to the HER, which produces fuel relatively simple to implement. Typically, electrocatalysts (IrO_2, RuO_2, Ni oxides, etc.) are used in conventional electrochemical cells for water splitting to reduce overpotential and Tafel slope and maximize energy conversion efficiency. However, the onset potential at an electrocatalyst is always higher with respect to the theoretical redox potential of water oxidation (1.23 V vs. RHE) due to the thermodynamic energy barriers of this electrochemical reaction [7].

Photoreduction of excessive CO_2 emissions with greenhouse effects into combustible CO and CH_4. However, due to the strong linear structure of CO_2, obtaining a highly efficient CO_2 reduction under visible-light irradiation is difficult. Furthermore, an effective CO_2 reduction photocatalyst should have appropriate redox potentials, high charge separation efficiency, and a long lifetime. The range of light absorption can be expanded using 1D/2D heterostructured photocatalysts, which can also affecting separation and migration of photoinduced charge carriers [4]. For instance, through a spatial distribution heterojunction made possible by the in situ growth of $ZnIn_2S_4$ (ZIS) nanosheets on g-C_3N_4 micro tubes (TCN), highly efficient charge separation, rapid interfacial charge transfer, strong light absorption, and larger CO_2 adsorption are required for photocatalytic CO_2 reduction. The vertically aligned $ZnIn_2S_4$ nanosheets on both the outer and inner surfaces of TCN generate a hierarchical core/shell structure, which increases the conversion efficiency of CO_2 to 1453 and 863 μmol g^{-1} h^{-1} of CO and H_2, respectively, which are 4.3 and 5 times higher than those of a bulk CN/ZIS heterojunction. Many reports such as the conversion of CO_2 to CH_4 or CO and H_2 production are available in these contexts. For instance, it is critical to collect solar energy during the day and then use it with a high conversion efficiency at night. Because phase-change materials have high phase-change enthalpies, acceptable phase-change temperatures, and modest volume expansion/shrinkage, they are frequently employed for solar-thermal energy storage [8].

1.2 Fundamental Processes and Mechanism in Photocatalytic Overall Water Splitting

Water splitting is a thermodynamically unfavorable reaction that needs more energy than 1.23 eV to proceed. Because every oxygen molecule needs four holes, water oxidation is more difficult than H_2 reduction in an artificial manner and five times slower than the hydrogen evolution process. There are several crucial phases involved in photocatalytic water splitting. Understanding the mechanism of each phase and identifying the important parameters impacting the process efficacy are critical when creating photocatalysts for this process. H_2 is a carbon-free, clean fuel with a high specific enthalpy [9]. Previously, roughly 95% of the world's hydrogen fuel was mostly generated from natural gas, but now, by combining methane with the steam utilizing fossil fuel, hydrogen and carbon dioxide CO_2 may be produced. As a result, hydrogen generation from methane is a fossil fuel product rather than a sustainable energy source. The key to utilizing hydrogen generation is to develop an efficient H_2-preparation technology that does not require the use of fossil fuels. The method using photocatalytic materials to produce hydrogen has been demonstrated since 1972 according to the following reaction equation,

$$2H_2O \rightarrow 2H_2 + O_2 \tag{7}$$

$$\Delta E^\circ = 1.23\,\text{ev},\ \Delta G^\circ = +237.2\,\text{kJ/mol}$$

This photocatalytic process is driven by a thermodynamic "uphill reaction" of solar energy with a large change of $\Delta G^0 = 1.23$ eV (237.13 kJ mol^{-1}) per photon.

There are three major steps to an overall photocatalytic split water for a semiconductor photocatalyst:

(1) the absorption of photons with the right energy on a semiconductor's surface, which can promote electron transfer from the valence band position to the conduction band position and generate electron–hole pairs (e^-/h^+);
(2) The separation and migration of charge carriers in a brief period of time on the material's surface; and
(3) The impingement between the corresponding free electrons and holes on the system surface to complete the water splitting reaction. On the other hand, the recombination of the free electrons and the excited holes may readily happen in the second step, giving rise to a poor photocatalytic activity of the nanocrystal surface. Therefore, even at a low solar energy flux density, the appropriate bandgap and the matching levels of the conduction and valence bands might facilitate light absorption on the surface of the semiconductor photocatalyst. In order to drive the electrons to the empty CB position from the filled valence band position of the semiconductor photocatalyst, the absorbed photon must have a higher energy than the corresponding bandgap of the semiconductor [10].

Subsequently, the excited electrons and holes pair (e^-/h^+) is formed according to the following expression,

$$\text{Photon activation: Semiconductor} + hv \rightarrow e^- + h^+,$$
$$\text{Oxygen adsorption}: (O_2)_{ads} + e^- \rightarrow O_2^{\cdot -},$$
$$\text{Water ionization}: H_2O \rightarrow OH^- + H^+,$$
$$\text{Superoxide protonation}: O_2^{\cdot -} + H^+ \rightarrow HOO^{\cdot}.$$

The bandgap energy should be more than 1.23 V from a thermodynamic perspective, and the band positions should be below 1.23 V versus normal hydrogen electrode (NHE) to share electrons from water O^{2-} anions and above 0.0 V versus NHE to decrease water H^+ ions. After light has been absorbed, electrons are moved to the conduction band and holes are simultaneously formed at the valence band. In order to carry the redox processes, photogenerated charge carriers must recombine in the bulk or on the surface. In the presence of metal particles with a larger work function than the oxide semiconductor, excited electrons may be pumped in order to speed up charge separation and convert H^+ ions into atomic hydrogen, which will subsequently mix with other molecules to form H^{2+}. At the energy level of the valence band, electrons are abstracted from O_2- anions of water to close the cycle. Accordingly, the reaction products desorb from the catalyst surface and are transferred to the medium to complete the overall process [11].

The fundamental concept of photocatalytic water splitting in semiconductor photocatalysis is depicted in Fig. 3. Absorption of photons creates pairs of electrons and holes in the first step (I) which will be recombine or separate and move to the reaction sites (II). Finally, water molecules react with electrons or holes to produce H_2 or O_2 (III). The highest occupied molecular orbital (HOMO, top of the valence band in a periodic crystal perspective) is excited to the lowest unoccupied molecular orbital (LUMO, bottom of the conduction band), leading to the generation of excitons, or excited electron–hole pairs. This step indicated that incident photon gives enough energy above the optical gap (also known as the absorption onset) of the irradiated photocatalyst. The excess energy known as the exciton binding energy, is converted into free electrons and holes (where "free" refers to non-bonded excitons). The band gap is defined as the optical band plus the exciton binding energy. At the same time, the potential of the CB should be more negative than the reduction potential of H_2O/H_2 (0 V vs. NHE, pH = 0), while the VB should be more positive than the oxidation potential of O_2/H_2O (1.23 V vs. NHE, pH = 0). Therefore, the theoretical minimum band gap for water splitting is 1.23 eV. The width of the band gap and the levels of the CB and VB are extremely important for semiconductor photocatalysts, which are essential for the second and the third step. These free charge carriers can drive redox reaction in photocatalytic water splitting, but also reform excitons in a process commonly referred to as electron–hole recombination [2] (Scheme 1).

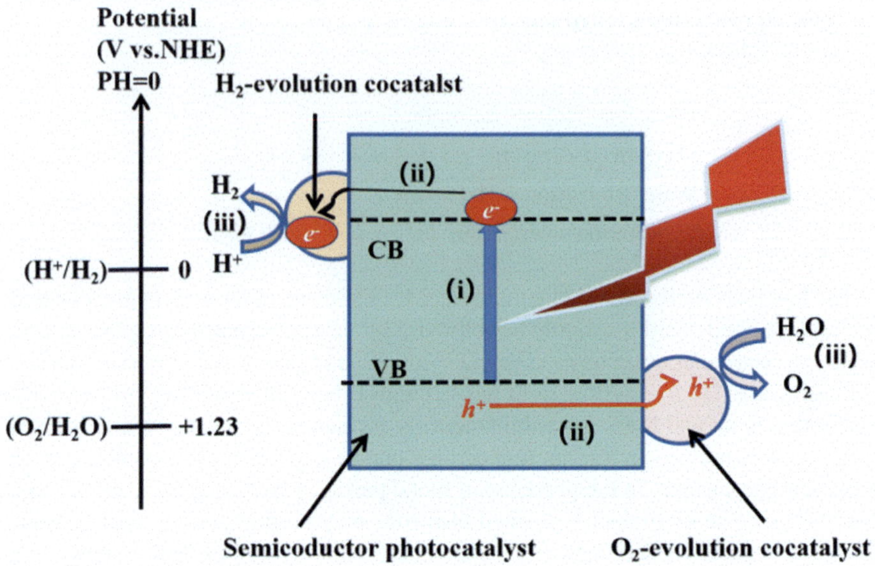

Fig. 3 Schematic diagram of the three major steps of the overall photocatalytic water splitting on a semiconductor photocatalyst

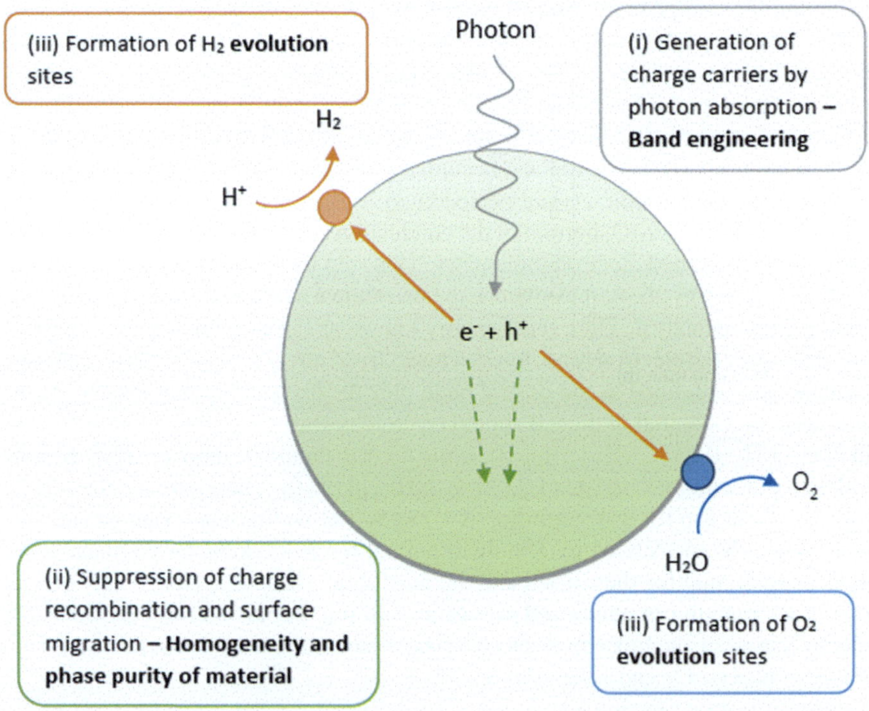

Scheme 1 Schematic illustration of three primary steps of photocatalytic water splitting [12]

1.3 Fundamental of Photocatalysis Based on Semiconductors

In general, the photocatalyst is activated by visible light, which leads to the generation of electron–hole pairs that are formed by the migration of generated electrons from the valence band (VB) to the conduction band (CB) of a semiconductor photocatalyst [13]. In a crystal, the valence band VB, the highest occupied band, the conduction band CB, and the lowest empty band are separated by a band gap in the so called region of forbidden energies. The bandgap model describes a photocatalytic system based on a semiconductor in which these two bands are separated by a band gap, a region of forbidden energies [14]. An electron that is excited to the CB leaves a hole (h^+) in the VB when the incident energy is equal to or larger than the band gap of the semiconductor, as depicted in Fig. 4.

Lower level of CB that is located at a more negative potential than the electrochemical potential of the desired reaction is the photoexcited electron used in a reduction reaction with an electron acceptor. For instance, the reduction of protons to hydrogen, production of an O_2-ion radical, and CO_2 reduction. The photo-generated hole in the semiconductor VB can also engage in an oxidative reaction with an electron donor with oxidation potential that is more negative than the maximum VB. In systems designed for the production of green fuel and other environmental applications, the semiconductor properties is the most crucial part (Fig. 5).

A wide range of semiconductors are now being synthesized and investigated in order to increase their performance, while novel architectures are being prepared and developed at the same time. The photocurrent density generated by the photocatalyst device is the performance benchmark; it should be greater than 10 mA/cm^2 in order to be useful in an industrial application. However, the device stability over a long period of time and continuous operation are required. Semiconductors can be categorized as follows in this context: (i) single crystal versus nanostructured and (ii) narrow against wide band gap. Single crystals (Si, GaAs and GaP) feature a high diffusion length and high mobility, while nanostructured materials exploit the small distance necessary to achieve a highly efficient rate of charge transfer (e−/h^+). A large band

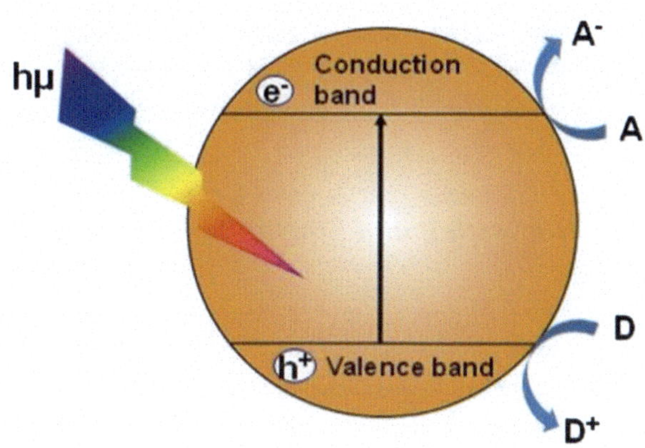

Fig. 4 Schematic illustion of electron–hole generation after adsorbed photon energy (hv) in an inorganic photosystem [14]

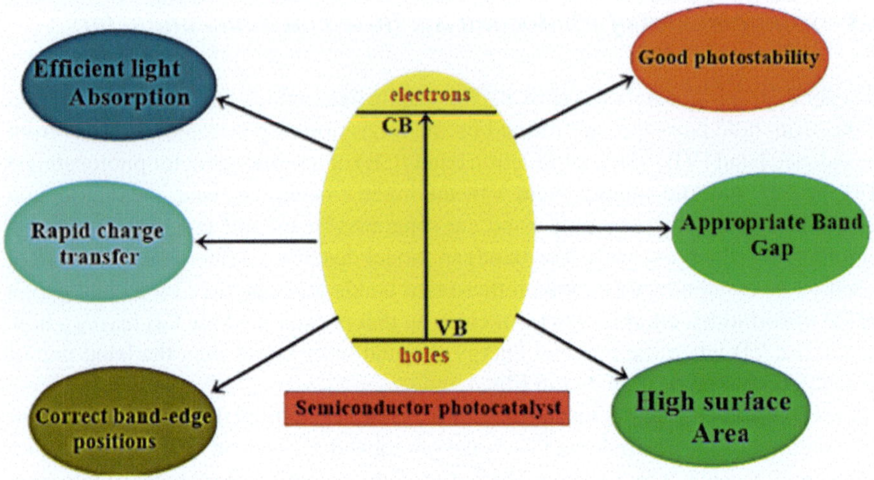

Fig. 5 The foremost properties of a noble photocatalyst

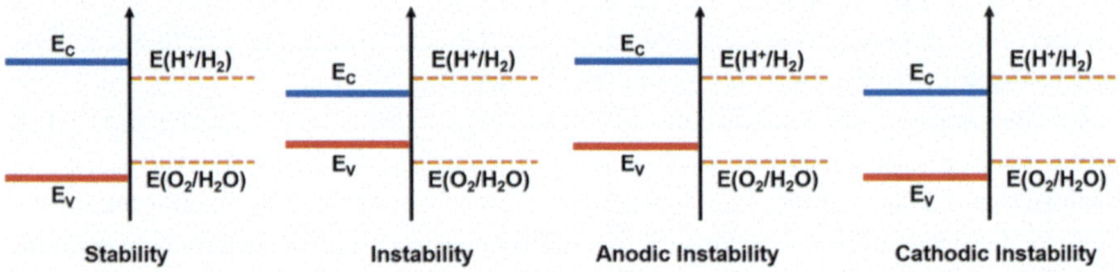

Fig. 6 Comparison in stability of semiconductor for the water splitting process's redox potential

gap (Eg) absorbs a narrow portion of the light spectrum, leading to poor absorption, whereas a tiny band gap absorbs a greater portion of the spectrum. Additionally, a wider band gap tends to resist photocorrosion, whereas a narrower band gap is more prone to it. The Ec and Ev bracket are the E(H/H_2) and the E(O2/H2O), respectively, which sums up the optimal conditions for semiconductor stability. Photocorrosion will occur due to several issues [15]. Figure 6 displays the stability of semiconductors in relation to the redox potential of water splitting.

1.4 Harvesting and Storing Solar Energy

The ultimate solution to the ever-increasing global energy need is to harvest electricity directly from photon energy utilizing photovoltaic (PV) systems. For improved usage of the copious but occasionally available sunlight, simultaneous solar energy conversion and storage is gaining a lot of attention. Photoelectrodes drive non-spontaneous reversible redox reactions in solar-powered redox cells (SPRCs), which can give energy via the related reverse reactions. This is a cost-effective and promising method for direct solar energy collection and storage. However, the lack of photoelectrodes having both high conversion efficiency and high durability becomes a bottleneck that hindered practical applications of SPRCs [16].

Due to its potential to address the world energy crisis, energy harvesting technology with a clean and renewable energy source has attracted a lot of interest. Solar energy is the most attractive among the renewable energy sources as compared to wind, geothermal, hydropower, and many others, since it can provide sustainable power by using hydrogen gas as an energy carrier. Practically, substantial advancements have been made in the photoelectrochemical (PEC) water splitting method of producing hydrogen from water using solar energy. The photoelectrodes (photoanode or photocathode), electrolyte, and light source make up the majority of the PEC water splitting system. Generally, light-absorbing semiconductors have been used as the photoelectrode and the electrode materials are the main research topics in the field of PEC water splitting [17] (Fig. 7).

Solar and wind energy can be distributed locally via the off grid system, thus reducing the need for costly transmission lines while also improves energy security. At larger scales, it is possible to harvest solar energy for local usage. Utility scale PV can enable micro-grids that can deliver dispatchable electricity to nearby towns when combined with other energy sources (such as wind, chemical fuels, and hydropower) and energy storage devices. Due to the high cost of maintaining transmission lines, this strategy can be effective, especially for isolated communities. An alternative for families to install PV modules on their rooftops and use batteries to store energy for use later, in the event that solar generating is no longer feasible (i.e., load shifting of daytime generation to night-time).

However, the solar-powered gadgets or appliances that can be used by following procedures:

Fig. 7 Solar Harvesting energy system (*Source* Linear Technology)

(i) directly gather solar energy and.
(ii) electrochemically store that energy is another intriguing potential.
(iii) supply electricity to power a load.

New opportunities such as low power electronics applications, remotely powered sensors, electrochromic smart windows, remote antenna tracking devices, and self-charging wearable electronics may arise from the seamless integration of solar powered gadgets. The terms photorechargeable energy storage systems, hybrid solar energy conversion/harvesting and storage systems, and solar batteries have all been used to refer to these integrated solar energy conversion and storage systems. Since the early 1970s, there have been reports of hybrid systems that can power electronics while also collecting solar energy and storing it electrochemically [18].

2 Photocatalysts for Oxygen Evolution

Numerous photocatalysts (e.g., ZnO, CdS, WO_3, and $SrTiO_3$) have been intensively researched for water splitting applications since the pioneering work by Fujishima and Honda on water splitting to create H_2 and O_2 under UV irradiation based on TiO_2 photoelectrode. Despite extensive investigation, majority of the photocatalysts produced are not suitable for visible-light-induced water splitting due to inadequate light utilization or photostability. Furthermore, the high recombination rate of photoexcited electron–hole pairs generated by an unfavorable band structure severely limits their practical applications [4]. Alternatively, photocatalysts with a band gap of approx. 1.23 eV and no photocorrosion are suitable for overall water splitting. To reduce the recombination of photogenerated electrons and holes during water splitting, high crystallinity and smaller particle size are preferred.

As water splitting catalysts, metal oxides, sulfides, nitrides, and phosphates with d^0 and d^{10} metal cations have been used. Perovskite materials made of Group I and Group II metals, as well as some lanthanides, can be utilized to catalyze photochemical water splitting. Figure 8 summarizes the band structure of various types of semiconductors in relation to the redox potentials of water splitting. To improve solar energy efficiency, modification of photocatalysts by doping with some transition metal cations such as Ni^{2+}, Cr^{3+}, and V^{5+} can help to increase the visible light response. To inhibit the energy decreasing backward reaction of water splitting and increase the hydrogen production yield, suitable cocatalysts including RuO_2, NiO, Au, and Pt were widely used.

2.1 *Tio_2-Based Photocatalysts*

Since the first report by Fujishima and Honda in the late twentieth century, the TiO_2-based photocatalyst has emerged as a viable material for renewable fuel production

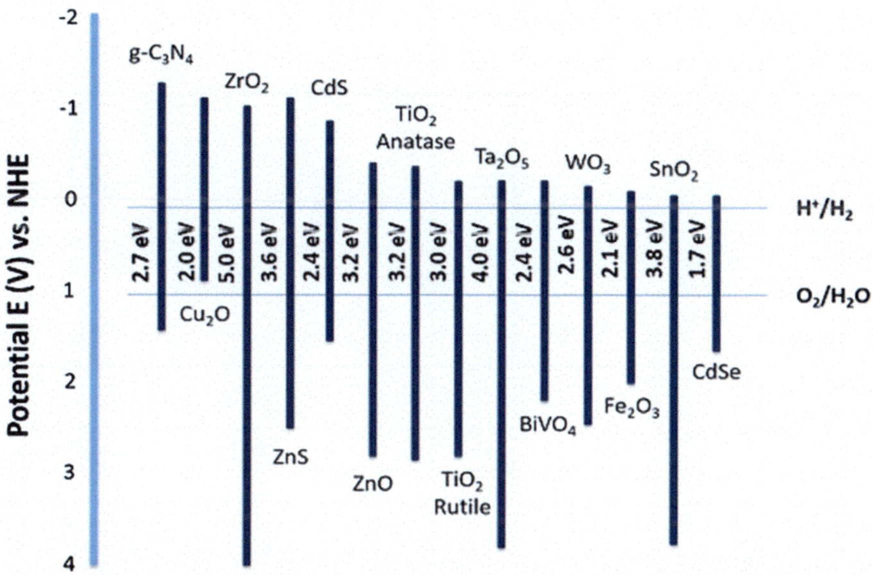

Fig. 8 Various band structures of semiconductors with the redox potentials for water splitting

and environmental remediation. As a result of its unique qualities (e.g., remarkable electrical properties, stability, eco-friendliness, and cost-effectiveness), TiO_2-based photocatalysts have been the most researched materials for photocatalytic applications in a recent study [13]. Improvements in three different pathways have been prioritized in the development of TiO_2-based photocatalysts: (i) the development of visible light-active TiO_2; (ii) the enhancement of photogenerated electron–hole separation using anatase–rutile homojunction; and (iii) the use of TiO_2-based nanocomposite. The gradient doping with nonmetallic heteroatoms in TiO_2 has been developed to improve the electronic structure with extensive efforts in elemental doping. When photons with energy ($E = hv$) equal to or greater than the band gap (3 eV for rutile and 3.2 eV for anatase phase) are absorbed by a semiconductor such as n-TiO_2, the photoactivity of the semiconductor is initiated, and a small percentage of electrons and holes are formed [19]. Small fraction of charge carriers can only reach the surface, which react with adsorbed molecules or ions, depending on the density of states (Fig. 9).

Despite its chemical stability in electrolyte solutions, n-TiO_2 has a number of drawbacks, and is able to only absorb 3% of solar irradiation. In order to improve the usability of n-TiO_2 as a water splitting catalyst, the improvement of its photosensitivity in the UV region and simultaneous shift in the spectral response to the visible region (400 nm–800 nm) have been performed. The best approach to meet this challenge is by doping n-TiO_2 with different elements (metal and nonmetals) and thus decrease its photo threshold energy. Transition metals have been used intensively as impurities dopands for n-TiO_2. Mixed results were reported for improving

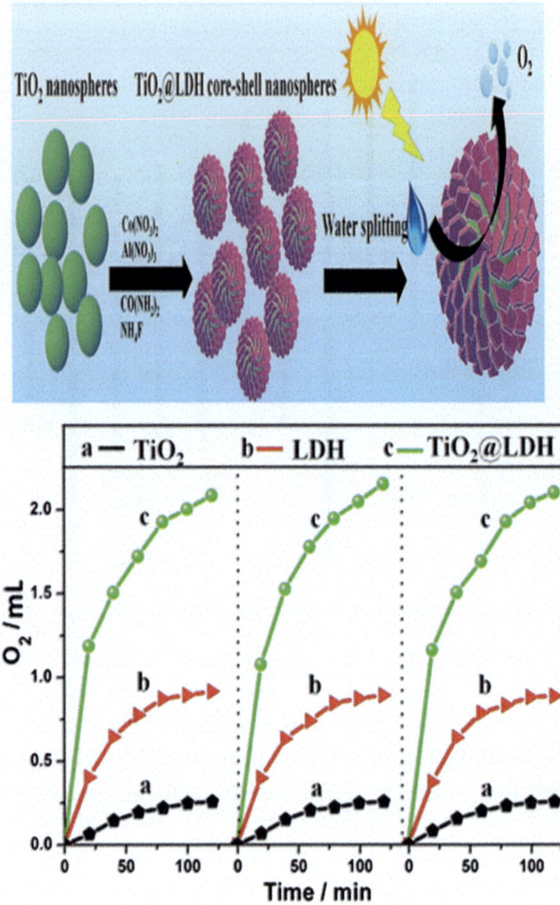

Fig. 9 **a** Schematic illustration of TiO_2@CoAl-LDH core–shell nanospheres toward O_2 from splitting of water (I), volume of O_2 generation as function of irradiation time ($\lambda > 200$ nm) in three consecutive cycles for a) TiO_2 nanospheres; **b** CoAl-LDH nanoplatelets; **c** TiO_2@LDH nanospheres, respectively (II)

the spectral response of metal doped Titania [20]. Dou et al. used a simple two-step method to make TiO@CoAl-LDH core–shell nanospheres: hydrothermal synthesis of TiO_2 hollow nanospheres followed by in situ growth of CoAl LDH shell (Fig. 9a). Under visible light illumination, the resultant material has an extraordinary high photocatalytic activity for oxygen evolution from water splitting, with O_2 generation rates of 2.34 and 2.24 mmol h^{-1} g^{-1}.

The broad spectrum absorption and strong electronic coupling between TiO_2 core and CoAl-LDH shell allow for improved solar energy and rapid electron–hole separation, resulting in significant photocatalytic activity toward oxygen generation. Furthermore, the TiO_2@CoAl-LDH photocatalyst produced has good recyclability and stability. TiO_2@CoAl-LDH nanospheres were prepared via two-step hydrothermal synthesis with TiO_2 hollow nanospheres as the core and in situ growth of CoAl-LDH as the shell. Figure 10 shows the XRD pattern of the resulting TiO_2@CoAl-LDH, with pristine TiO_2 and CoAl-LDH as reference samples. To better understand the interaction between the CoAl-LDH shell and TiO_2 core in Fig. 11, a narrow-scan X-ray photoelectron spectroscopy (XPS) as shown in Fig. 12 was used. For the CoAl-LDH sample, the binding energies at 781.3 and 797.3 eV correspond to Co $2p_{3/2}$ and Co $2p_{1/2}$, respectively. The appearance of satellite and 803.3 eV implies the presence of a high-spin divalent state of Co^{2+} in this sample. For the TiO_2@LDH nanospheres, however, the binding energies of Co $2p_{3/2}$ and Co $2p_{1/2}$ shift to 781.7 and 797.6 eV, respectively (Fig. 12a).

The binding energies of Ti $2p_{3/2}$ and Ti $2p_{1/2}$ of TiO_2@LDH nanospheres demonstrate a negative shift compared with those of pristine TiO_2 microspheres (Fig. 12b). In addition, the increased and decreased binding energies of Co 2p and Ti 2p indicate an obvious electron transfer from the LDH shell to the TiO_2 core. The strong electronic coupling between LDH and TiO_2 would probably accelerate the electron–hole separation [21]. Water-splitting on TiO_2: mechanism, reduction, and oxidation are fundamental chemical processes involved in water splitting. Previously, in order to produce H_2 and O_2, the semiconductor positions CB and VB must both fall within the potential windows of water reduction and oxidation potential, respectively. When light strikes TiO_2, electrons in the VB travel to the CB and holes are created. Excited electrons promote the production of hydrogen, whereas holes aid in the production of oxygen (Fig. 13).

The valence band of TiO_2 is more positive than E^0oxid of O_2/H_2O (1.23 V vs. NHE, pH = 0), while the conduction band is more negative than E^0 red of H^+/H_2 (0 eV vs. NHE, pH = 0). However, there are two fundamental drawbacks to TiO_2 materials. Firstly, rapid charge carrier recombination cause undesired energy to be released. Secondly, the poor visible light harvesting ability such as TiO_2 can only be

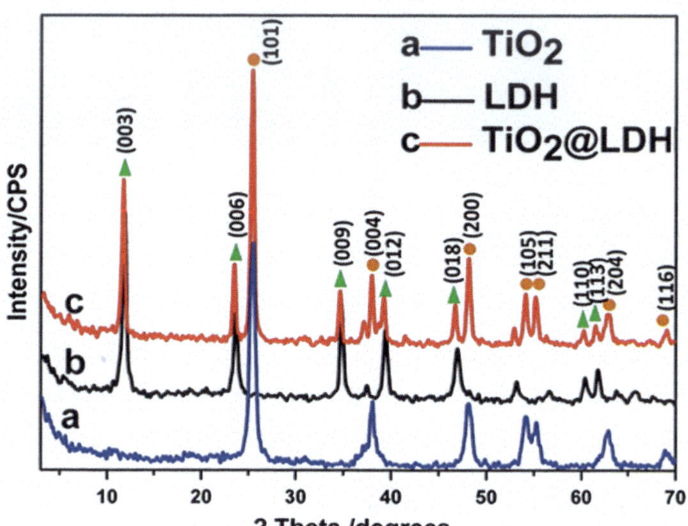

Fig. 10 XRD pattern of TiO_2@CoAl-LDH, with pristine TiO_2 and CoAl-LDH [21]

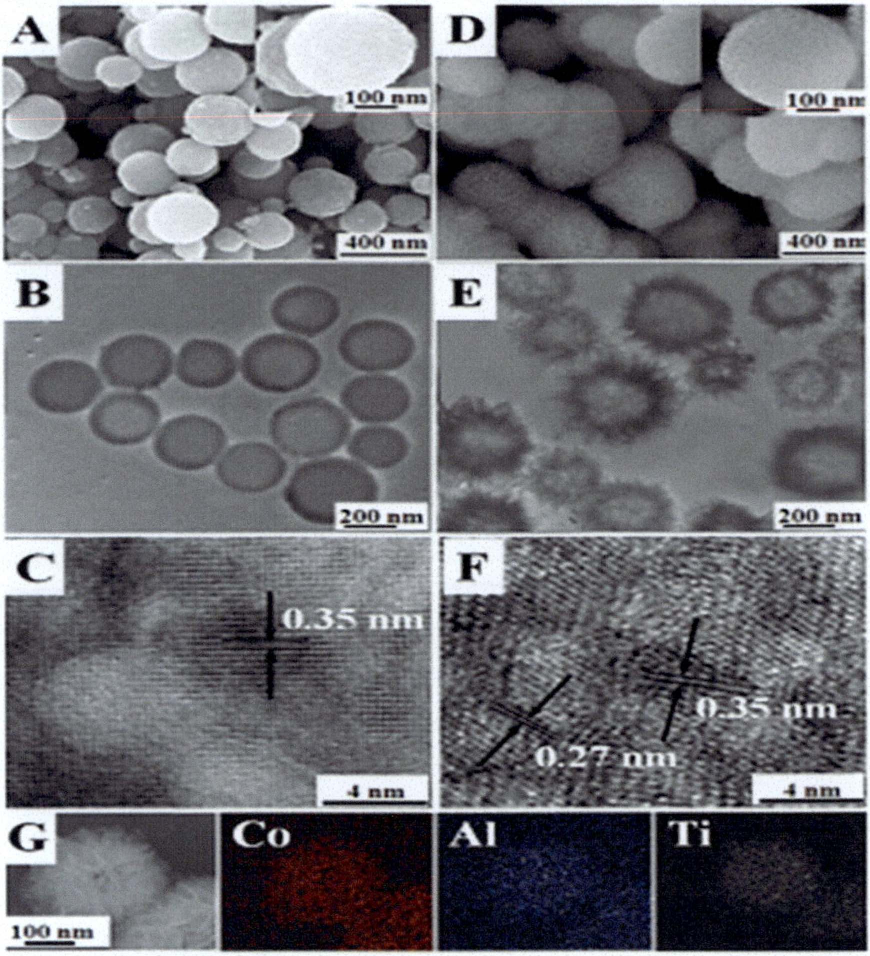

Fig. 11 **a** SEM, **b** TEM, and **c** HRTEM image of TiO_2 hollow nanospheres, **d** SEM, **e** TEM, and **f** HRTEM image of TiO_2@LDH core–shell nanospheres. g) SEM image and EDX mapping for an individual TiO2@LDH core–shell nanosphere [21]

excited by UV light due to its wide band gap of 3.0–3.2 eV, which only covers 5% of the solar spectrum. To enable visible light harvesting and prevent photogenerated electron–hole pairs from recombination, proper modification is required [22].

2.2 $BiVO_4$-Based Photocatalysts

Bi-based oxides photocatalysis ($Bi_4V_2O_{11}$, Bi_2O_3-TiO_2, Bi_2WO_6, Bi_2O_3, Bi_2WO_6/Bi/Bi_2O_3, Bi_2O_3/Bi_2WO_6 [23]) is non-toxic, chemical corrosion resistant,

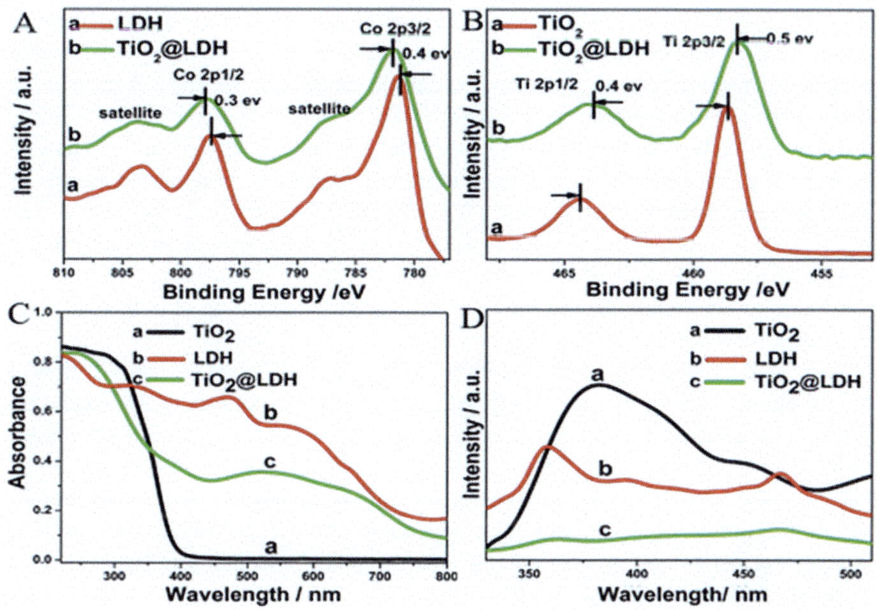

Fig. 12 **A** Co 2p and **B** Ti 2p orbital XPS spectra: **a** LDH nanoplatelets and **b** TiO_2@LDH nanospheres. **C** UV–vis DRS and **D** photoluminescence spectra: **a** TiO_2 nanospheres; **b** LDH nanoplatelets; and **c** TiO_2@LDH [21]

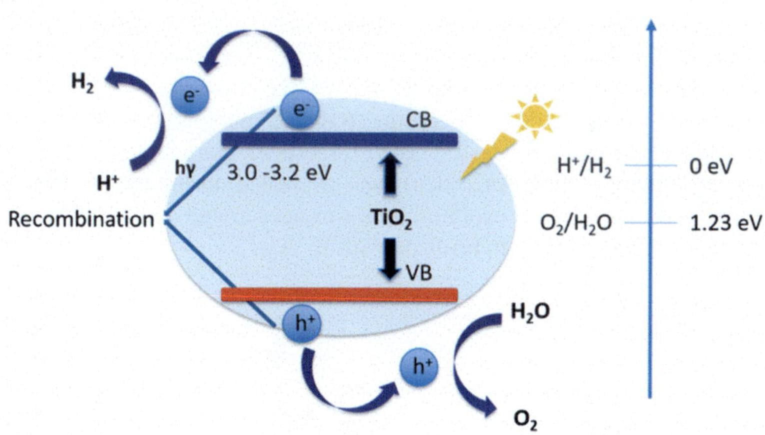

Fig. 13 Bandgap of TiO_2 taken from references (Copyright Elsevier License Number 5312530861756)

and visible light harnessing and is one of the most essential characteristics of bismuth vanadate, which makes it an effective photocatalyst. Because the energy from light absorption is used to trigger green photocatalytic processes, the latter is critical for prospective uses in renewable technology. Despite the fact that further research is needed to assess the benefits of $BiVO_4$, bismuth vanadate and its derivative materials are promising for environmental applications [24]. Due to its moderate bandgaps of 2.3–2.5 eV, excellent VB level, good photochemical stability and low-cost, the n-type semiconductor bismuth vanadate ($BiVO_4$) has attracted a lot of attention for the past few years for photocatalytic water oxidation. Monoclinic scheelite is $BiVO_4$'s most common crystalline phase, with a crystal structure made up of deformed VO_4 tetrahedrons and BiO_8 dodecahedrons as depicted in Fig. 14a–c.

Shang et al. developed a photocatalytic water oxidation system using a hybrid artificial photosynthesis system. The light harvester $BiVO_4$ and the water oxidation catalyst is a transitional metal complex ($M(dca)_2$, M = Co, Ni, DCA: dicyanamide) and the sacrificial electron acceptor is $S_2O_8^{2-}$ [25]. During the incorporation of $M(dca)_2$ in the system, the oxygen evolution activity increases. Under visible light irradiation, photocatalytic studies for oxygen evolution in $BiVO_4/M(dca)_2$ hybrid systems were performed in a completely aqueous solution containing $Na_2S_2O_8$ as a sacrificial electron acceptor ($\lambda > 420$ nm).

According to Fig. 15a, $BiVO_4/Co(dca)_2$ and $BiVO_4/Ni(dca)_2$ exhibited a higher performance than pure $BiVO_4$, which shows the catalytic role of $M(dca)_2$ in the water oxidation steps. Compared to pure $BiVO_4$, the photocatalytic oxygen evolution rate of 252.2 μmol/(h·g) during 6 h of reaction, the $BiVO_4/Co(dca)_2$ and $BiVO_4/Ni(dca)_2$ systems exhibited higher photocatalytic performances, with oxygen evolution rates of 508.1 and 297.7 μmol/(h·g) and TONs (vs. $M(dca)_2$) of 3.46 and 2.03, respectively. The $BiVO_4/Co(dca)_2$ had a photocatalytic oxygen evolution rate of 1.7 times higher than the $BiVO_4/Ni(dca)_2$ system. According to surface photovoltage (SPV) analysis and subsequent electrochemical testing, $BiVO_4/Co(dca)_2$ has a higher surface charge transfer. The stability of $BiVO_4/Co(dca)_2$ was evaluated during the 30-h photocatalytic experiment. The oxygen evolution rate was accelerated and the photocatalyst was stable when $Co(dca)_2$ was hybridized with $BiVO_4$.

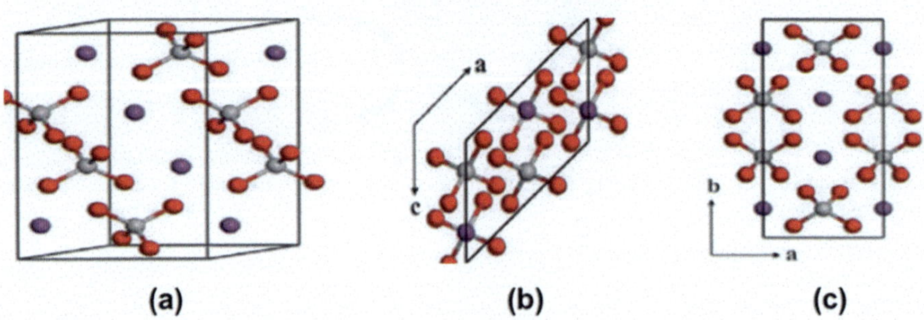

Fig. 14 **a** $BiVO_4$ crystal structure, **b** vertical view and **c** side view

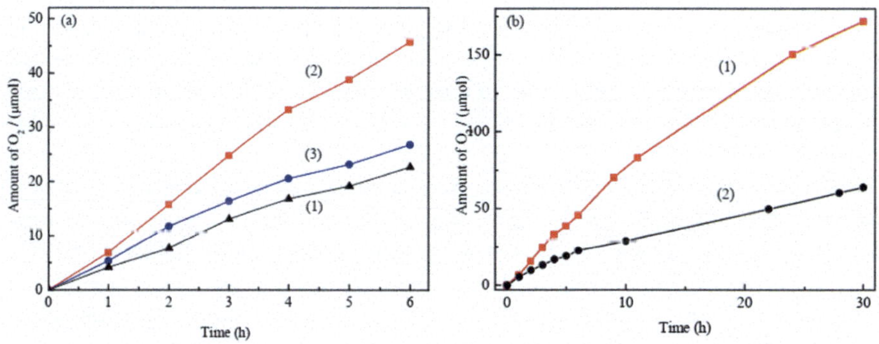

Fig. 15 **a** $BiVO_4$ Photocatalytic O_2 evolution 6 h for 1, $BiVO_4$/Co(dca)$_2$ 2, $BiVO_4$/Ni(dca)$_2$ (3) systems, **b** photocatalytic oxygen evolution for 30 h for the $BiVO_4$/Co(dca)$_2$ [25]

From Fig. 16b, $BiVO_4$/Co(dca)$_2$ had a very steady oxygen evolution rate of more than the 30 h photocatalytic reaction, yielding 171.95 mol of oxygen (TON = 13, vs. Co(dca)$_2$). $BiVO_4$/Co(dca)$_2$ had a substantially higher stability than previously reported $BiVO_4$/molecular catalyst photocatalytic systems that had activity less than 10 h. As a water oxidation catalyst, Co(dca)$_2$ is found to be more effective than Ni(dca)$_2$. The M(dca)$_2$ engineered $BiVO_4$/electrolyte interface energetics, as well as the M(dca)$_2$ catalyzed surface water oxidation, resulting in improved photocatalytic performance. The energy barrier for hole transfer to the surface of $BiVO_4$ from the bulk is reduced, thus promoting water oxidation kinetics.

The interfacial energetics in terms of interfacial band bending were investigated further using Mott-Schotky (M-S) measurements. The energy band of $BiVO_4$ bends upward at the $BiVO_4$/electrolyte contact because it is a typical n-type semiconductor. As a result, for water oxidation, there is a pushing force for hole transfer from the bulk to the surface. As indicated in Fig. 16b, the flat band potential of $BiVO_4$ was negatively shifted by 78 and 49 mV when Co(dca)$_2$ and Ni(dca)$_2$ were added, respectively.

As can be seen in Fig. 16d, a higher band bending lead to increase in the driving force for hole transfer, which promotes the water oxidation cycle. The Co(dca)$_2$ engineered $BiVO_4$/electrolyte interface exhibited a higher band bending than the Ni(dca)$_2$ engineered interface, as evidenced by the higher cathodic shift of the flat band potential in the $BiVO_4$/Co(dca)$_2$ system. This implies that the energetics of the Co(dca)$_2$ designed $BiVO_4$/electrolyte interface favored more efficient interfacial hole transfer. Electrochemical impedance spectroscopy (EIS) analysis indicates the charge of transfer behaviors at the $BiVO_4$/electrolyte interface. The semicircle diameter of the $BiVO_4$/M(dca)$_2$ Nyquist plot was smaller than that of pure $BiVO_4$ as can be seen in Fig. 16c. Smaller interfacial resistivity for charge carrier transfer with $BiVO_4$/electrolyte interface energetics engineered by M(dca)$_2$. The equivalent circuit for the results in Nyquist plots is shown in the Fig. 16c. Three sample charge transfer resistances (Rct) were in the following order: $BiVO_4$/Co(dca)$_2$ (507 Ω), $BiVO_4$/Ni(dca)$_2$ (1747 Ω), and $BiVO_4$ (5130 Ω). This revealed that the

$BiVO_4/Co(dca)_2$ system had the best OER kinetics, which was consistent with the photocatalytic performance. The inclusion of $M(dca)_2$ considerably promotes the photogenerated hole transfer from $BiVO_4$ to the electrolyte, while $Co(dca)_2$ provides interfacial charge transfer and thereafter catalyzed the OER more efficiently [25] (Fig. 17).

Wei et al. prepared $ZnO/BiVO_4$ nanoarchitectures to show the multi-electric field modulation technique for charge separation efficiency [26]. A seeded growth technique was applied for the directed development of ZnO photocatalyst on $BiVO_4$ by combining facet engineering with interfacial defect modification, resulting in a remarkable oxygen evolution rate of 68 mol h^{-1} (step i-iii in Fig. 18a). The electrical structure of 3-D nanoarchitectures was modified in the final phase (step iv in Fig. 18a) by adjusting the calcination atmospheric environment, resulting in interfacial oxygen vacancy-rich nanoarchitectures. Well-aligned nanorod arrays were successfully constructed onto the top (010) plane surfaces and isosceles tapezoidal (110) sides of $BiVO_4$ decahedrons, as evidenced by XRD and morphological studies as shown in Fig. 18b–d. From Fig. 19a, it can be observed that the O_2 evolution activities for all Ce-$BiVO_4$ samples are significantly improved compared to pristine $BiVO_4$.

The Ce-$BiVO_4$(C) sample showed the most O_2 evolution activity. The amount of O_2 evolution after 4 h of irradiation is approximately 4 times higher than that of

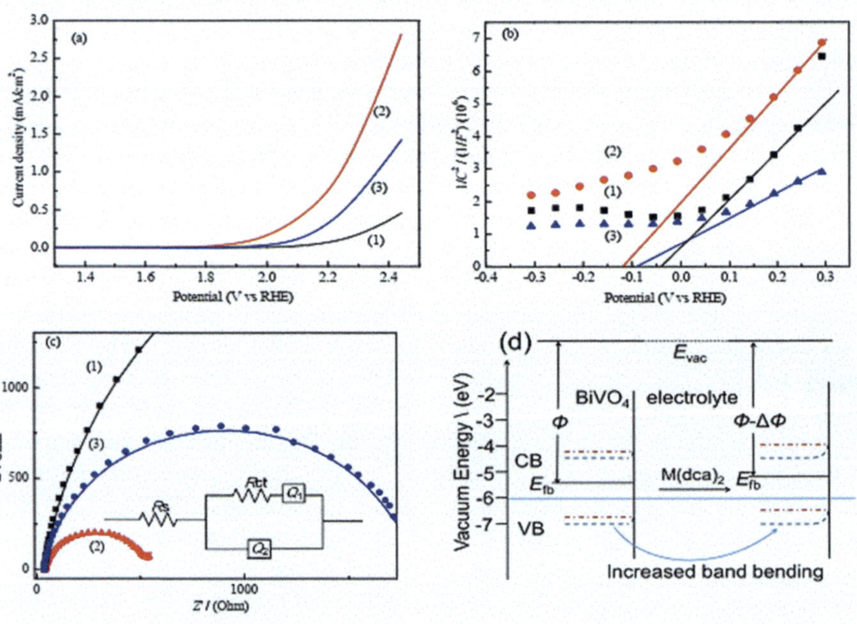

Fig. 16 $BiVO_4$ Electrochemical test results (1), $BiVO_4/Co(dca)_2$ (2), and $BiVO_4/Ni(dca)_2$ (3), **a** LSV curves; **b** Mott-Schottky plots; **c** Nyquist plots with the inset **d** structural band of the $BiVO_4$/electrolyte interface with or without the addition of $M(dca)_2$ [25]

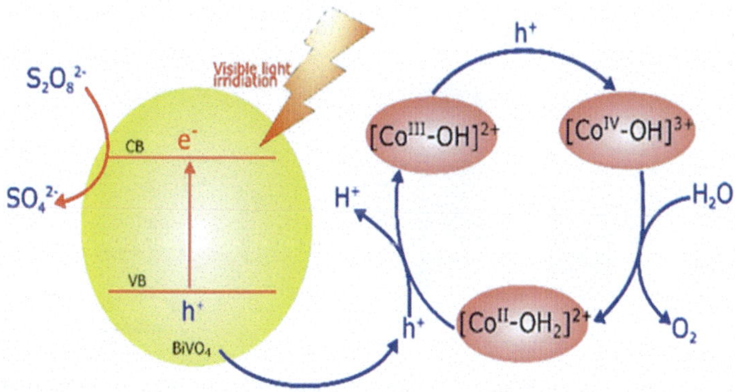

Fig. 17 $BiVO_4/Co(dca)_2$ reaction pathway of proposed photocatalytic water oxidation [25]

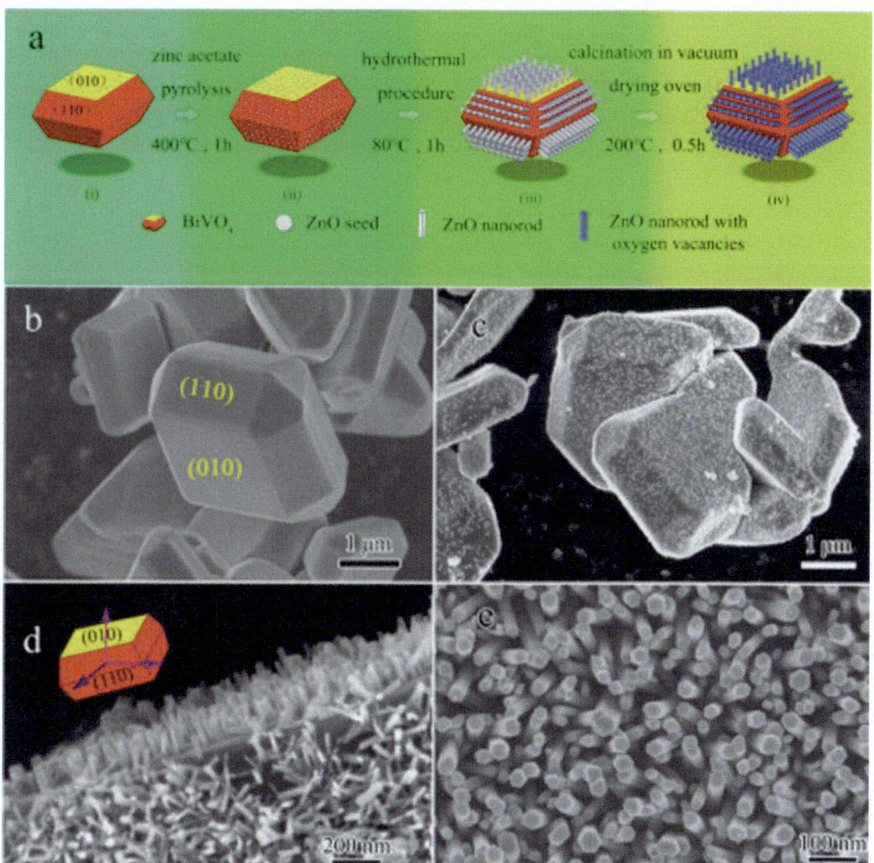

Fig. 18 **a** Preparation of $ZnO/BiVO_4$ with modulated interfacial structures and corresponding samples SEM images captured at different angle [26]

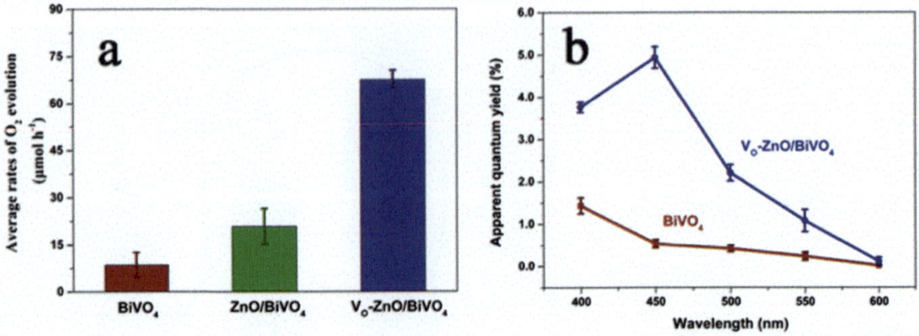

Fig. 19 **a** $BiVO_4$, $ZnO/BiVO_4$ and Vo-$ZnO/BiVO_4$ Photocatalytic water oxidation under irradiation of visible light ($\lambda > 420$ nm); **b** AQY of $BiVO_4$ and Vo-$ZnO/BiVO_4$ adopted from reference [26]

pristine $BiVO_4$. Under visible light irradiation, the photocatalytic oxygen evolution rates of $CeBiVO_4$ (C) and pristine $BiVO_4$ were evaluated (Fig. 18b). After 4 h irradiation, $CeBiVO_4$ (C) also displays a higher photocatalytic activity than $BiVO_4$ (about 4 times). According to the result, the substitution of Ce^{3+} ions in $BiVO_4$ for Bi^{3+} successfully enhances the photocatalytic activity.

The charge transfer resistance and efficiency of separation of Ce-$BiVO_4$ and pristine $BiVO_4$ were investigated using EIS. As demonstrated in Fig. 20c, Ce-$BiVO_4$ (C) has a smaller arc radius under the UV–Vis region than $BiVO_4$, implying that Ce-$BiVO_4$ (C) has a lower resistance and faster interfacial transfer of charge carrier than $BiVO_4$. The BET surface areas of pure $BiVO_4$ and Ce-$BiVO_4$ were determined to be 11.19 and 14.59 m^2g^{-1}, respectively, to better understand the major reason for photocatalytic activity (Fig. 21).

2.3 WO_3-Based Photocatalysts

Photocatalysts for O_2 evolution under visible light irradiation in the presence of suitable reagents are WO_3 that cause the potential at the top of the VB to be significantly higher than the potential of O_2 evolution under visible light irradiation. Another effective photocatalyst for O_2 evolution driven by visible light is $BiVO_4$. $BiVO_4$ has a band gap of 2.4 eV, which is smaller than WO_3, and also has the VB edge potential at +2.86 V versus RHE, which is sufficient for oxidizing water to create O_2 [27]. Dongsheng Li developed a deposition–precipitation procedure to prepare the Ag_3PO_4/WO_3 S-scheme heterojunction, and its photocatalytic activity was measured by monitoring water splitting and pollutant degradation under visible light. Consequently, pure Ag_3PO_4 and Ag_3PO_4/WO_3 heterojuction photocatalysts with an optimum ratio photocatalyst demonstrate an improved oxygen production photocatalytic activity at 306.6 $\mu molL^{-1}\ h^{-1}$ and 204.4 $\mu molL^{-1}\ h^{-1}$,

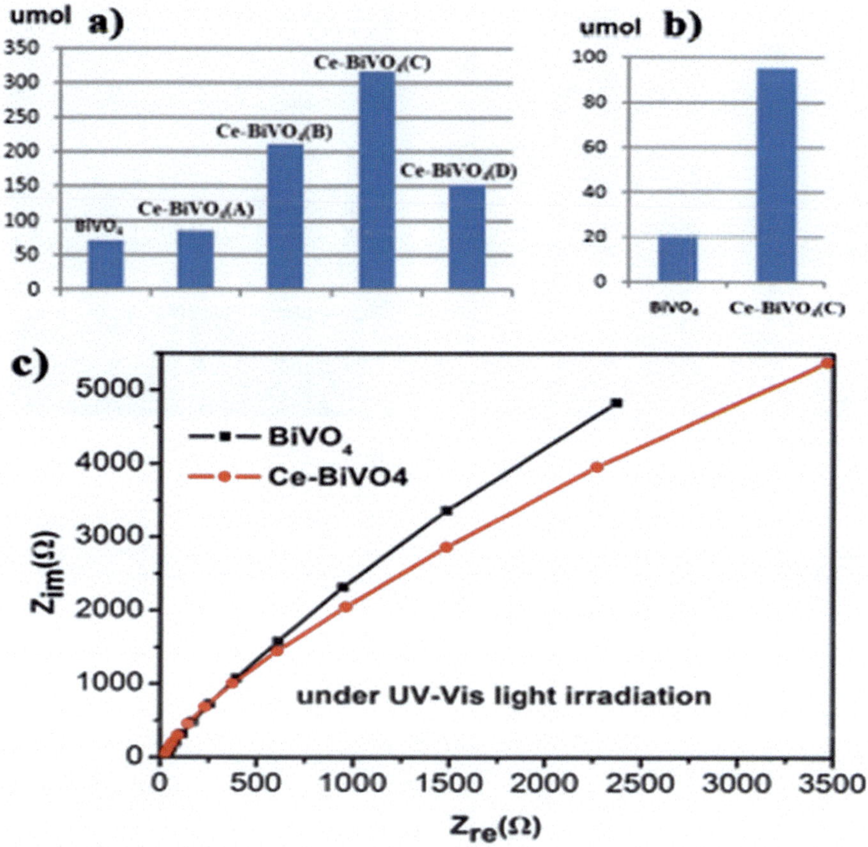

Fig. 20 **a** Photocatalytic O_2 evolution of $BiVO_4$ and Ce-$BiVO_4$ under irradiation of UV—vis light and **b** $BiVO_4$ photocatalytic O_2 evolution and Ce-$BiVO_4$ (C) in the irradiation of visible light. **c** EIS of $CeBiVO_4$ (C)/ITO and $BiVO_4$/ITO

respectively. Figure 22 demonstrate a putative S-scheme photocatalytic mechanism of Ag_3PO_4/WO_3 as indicated in detail according to a series of characterization, which supplied the heterojunction with high redox abilities to offer a tremendous photocatalytic process.

The Ag_3PO_4 has a well-defined spherical structure, smooth surface and diameter of 300 nm as shown in Fig. 23a. In contrast, the WO_3 structure was in a laminar form (Fig. 23b), whereas Ag_3PO_4 particles are tightly enveloped by WO_3 nanoparticles in Ag_3PO_4/WO_3 composites (Fig. 23c). TEM analysis of Ag_3PO_4/WO_3 composites (Fig. Schemed, e) revealed that the lattice plane spacing of 0.371 and 0.269 nm is indexed to (100) and (210) for WO_3 and Ag_3PO_4, respectively. Furthermore, interfaces between Ag_3PO_4 and WO_3 were clearly visible in the HRTEM image, demonstrating that WO_3 nanoparticles were firmly fixed on the surface of Ag_3PO_4. The EDS elemental mapping displayed that the composites are composed of Ag, P,

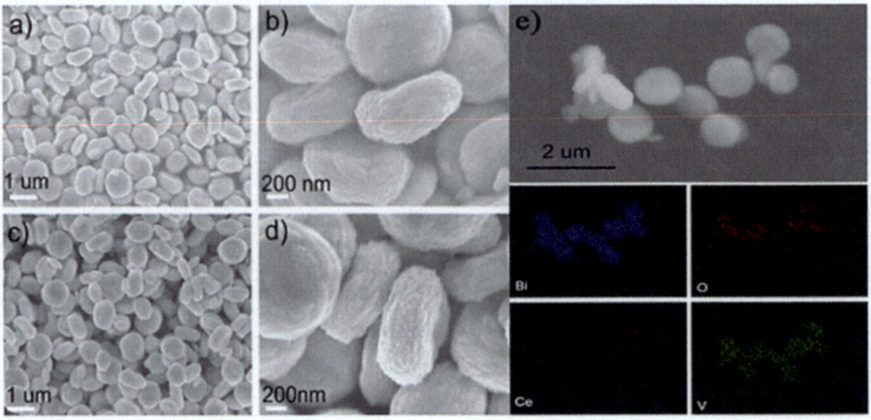

Fig. 21 Images of SEM **a**, **b**, pristine $BiVO_4$, **c**, **d** Ce-$BiVO_4$ (C), **e** Ce-$BiVO_4$ and its respective element mapping analysis

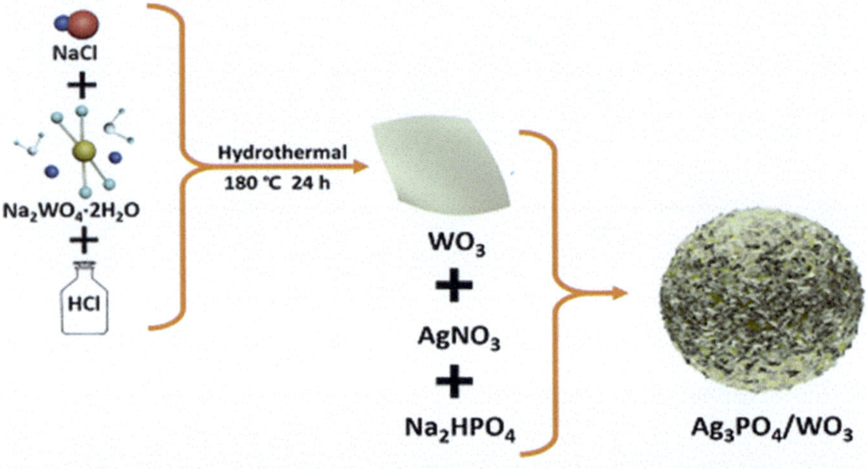

Fig. 22 Synthesis route of Ag_3PO_4/WO_3 heterojunction (Copyright Elsevier License Number 5319270917223)

W, and O elements with uniform elemental distributions (Fig. 23f), which further confirm that the WO_3 nanoparticles are distributed equally.

Figure 24a displayed the XRD profiles of the Ag_3PO_4/WO_3 heterojunction. The diffraction peaks of Ag_3PO_4 and WO_3 clearly match those of cubic Ag_3PO_4 (JCPDS card no.06–0505) and hexagonal WO_3 (JCPDS card no.33–1387), respectively. All Ag_3PO_4/WO_3 samples showed the presence of Ag_3PO_4 diffraction peaks after incorporation of WO_3, thus confirming that the phase structure of Ag_3PO_4 is not influenced by WO_3 loading. When the amount of WO_3 in Ag_3PO_4/WO_3 composites is less than

Fig. 23 **a–c** SEM images of pristine Ag_3PO_4, WO_3 and Ag_3PO_4/WO_3; TEM and HRTEM images of Ag_3PO_4/WO_3 (**d**, **e**); the elemental mapping images of Ag_3PO_4/WO_3 (Copyright Elsevier License Number 5319270917223)

5%, negligible diffraction peaks of WO_3 were observed, which can be ascribed to low loading and crystallization degree of WO_3.When WO_3 loading is higher than 5%, the diffraction characteristic peaks of WO_3 appear close to 22.89° and 28.38°, and gradually become stronger with the increase in WO_3 loading.

XPS analysis in the Ag_3PO_4/WO_3 heterojuction clearly indicates the presence of Ag, P, O, and W elements that can be seen in Fig. 24b. A doublet signal of binding energy centered at 373.6 and 367.7 eV correspond to Ag 3d5/2 and Ag 3d3/2, respectively. Figure 25b showed that the lattice oxygen atoms are represented

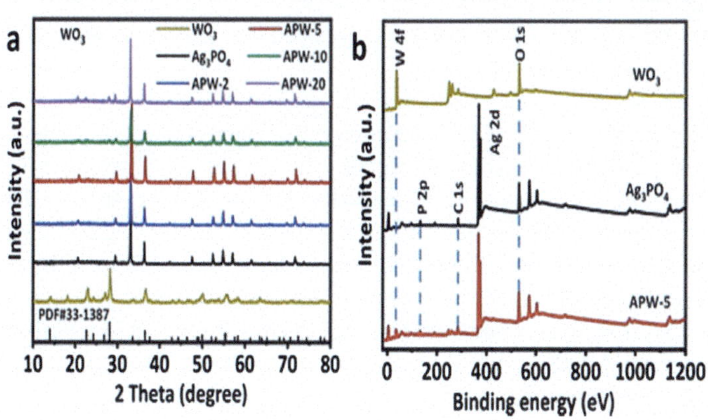

Fig. 24 **a** XRD patterns of pristine Ag_3PO_4, WO_3, Ag_3PO_4/WO_3 heterojunction, **b** XPS spectra of Ag_3PO_4/WO_3 (Copyright Elsevier License Number 5319271133782)

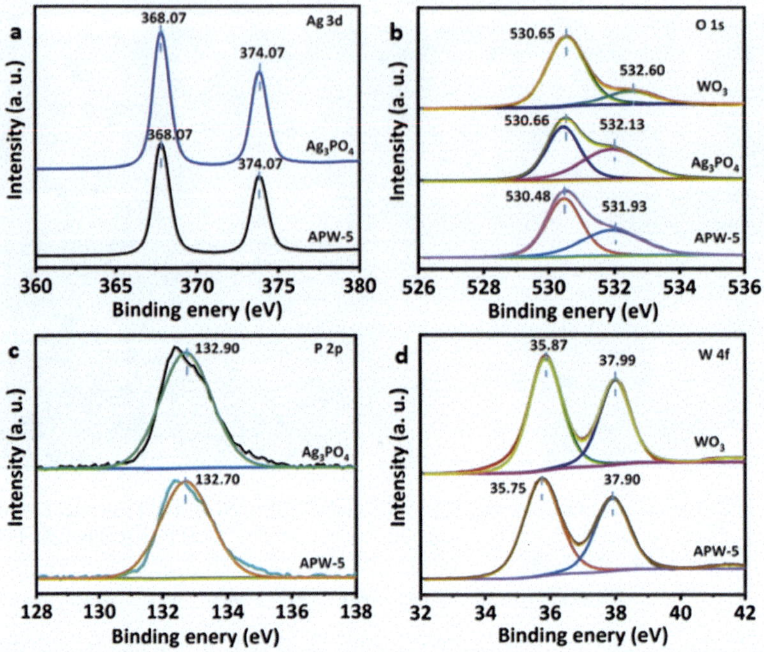

Fig. 25 XPS spectra of Ag_3PO_4/WO_3: **a** Ag 3d, **b** O 1 s, **c** P 2p, and **d** W 4f. (Copyright Elsevier License Number 5319271133782)

by two peaks in the O 1 s spectra located at 530.4 eV and 532.1 eV, which attributed to the O^{2-} and hydroxyl group OH^-, respectively. Narrow scan of P 2p in Fig. 25c revealed a single broad peak at 132.5 eV, which can be ascribed to P^{5+} of Ag_3PO_4. W 4f5/2 and W 4f7/2 of W^{6+} in WO_3 correlated to the peaks at 37.6 and 35.5 eV as shown in Fig. 25d. The binding energies of W 4f, O 1 s, and P 2p in the Ag_3PO_4/WO_3 heterojunction was slightly shifted when compared to pure WO_3 and Ag_3PO_4, which may due to instrument detection error.

The densely wrapped WO_3 nanoparticle on the surface of Ag_3PO_4 is favorable for promoting the separation of photoexcited e^-/h^+ pairs while also relieving photocorrosion and protecting Ag_3PO_4 particles. The composite photocatalyst uses the S-scheme mechanism to demonstrate excellent photocatalytic activity for oxygen synthesis, with up to 306.6 $molL^{-1}$ h^{-1} ($AgNO_3$ as sacrificial agent). As demonstrated in Fig. 27, a putative s-scheme transfer of charge mechanism of Ag_3PO_4/WO_3 has been postulated. The difference in Fermi energy between Ag_3PO_4 and WO_3 lead to free electrons which migrate from Ag_3PO_4 to WO_3, while a built-in electric field guiding from Ag_3PO_4 to WO_3 was formed. Both Ag_3PO_4 and WO_3 can create electron–hole pairs when exposed to visible light (Fig. 26).

The photogenerated electrons in the CB of WO_3 will be transferred to Ag_3PO_4 under the internal electric field force and recombine with the photoinduced holes

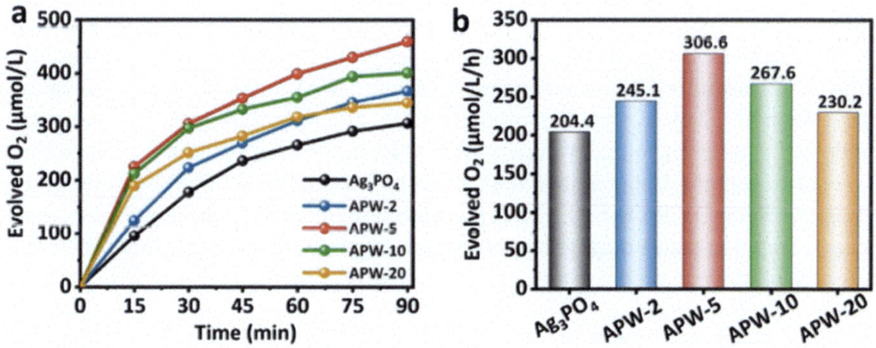

Fig. 26 Photocatalytic O_2 evolution activity for different samples (**a**) and generation rate (**b**) of the samples (Copyright Elsevier License Number 5319271278672)

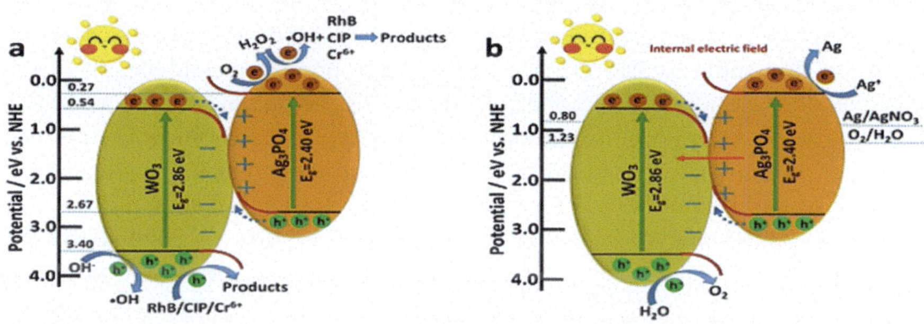

Fig. 27 S-scheme mechanism of photocatalytic degradation and O_2 evolution of Ag_3PO_4/WO_3, adopted from reference [28]

in the VB of Ag_3PO_4. This S-scheme arrangement, effectively suppressed photo-induced e^-/h^+ pair recombination while maintaining high redox abilities, hence enhancing its photocatalytic performance. The O_2 adsorbed on the surface of Ag_3PO_4 could not be converted to O_2 by electrons because the CB position of Ag3PO4 (0.27 V) could not surpassed above 0.3 However, considering that Ag_3PO_4 has a more negative E_{CB} position compared with $E(O_2/H_2O_2)$ (0.68 V vs. NHE), the CB electrons in Ag_3PO_4 can reduce O_2 into H_2O_2, and the generated H_2O_2 can then react with electrons to generate .OH.

Due to its deep VB position (3.40 V), which is more positive than the potential of E(OH/OH–) = 2.7 V, the holes accumulated in the VB of WO_3 have enough oxidation capability to oxidize OH– into OH [29]. In addition, the water splitting to generate O_2 reaction in Fig. 27b was based on a similar charge carrier transfer method.

Wei et al. developed and synthesized WO_3 catalysts with oxygen vacancies for effective photocatalytic oxygen evolution reactions in another study. The electron trapping state was found to have a key role in the charge transfer mechanism for improving the oxygen generation reaction of photocatalytic activity. The unique insights can be used to provide guidelines for designing the effective oxygen evolution catalysts for advanced renewable energy schemes. WO_3 containing oxygen vacancies (Ov) in Fig. 28 was obtained through an annealing process employing dicyandiamide and a tungsten source, whereas WO_3 was synthesized without dicyandiamide as a control sample. Using a photodeposition approach, Ov-WO3-Pt was created by depositing Pt on Ov-WO_3-600. The defect structure of Ov-WO_3 generates additional electron trapping states during the photocatalytic process, reducing direct recombination of photogenerated carriers and further enhancing the photocatalytic oxygen evolution performance.

The architecture of Ov-WO_3-Pt was also confirmed using a high-angle-annular-dark-field scanning transmission electron microscopy (HAADF-STEM) and EDX mapping (Fig. 28c), which showed that Pt particles were evenly distributed on Ov-WO3-600 upon photodeposition. Three main points can be used to sum up the Ov-WO_3 photocatalytic oxygen evolution system mechanism as described in Fig. 29. First, the existence of oxygen vacancies makes it easier for the trap states to develop, which can trap some excited electrons to partially prevent recombination of the photogenerated carriers [30]. Secondly, by increasing the lifetime of the active electrons, this process can boost its photocatalytic activity. The photo-deposited Pt can function as an active site to acquire trapped state electrons and interact with electron sacrificial materials, thus boosting the photocatalytic oxygen evolution process.

Third, some thermodynamic variables including absorption of light and ground-state OER onset potential, can be used to estimate the performance of Ov-WO_3

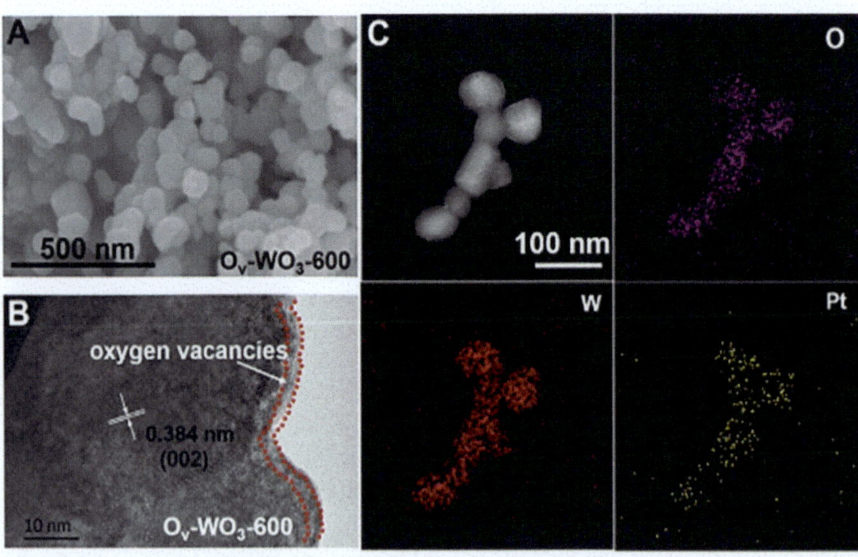

Fig. 28 **a** Image of SEM of Ov-WO_3-600; **b** HRTEM for Ov-WO_3-600; **c** HAADF-STEM image and EDX elemental mapping images of Ov-WO_3-Pt [31]

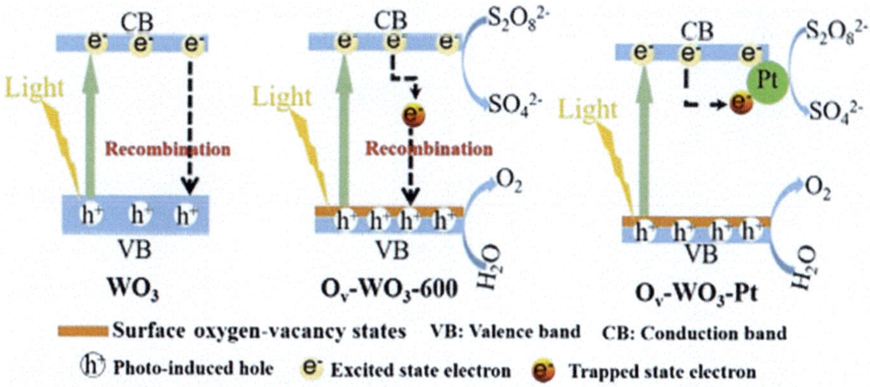

Fig. 29 Schematic illustration of photoinduced electron transfer mechanism of WO_3, Ov-WO_3-600, and Ov-WO_3-Pt charge recombination [31]

samples during photocatalytic oxygen evolution. In summary, Ov-WO_3 outperformed pristine WO_3 in terms of photocatalytic oxygen evolution, and Pt can further increase photocatalytic oxygen evolution activity. To determine the function of oxygen vacancies and cocatalysts in the photocatalytic oxygen evolution performance of WO_3, the photocatalytic oxygen evolution performance of Ov-WO_3 series samples, Ov-WO_3-Pt, and WO_3 were investigated. As the photocatalytic oxygen evolution results shown in Fig. 30a, Ov-WO_3-600 prepared at 600 ºC exhibited superior photocatalytic oxygen evolution performance compared with Ov-WO_3-650 and Ov-WO_3-550, likely due to the poor crystallinity of Ov-WO_3-550 and the large particle size of Ov-WO_3-650. The average O_2 evolution rate of Ov-WO_3-600 reaches 683 μmmol h^{-1} g^{-1} with $Na_2S_2O_8$ as a sacrificial agent under full-spectrum irradiation (Fig. 30a), while WO_3 only exhibited 159 μmmol h^{-1} g^{-1} under the same condition. The photocatalytic oxygen evolution performance of Ov-WO_3-600 was 4.3 times higher than that of pristine WO_3 without oxygen vacancies.

2.4 Fe_2O_3-Based Photocatalysts

Hematite (-Fe_2O_3) is regarded as a potential material for various applications since it is a typical p-type semiconductor with abundant, environmental friendly, low-cost and appropriate energy band structure [32]. Recently, Fe_2O_3 was widely use in field-effect transistors, sensors, catalysis, especially OER, and lithium ion batteries. According to earlier studies, Fe_2O_3 with a sufficient band gap (1.9–2.2 eV) can reach a solar conversion efficiency of 16.8%. However, the lifespan of photogenerated carriers in Fe_2O_3 is only of 10–12 s due to a short carrier diffusion length of 2–4 nm. Even though Fe_2O_3 has the potential to oxidize water due to its valence band position, the OER kinetics is rarely investigated. Numerous approaches have

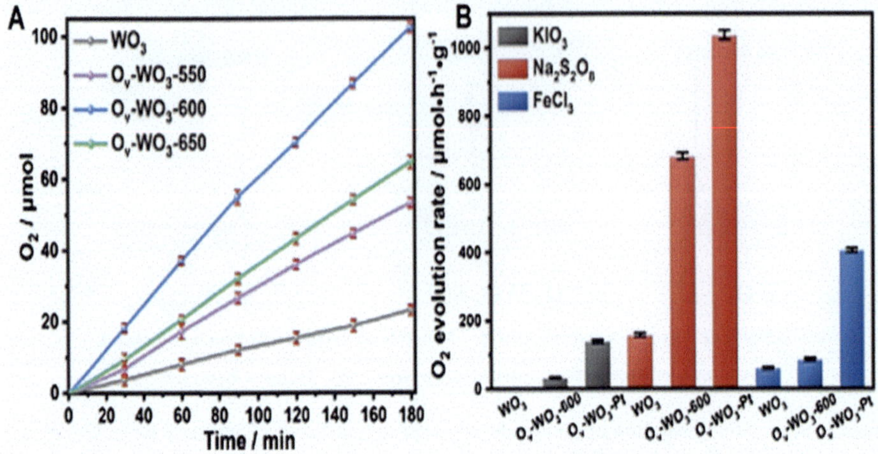

Fig. 30 **a** Photocatalytic OER for a series of Ov-WO_3 samples and WO_3 in the irradiation in an aqueous solution, **b** O_2 evolution rates of Ov-WO_3-600, Ov-WO_3-Pt, and WO_3 [31]

been used to address these problems, including chemical doping, designing new nanostructure, the combination with cocatalysts, and composites with other semiconductors [33]. Low cost, thermal stability, and unique characteristics of hematite Fe_2O_3, an n-type semiconductor with a narrow band gap (2.2 eV), has been widely used in visible-light-driven photocatalysts. Similar to other semiconductors, Fe_2O_3 has short carrier lifetimes and short hole migration distance. One effective way to address the aforementioned problems is through conjugation of coupling [34].

Cocatalyst integration is one of the most successful technique. For instance, RuOx and CoOx, are the most efficient OER cocatalysts. The use of the cocatalyst not only increases the photogenerated e–h pair separation efficiency, but it also gives the active sites a low molecular activation ability. Accordingly, the cocatalyst can be incorporated to improve the semiconductor carrier lifespan and carrier diffusion length as well as the reaction kinetics. Li et al., designed the hexagonal pure α-Fe_2O_3 and CoOx/α-Fe_2O_3 composites through the thermolysis approach as shown in Fig. 31a. The addition of $FeCl_3$. $6H_2O$ and CH_3COONa is a solvothermal process, where hexagonal α -Fe_2O_3 was successfully synthesized on CoOx to investigate the photocatalytic O_2 evolution activity. The CoOx as the holes acceptor was incorporated on the surface of hexagonal α-Fe_2O_3, which contains Co^{2+} and Co^{3+} species. Aiming to enhance the OER performance, the photocatalytic of α-Fe_2O_3 was dramatically enhanced after incorporating the CoOx cocatalyst. The increase in activity results from efficient charge separation and high hole transfer kinetics.

According to Fig. 31b–j, it has been discovered that pure hexagonal α-Fe_2O_3 has a hexagonal structure and exhibits a uniform size. The hexagonal α-Fe_2O_3 has a thickness of 15 nm and a diameter of 200 nm (Fig. 31b, c). The shape of α-Fe_2O_3 altered slightly after the thermolysis reaction, and the thickness of hexagonal α-Fe_2O_3 is up to 27 nm, which may have been caused by the calcining process, according to

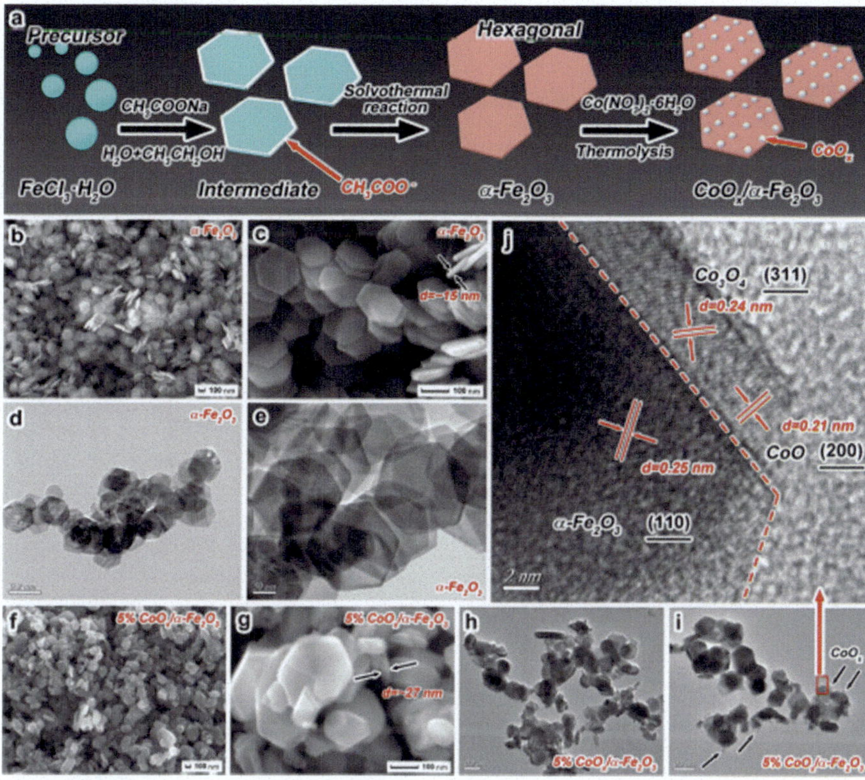

Fig. 31 **a** synthetic processes of the hexagonal α-Fe_2O_3 and CoOx/α-Fe_2O_3 composites, (**b**, **c**) and (**f** and **g**) SEM images of hexagonal α-Fe_2O_3 and 5% CoOx/α-Fe_2O_3 composite, (**d**, **e**) and (**h**, **i**) TEM images of α-Fe_2O_3 and 5% CoOx/α-Fe_2O_3 composite, **j** HRTEM image of 5% composite of CoOx/α-Fe_2O_3

Fig. 31f, g. The CoOx particles arrangement on the hexagonal α-Fe_2O_3 surface is clear from Fig. 31h, i. The HRTEM image (Fig. 31j) demonstrates that CoOx was integrated on the hexagonal -Fe_2O_3 surface. CoOx mostly consists of Co_3O_4 and CoO with lattice spacing of 0.24 and 0.21 nm, respectively, which can be ascribed to the (311) plane of Co_3O_4 and CoO. The (110) plane is visible in hexagonal α-Fe_2O_3 with a lattice spacing of 0.25 nm.

According to the XRD pattern, the sample crystal structures correlated to hematite as shown in Fig. 32a. The hexagonal CoOx/ α -Fe_2O_3 composites and Fe_2O_3-based materials exhibit similar XRD pattern. Low loading amount of CoOx in composites cause the negligible XRD peaks observed for CoOx. All the peaks in the XRD pattern can be correlated with the individual crystal planes of α-Fe_2O_3 (JCPDS No. 33–0664). The chemical structures were further examined using FTIR and Raman spectra (Fig. 32b, c).

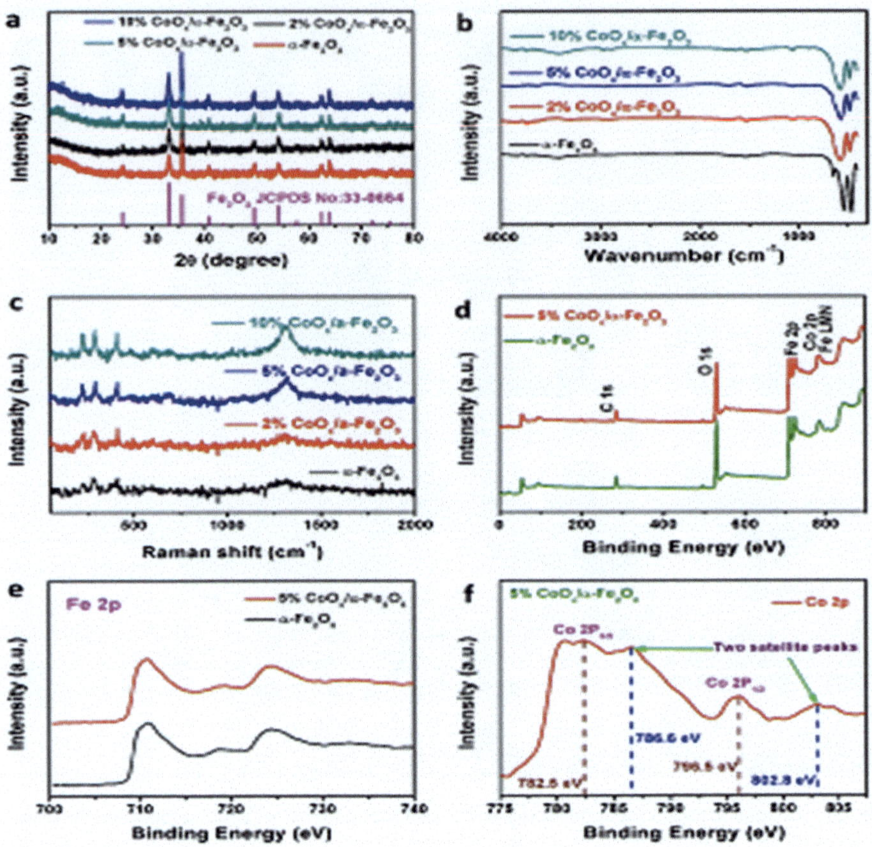

Fig. 32 **a** XRD pattern, **b** FT-IR spectra, **c** Raman spectra, **d** XPS spectra and **e**, **f** the narrow scan XPS of Fe 2p and Co 2p of the CoOx/α-Fe_2O_3 composites

The peaks in Fig. 32b between 400 and 750 cm^{-1} were ascribed to the natural lattice vibrations of α-Fe_2O_3. For the hexagonal α-Fe_2O_3, the peak at 650 cm^{-1} corresponds to the vibration mode (A) that is polarized along the c axis, while the peaks at 532 and 460 cm^{-1} were attributed to the vibration mode (B) that is perpendicular to the c-axis. The intensity of A mode is much weaker than that of B mode.

Since the thickness of the hexagonal α-Fe_2O_3 increases after calcination, as illustrated in Fig. 32c, g, two B mode peaks shift to a higher wavenumber, which can be attributed to an increase in the c/a aspect ratio. Furthermore, the two peaks at 650 and 532 cm^{-1} appeared to combine into a single broad peak at 575 cm^{-1}. The Raman spectra were analyzed at room temperature with a laser at a wavelength of 532 nm as shown in Fig. 32. There are no additional iron oxide in the samples, such as magnetite or maghemite implying high purity of α-Fe_2O_3. The peaks at 223 and 494 cm^{-1} are attributed to A 1 g, and those at 285, 413, and 608 cm^{-1} correspond to Eg. From Fig. 3d, it is clear that the composite of CoOx and α-Fe_2O_3 has four components instead of two that are present in maghemite α-Fe_2O_3: Fe, Co, O, and C [35].

For the solid-state reaction of Fe_2O_3/g-C_3N_4 samples under nitrogen conditions, Kong et al. proposed a one-step efficient, easily repeatable, and straightforward method. Solid-state reaction using layered glucose-coated FeOOH and g-C_3N_4 as the precursors for the visible-light-driven photocatalytic water oxidation process resulted in the creation of a unique layered Fe_2O_3/g-C_3N_4 heterojunction sample. Layered FeOOH and C-C_3N_4 were completely pulverized before calcined at 580 °C. Water vapor disrupted the van der Waals force between layered g-C_3N_4 during this process, leading to exfoliation and a specific edge carbonization as a result of the dehydration effect. Subsequently, FeOOH transform into α-Fe_2O_3. Layered and carbonized g-C_3N_4 with loading α-Fe_2O_3 showed an enhanced visible-light-driven water oxidation performance without addition of cocatalysts.

Typically, g-C_3N_4 was obtained from the thermolysis of urea at 550 °C. Fe_2O_3/C–C_3N_4 sample gives a relatively low XRD signal intensity compared to Fe_2O_3/C_3N_4, suggesting the strong interaction between FeOOH and g-C_3N_4. Furthermore, the XRD peaks of g-C_3N_4 and α-Fe_2O_3 can be easily detected without any impurities. In the case of g-C_3N_4, two peaks loaded at 2θ of 13° and 27.4°, are assigned to (100) and (200) respectively, which correspond to the Fe_2O_3/C_3N_4 sample. A wide peak centered at 2θ of 21° detected for Fe_2O_3/C–C_3N_4 can be attributed to the amorphous carbon due to carbonization. The Fe_2O_3/C_3N_4 sample layered structure is shown in the SEM image, as illustrated in Fig. 34a. The examined sample 2-D layered structure is relatively thin when compared to the fresh, thick g-C_3N_4 and FeOOH. TEM result shown in Fig. 34b strongly indicates the layered g-C_3N_4 with incorporation of 2-D α-Fe_2O_3 (Fig. 33).

The amorphous components in the edge region can be clearly seen, where inter-planar crystal separation of 0.27 nm was obtained. Figure 34c shows the HRTEM

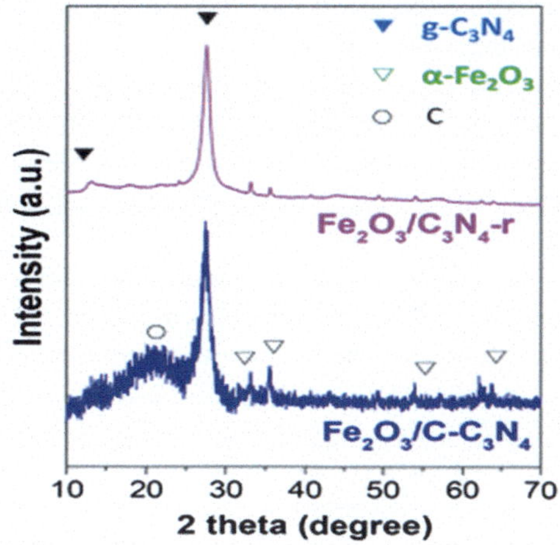

Fig. 33 XRD pattern for Fe_2O_3/C–C_3N_4 and Fe_2O_3/C_3N_4 composites

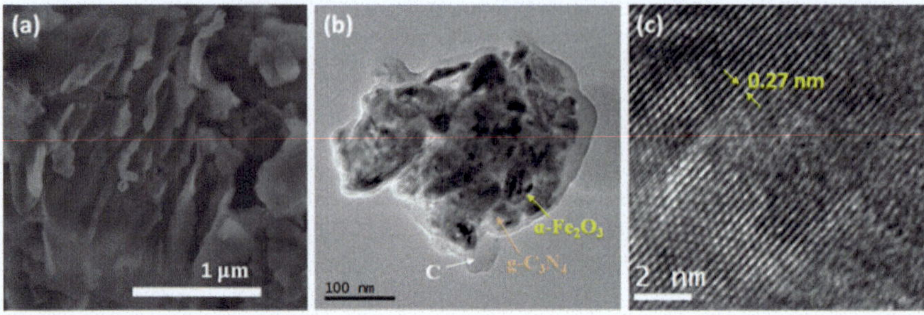

Fig. 34 **a** SEM, **b** TEM, and **c** HRTEM images of the Fe_2O_3/g-C_3N_4 sample

image of (104) plane of α-Fe_2O_3. A novel Fe_2O_3/C_3N_4 sample with the characteristics shown above should demonstrate visible-light-driven photocatalytic activity. A remarkable electrocatalytic performance toward water oxidation to oxygen evolution has recently been verified for g-C_3N_4 with some conductive crystalline carbon. In order to prepare a Z-scheme heterojunction structure and achieve significant photocatalytic water splitting activity, the synthesized investigated Fe_2O_3/C_3N_4 sample should have suitable conduction and valence band level. In order to complete the multistep photocatalytic water oxidation cycle, four electrons or holes are required. $AgNO_3$ solution (0.01 M) and LED lamp (420 nm) were initially utilized as the electron scavenger and visible light for water oxidation to oxygen evolution, in order to examine the photocatalytic performance of the Fe_2O_3/C_3N_4 sample. As reference samples, the pristine of g-C_3N_4 and α-Fe_2O_3 were also evaluated. All the samples exhibit photocatalytic water oxidation performance with discernible variations in oxygen evolution, as shown in Fig. 35a. Normal reaction rates for pristine g-C_3N_4 and α-Fe_2O_3 are around 0.7 and 1.4 mol/h, respectively, and these values indicate a rather modest activity.

The increased oxygen evolution for the Fe_2O_3/C_3N_4 heterojunction samples and Fe_2O_3/g-C_3N_4 at 7.3 and 22.3 μmol/h are about 10 and 30 times higher, respectively, than those for g-C_3N_4 as shown in Fig. 35a. This indicate that the heterojunction structure can facilitate photogenerated carrier separation, migration, and subsequent reaction. Further demonstrating the critical role of strong contact and amorphous carbon for further induced electron–hole pairs, the Fe_2O_3/g-C_3N_4 sample exhibits around three times higher photocurrent density than the reference Fe_2O_3/C_3N_4-r (Fig. 35b) [36].

Figure 35b shows the polarization curves of the Fe_2O_3/C_3N_4 and Fe_2O_3/C_3N_4 samples under potentials ranging from 0.4 to 1.6 V vs Ag/AgCl. The electrocatalytic water oxidation for the reference Fe_2O_3/C_3N_4 sample starts off mildly and occurs near 0.9 V (overpotential). In contrast, the Fe_2O_3/C_3N_4 sample operates at 0.4 V with a higher photocurrent density value than the reference sample. The increased overpotential of water oxidation shows that charge carrier easily transfers the catalyst/solution interface. The presence of edge carbon in the Fe_2O_3/C_3N_4 sample promotes the migration of the interface. Based on EIS analysis, Fe_2O_3/C_3N_4

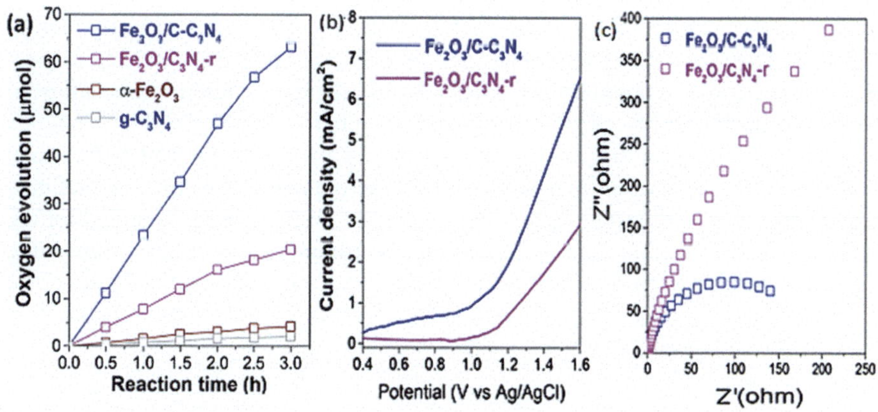

Fig. 35 **a** Oxygen evolution activities ($\lambda = 420$ nm), **b** Polarization curves of the $Fe_2O_3/C-C_3N_4$ and Fe_2O_3/C_3N_4 sample. **c** Nyquist plots of the Fe_2O_3/g-C_3N_4 and Fe_2O_3/C_3N_4 samples

and Fe_2O_3/C_3N_4 as seen in Fig. 35c, the semicircle radius of the Fe_2O_3/C_3N_4 sample is particularly smaller than that of the Fe_2O_3/C_3N_3, indicating a lower value for the overall average resistance and supporting the polarization results. To assess the carrier separation, the time-resolved PL spectra were carried out as shown in Fig. 36. Pristine α-Fe_2O_3, 1% and 1.5% IrOx/Fe_2O_3 were also used as the reference samples. Bare α-Fe_2O_3 shows the longest fluorescence quenching, followed by Fe_2O_3/C_3N_4-r, 1% IrOx/Fe_2O_3, Fe_2O_3/C–C_3N_4, and 1.5% IrOx/Fe_2O_3.

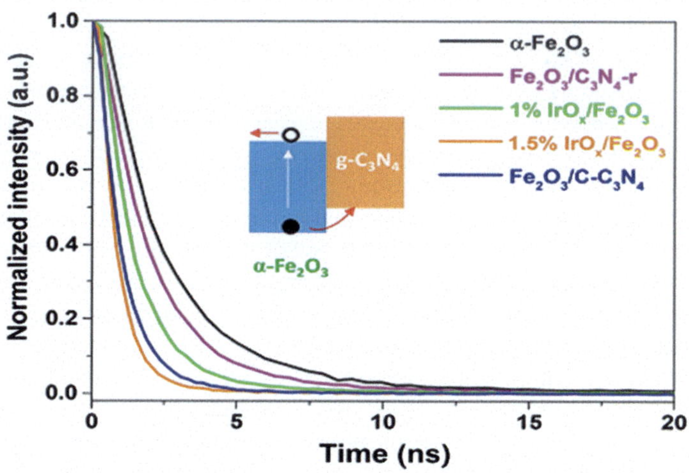

Fig. 36 Time-resolved photoluminescence (TRPL) spectra of Fe_2O_3/g-C_3N_4 and Fe_2O_3/C_3N_4; the excitation and emission wavelengths were 365 and 520 nm

Figure 36 shows the PL lifetimes of 24.5, 19.9, 14.1, 8.55, and 10.1 ns for bare α-Fe_2O_3, Fe_2O_3/C_3N_4, 1% $IrOx/Fe_2O_3$, 1.5% $IrOx/Fe_2O_3$, and Fe_2O_3/g-C_3N_4, respectively. For bare α-Fe_2O_3 and IrOx-loaded α-Fe_2O_3, the reduced lifetimes result from the good transfer ability of IrOx for photogenerated holes. Fe_2O_3/C_3N_4 exhibits a fluorescence lifetime that is shorter than that of bare α-Fe_2O_3 and longer than 1% $IrOx/Fe_2O_3$. This result indicates that the heterojunction structure can facilitate carrier separation. An intriguing and significant occurrence can be identified for the Fe_2O_3/C_3N_4 sample. Amorphous carbon and heterojunction both have a favorable impact on the separation of carrier and migration, as evidenced by fitted fluorescence lifetime, which is smaller than that of 1% $IrOx/Fe_2O_3$ and only slightly better than that of 1.5% $IrOx/Fe_2O_3$ [36].

3 Strategies for Enhancing Photocatalytic Oxygen Evolution

An appealing, clean, and efficient way to use solar energy to carry out chemical reactions is through solar-driven photocatalytic reactions. Five independent processes can be used to breakdown the full heterogeneous photocatalysis in liquid phase process: (1) Reagents are transferred from the liquid phase to the catalyst surface; (2) Reactants are adsorbed; (3) Reaction occurs in the adsorbed phase; (4) Products are desorped; and (5) Products are removed from the solid/liquid interface. Instead, the photocatalytic reaction in step (3) can be broken down into four stages: (a) light absorption followed by the separation of the electron–hole couple; (b) reagent adsorption; (c) redox reaction; and (d) desorption of products.

The redox process takes place in the semiconductor's electronic structure, which is made up of two bands. Electrons gain energy from the photons and are thus promoted from VB to CB when the semiconductor is activated by photons with energy equal to or higher than the band gap energy level ($hv \geq Eg$). Depending on the redox potential (or energy level) of the substrate, electrons and holes that move to the surface of the semiconductor can, reduce and oxidize the reactants adsorbed on the semiconductor surface. The relationship between the energy levels of the catalyst and substrate affects the photocatalyst's capacity to catalyze a given process. The adsorbed molecule undergo reduction if its potential for reduction exceeds that of the photoelectrons, or it can undergo oxidation if its potential for reduction falls below that of the photohole. The inherent band structure of the semiconductor photocatalyst governs its electronic transport characteristics, including its electron energy-level transition. The light absorption, redox potential, charge-carrier mobility, and the ability to change the band structure all contribute to the photoactivity [37].

3.1 Cocatalysts Loading

Photocatalyst systems is essential with cocatalyst loading. The formation of junctions with semiconductors, cocatalysts can offer active sites for the catalytic redox processes on the surface, improving the efficiency of charge separation. Along with acting as an OER catalyst, CoOx also formed a heterojunction with the semiconductor, facilitating charge separation. By altering the reaction pathways, cocatalysts can also lower the overpotential. Dual cocatalyst loading has been demonstrated to be a highly successful strategy for fostering CST [7].

O_2 development from photocatalytic water oxidation uses the oxidation cocatalyst functions. RuO_2, CoOx, IrOx, and other examples are typical inorganic cocatalysts for water oxidation. Many oxide photocatalysts do not sufficiently have a negative CB level to overcome the overpotential and provide the dynamic driving force for hydrogen evolution. Cocatalysts with more active sites, such PtOx, RuO_2, and Ni/NiO, typically only need a slight overpotential. These cocatalysts can also help improve charge separation in a photocatalyst by forming Schottky junctions with semiconductors. Cocatalysts have the ability to improve both water oxidation and reduction.

PtOx, IrOx, and CoOx are the most often used cocatalysts for the formation of oxygen in the half water oxidation reaction. However, VBs of cocatalysts are often significantly more positive than the redox potential of O_2/H_2O, whereas loading cocatalysts on oxide photocatalysts are not necessary for photocatalytic O_2 development (1.23 V vs. SHE). However, the 4-hole oxidation of water to produce one molecular O_2 is a more difficult reaction than the 2-electron reduction of H_2O to produce one molecular H_2 [52]. Hence, the active spots on cocatalysts for this reaction are extremely beneficial to speed up the O_2 evolution. Lower light absorption of the photocatalysts would result from the more active sites induced by cocatalysts. Thus, optimizing the amount of loading cocatalysts should be considered to obtain the maximum activity of water splitting under light irradiation.

Single atom catalysts (SACs) also have a tremendous potential for achieving high activity and selectivity owing to their homogeneous and well-defined single-atom dispersion. For instance, a simple single-step method for stabilizing solitary metal atoms like Pt, Pd, Rh, or Ru on TiO_2 led to a 6–13-fold increase in photocatalytic activity compared to metal clusters loaded on TiO_2 using the conventional method, as well as improving the stability of the photocatalysts [2].

Oxidation reaction calls for thermodynamic and kinetic parameters during the photocatalytic water splitting process. It can lower the activation energy and improve photocatalytic activity, because small loading of cocatalysts can offer the active sites and trap the photogenerated charge carriers. Therefore, it is required to use the suitable cocatalyst loading technique on the surface of the semiconductor. For the O_2 evolution reaction, the cocatalysts can be divided into the noble metals (e.g., Ag, Ru, Rh, Ir), noble metal oxides (e.g., RuO_2, RhO_2, IrO_2), and transition metal oxides (TMOs) (e.g., MnO_2, Co_3O_4). Pt is the noble metal that is most frequently utilized to reduce overpotential and to encourage charge separation by accumulating electrons.

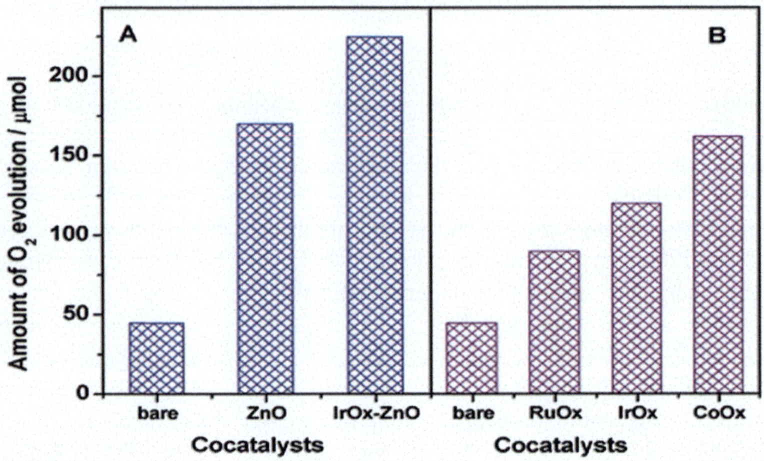

Fig. 37 Photocatalytic performance of O_2 evolution on $Zn_{2-x}GeO_{4-x-3y}N_{2y}$ loaded with different cocatalysts: **a** coupling with ZnO and loading with IrO_x and **b** deposition of IrO_x, CoO_x, and RuO_x

The uniformly dispersed Pt species were clearly visible on the surface of WO_3 when the loading amount of Pt was 0.5 wt percent at formation and additional techniques, such as the development of specific microstructure, surface modification, and solid solution construction at 823 K via the impregnation method.

Figure 37a indicates the photocatalytic activity of O_2 evolution on the Zn_2-$_x$GeO_4-x-3yN_2y semiconductor denoted as ZGON loaded with various cocatalysts. Due to the formation of the solid solution phase junction, ZGON combined with ZnO exhibits enhanced O_2 evolution activity. IrOx can act as a cocatalyst to further boost the activity as shown in Fig. 37a. As demonstrated in Fig. 37b, the activity for O_2 evolution can be increased by depositing IrOx, CoOx, and RuOx cocatalysts on ZGON, but CoOx was shown to be the most effective. These cocatalysts can also be loaded into other systems to increase the activity of O_2 evolution. The metal-free boron oxynitride cluster has recently been discovered to be an oxidation cocatalyst that enhances the photocatalytic oxygen evolution performance. These findings suggest that in order to obtain high water splitting efficiency, a water oxidation cocatalyst is also essential [38].

3.2 Heterojunction Construction

To increase charge-separation effectiveness, many kinds of semiconductor heterojunction photocatalysts have been developed and produced over the past few decades. The majority of p-type semiconductors are unstable, and p–p heterojunctions are even more unstable, which has hampered the development of p–n heterojunction-based photocatalysts. The main obstacles to getting the requisite properties for a water splitting photocatalyst are design techniques and the construction of heterostructures using various functional materials.

Each coupling material's energy level must have almost identical overlapping band structures in order to create a heterojunction. When photo-generated charges move from one material to another at the heterojunction interface, the ability of the migrated holes and electrons to oxidize and reduce decreases, respectively. Holes transfer to a material's less positive valence band while electrons transfer to the less negative conduction band location, particularly in type I heterojunctions (Fig. 38a). As a result, after transfer through the heterojunction, the photoexcited charge separation improves but their redox ability declines. The materials involved in a heterojunction structure for water splitting must adhere to strict requirements, such as having a narrow band gap that facilitates the absorption of visible light and proper energy level matching that promotes good charge transfer and improved oxidation and reduction ability. For total water splitting under visible light irradiation, the fabrication of heterojunctions with the right combination of one H_2 evolving and one O_2 evolving photocatalyst is crucial [3].

Typically, water splitting cannot be accomplished with just one n-type semiconductor. For instance, the intrinsically short charge-diffusion lengths or limited charge-carrier mobility of several n-type semiconductors that visible light absorb and have

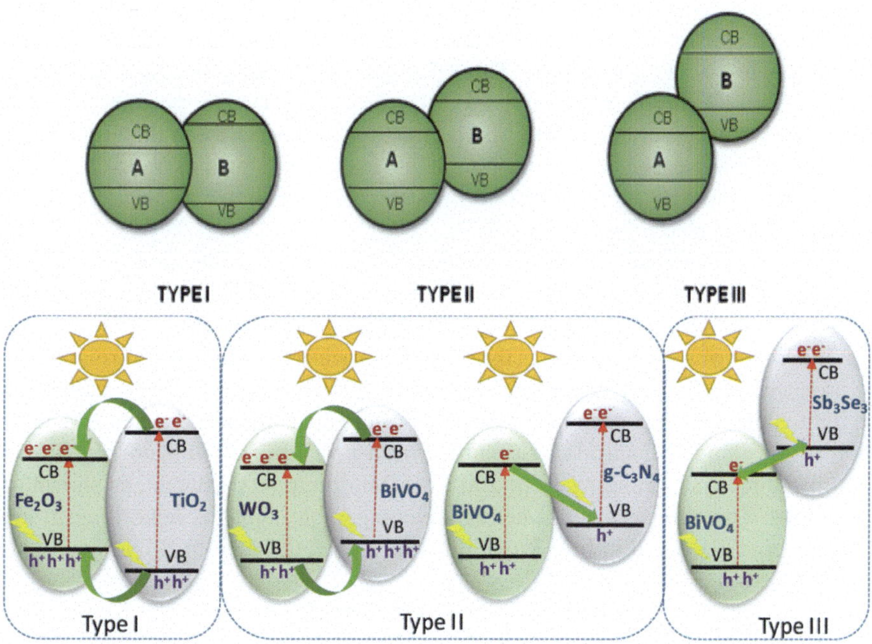

Fig. 38 **a** Alignment of Band in heterojunctions Type (I, II, and III) **b** Band alignment for of heterojunctions [3]

narrow band gaps, such as Fe_2O_3, Bi_2WO_6, and $SnWO_4$, results in ineffective CST. However, several n-type wide-band-gap semiconductors, such TiO_2 and ZnO, only respond to UV light despite having strong CST. Because the advantages of the beneficial qualities of the other components are exploited to overcome shortcomings, n–n heterojunction photocatalysts made of narrow- and wide-band gap semiconductors are of great interest. The amount of requirements that must be satisfied by a single photocatalyst can be decreased by creating a heterojunction out of two or more semiconductors. The top of the valence band in photoanode materials must be more positive than the oxygen evolution potential in order to participate in oxidation reactions, whereas a p-type semiconductor (photocathode) needs a conduction band position that is more negative than the hydrogen evolution potential. Then, to enhance charge separation, a modest external bias (electrical energy) can be applied to move electrons to the counter electrode. There are three primary types of heterojunction formations between binary dissimilar particles, according to band alignment. For type I heterojunctions, material A has a smaller band gap than material B. Holes and electrons transfer from material B to material A due to a more negative CB and a more positive VB of material B.

Heterojunction of Type I is made up of (two) semiconductors, with component B's CB being higher than component A's. Since component B has a lower VB than component A, holes and electrons will move to and collect on component A. Due to B's more negative CB location, a type II connection depends on the flow of photoexcited electrons from B to A. Because holes can move in the opposite way from the more positive VB of A to B, charge separation is generally effective and the photocatalytic activity is increased. The VB and CB locations of the third kind, Type III, differ significantly from those of Type II, providing a stronger pushing force for charge transfer. The semiconductor–liquid junction is created when the Fermi level (E_f) of a semiconductor electrode is brought into equilibrium with the redox potential of the electrolyte as a result of electron transfer that occurs when the electrode is submerged in an electrolyte solution [39].

Wang et al., designed a nanoporous MoO_3-x/$BiVO_4$ heterojunction photoanode to promote charge separation and oxygen deficient WO_3-x/$BiVO_4$ heterojunction photoanodes to boost the PEC performance. Using a standard three-electrode cell and a Xe 150 W lamp as the light source, the photoelectrochemical (PEC) performances of the produced films were assessed. A thermopile optical detector precisely calibrated the light intensity to be around 100mW cm^{-2} after being irradiated via an AM 1.5G filter. By varying the electrodeposition time from 1 to 5 min, BiOI nanosheet arrays with film thicknesses ranging from 200 to 600 nm may be produced, as shown in Fig. 39b–d. The nanosheets' diameters are comparable, though. According to the XRD patterns displayed in Fig. 39e, all peaks can be attributed to BiOI along with the FTO signals.

Interestingly, the nanosheet array structures can still be seen in the BOI-x films after hydrothermal treatment with a $(NH_4)6Mo7O_{24}.4H_2O$ solution (Fig. 39f–h). The BOI-x films have successfully been transformed to Bi_2MoO_6 based on the XRD patterns displayed in Fig. 1(i) (PDF#33–0208; the matching samples were designated as BMO-x). On an electrochemical workstation, photocurrent-potential curves were

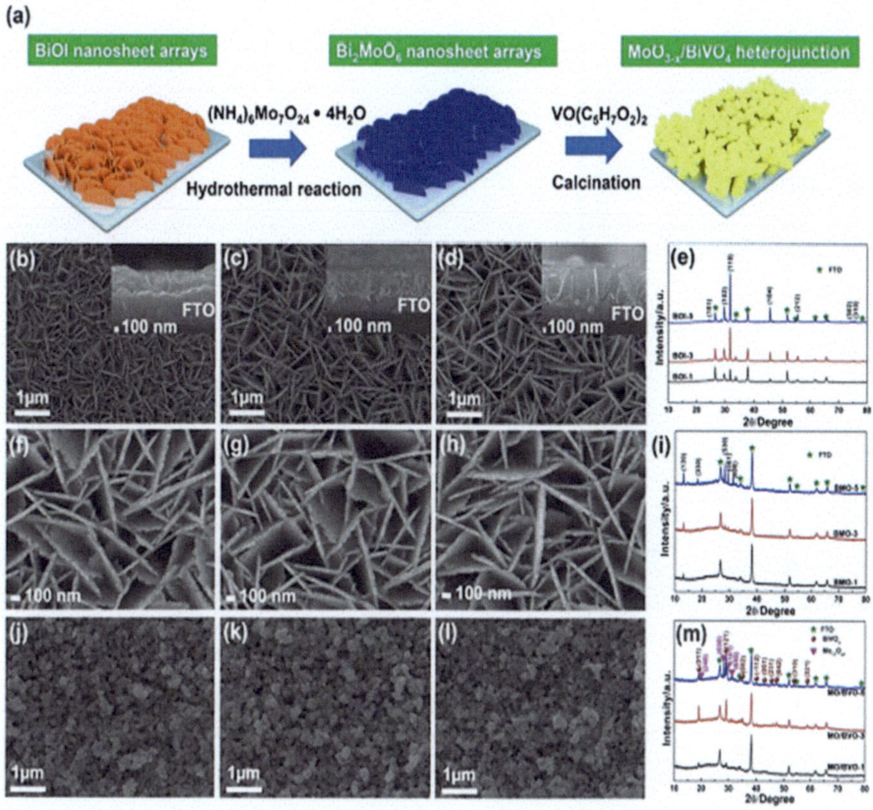

Fig. 39 **a** Schematic the preparation of MoO_3-x/$BiVO_4$ heterojunction. SEM images of **b** BOI-1, **c** BOI-3 and **d** BOI-5. insets in (**b–d**): **e** XRD patterns of BOI-1, BOI-3 and BOI-5. SEM images of **f** BMO-1, **g** BMO-3 and **h** BMO-5. **i** XRD patterns of BMO-1, BMO-3 and BMO-5. SEM images of **j** MO/BVO-1, **k** MO/BVO-3 and **l** MO/BVO-5. **m** XRD patterns of MO/BVO-1, MO/BVO-3 and MO/BVO-5

produced using LSV with a scan rate of 50 mV s-1 and a voltage window of 0.1–1.3 V versus RHE. Long-term stability of the photoanode was evaluated at 1.23 V versus RHE under AM 1.5 G illumination for 10 h.

All samples have much greater photocurrent densities when Na_2SO_3 is present, showing that oxygen evolution cocatalysts (OECs) are necessary to speed up OER at the photoanode/electrolyte interfaces. The resulting MO/BVO-3/FeOOH/NiOOH photoanode displays an improved photocurrent density of 4.81 mA cm^{-2}, which is noticeably higher.

The photocurrent densities of the MoO_{3-x}/$BiVO_4$/FeOOH/NiOOH films were measured in an airtight cell at 1.23 V versus RHE under AM 1.5 G illumination for 10 h, and the amount of gases was detected with a gas chromatograph to assess the overall water splitting performance. By adjusting the electrodeposition duration of the BiOI precursor films, the composition of MoO_{3-x} can be customized. With a high

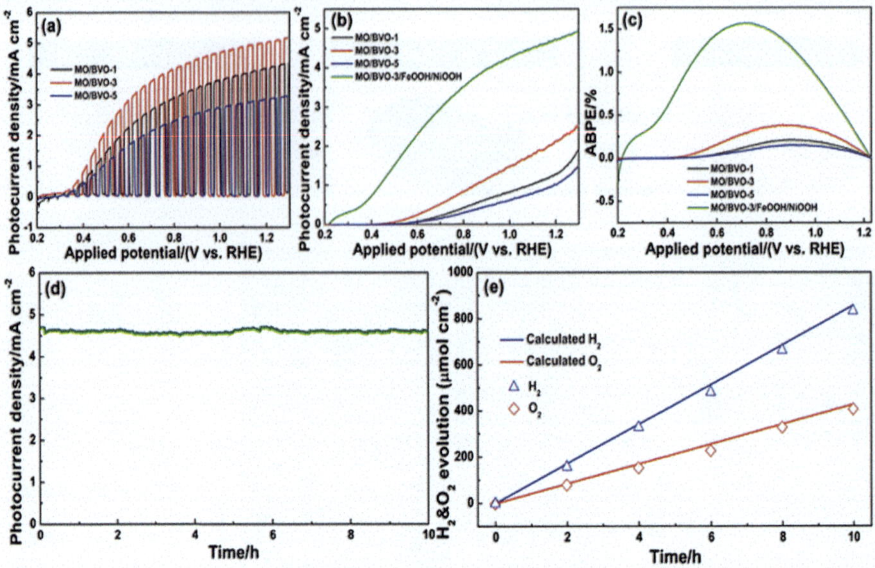

Fig. 40 **a** Photocurrent density versus potential of the MO/BVO-1, MO/BVO-3 and MO/BVO-5 photoanodes in a 1 M borate buffer electrolyte containing 0.2 M Na_2SO_3 under AM 1.5 G **b** Photocurrent density versus potential and **c** ABPE curves of the MO/BVO-1, MO/BVO-3, MO/BVO-5 and MO/BVO_3/FeOOH/NiOOH photoanodes in a 1 M borate buffer electrolyte under AM 1.5 G illumination. **d** J–t curve of the MO/BVO-3/FeOOH/NiOOH photoanode at 1.23 V versus in a 1 M borate buffer electrolyte. **e** Gas evolution from PEC water splitting of the MO/BVO-3/FeOOH/NiOOH photoanode at 1.23 V versus RHE

photocurrent density of 5.07 mA cm–2 at 1.23 V versus RHE under AM 1.5 G illumination and Na_2SO_3 as the hole sacrificial agent, the optimized $MoO_{3-x}/BiVO_4$ heterojunction photoanode exhibits a nanoporous shape that effectively promotes charge separation. By depositing FeOOH/NiOOH dual OECs to accelerate the surface reaction kinetic for OER, the optimized $MoO_{3-x}/BiVO_4$/FeOOH/NiOOH photoanode can achieve a photocurrent density of 4.81 mA cm^{-1} [40]. The ability to achieve visible light responsive photocatalytic water splitting with preferential band alignment is one of the most significant benefits of a heterojunction arrangement (Fig. 40).

3.3 Doping and Vacancy Formation

Another reasonably popular technique for extending the absorption response to longer wavelengths via incorporation of dopant. Additionally, it can be employed to control the photocatalyst development in order to stabilize more reactive facets and improve the photocatalytic activity. Nitrogen, sulfur, carbon, phosphorus, fluorine, and non-metal elements including Ag, Mg, Cu, Rb, Co, Mo, Au, Pt, Fe, Cr, and V are frequently used as dopants for broad bandgap oxide photocatalysts like TiO_2

and ZnO. The preparation of impurity states or defect states in the forbidden band as a result of doping or vacancy formation can cause the periodicity of a crystal atomic arrangement cause lattice distortion, hence widening the range of the light response.

Additionally, the separation of photogenerated charge carriers is done by improving the efficiency of the photocatalytic O_2 evolution. Doping is essential for altering the characteristics of catalyst materials. Since Asahi et al. announced nitrogen-doped TiO_2 in 2001, doping, particularly non-metal doping, has been thoroughly researched in photocatalysis [21]. Since 1982, it has been shown that TiO_2 can absorb more visible light when it contains a specific quantity of transition metals like Cr and Ru. Then, metallic doping was used to enhance semiconductors' photocatalytic activity. Non-metallic element doping (such as B and N) entered the mainstream at the beginning of this century because it allowed for more flexible tuning of photocatalytic activity. Currently, it is widely believed that the types of elements, doping techniques, concentration of dopant, and position of the dopant can all have an impact on the effect of metallic/nonmetallic doping for photocatalysts. The concept of gradient doping with nonmetallic heteroatoms in TiO_2 to improve the electronic structure has been developed based on the mature research works on elemental doping in TiO_2. N- and p-doping are the two basic kinds of doping.

Doping via foreign elements may alter a semiconductor band structure by adding intermediate donor or acceptor energy levels, and narrowing the band gap to absorb more light toward the visible region. Doping has an impact on the semiconductor lattice structure, which improves charge-separation effectiveness and carrier mobility in the semiconductor core. The band structure of TiO_2 is shown in the scheme along with the impact of non-metal doping, illustrating the effect of doping. Doping may also result in the introduction of extra color centers, as shown in portions in Fig. 41d, e. The origin of the visible-light absorption for nitrogen-doped TiO_2 has been explained. According to calculations using basic principles, the reduced bandgap of TiO_2 helps nitrogen-doped TiO_2 absorb visible light by combining O 2p states with N 2p states of the substituted nitrogen atoms in VB. Figure 41a–c highlight that three mechanisms causing this visible light absorption by nitrogen doping have persisted, particularly at the theoretical level where chemical state, dopant concentration, and various calculation techniques have been carefully taken into account. In independent dopant species, the visible-light absorption band originates from the decrease of TiO_2 after heat treatment or photostimulation procedure (Fig. 41d). TiO_2 with nitrogen doping cause band-to-band and shoulder tail visible-light absorption band properties of $TiO_{2x}N_x$ films prepared by sputtering technique on the TiO_2 plate in an Ar/N_2 mixed gas (Fig. 42a) [41]. Regardless of applying metal or non-metal doping, it is obvious that preparation procedures have a significant impact on band electronic structure and the subsequent properties of the doped TiO_2 visible-light absorption band. The existing localized states in the bandgap or the sensitizer attached to the semiconductor surface produce an absorption tail, whilst the narrowed bandgap might lead to steep and parallel absorption edges.

The important states of the CBs and/or VBs can be changed by modifying the compositions of the materials under research to produce full and long-range coupling between dopant states with spectral distributions. For this kind of coupling, the entire

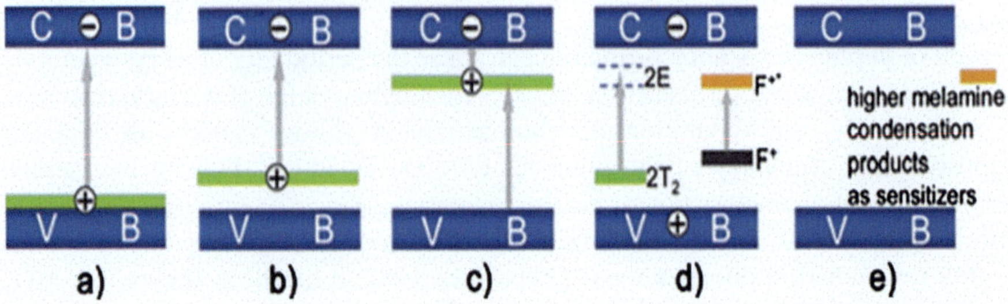

Fig. 41 The possible origin of visible-light absorption in doped anatase-type TiO_2 by non-metal doping: nitrogen species [7]

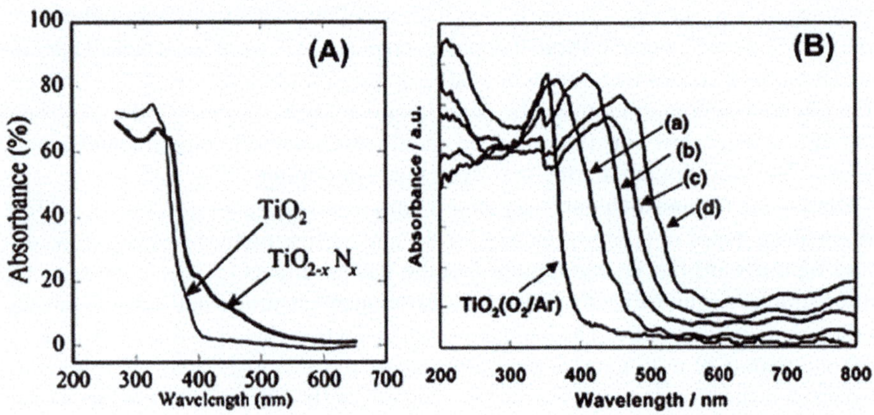

Fig. 42 **a** TiO_2xNx optical-absorption spectra and TiO_2 films. **b** UV–Vis absorption spectra of TiO_2 (O_2/Ar) and N–TiO_2(X) thin films

doped semiconductor matrix must have a uniform distribution of dopants at the atomic level without affecting the semiconductor crystal structure. The metal states of 3d and O 2p states make up the CBs and VBs of the majority of transition-metal oxide semiconductors.

The majority of doping techniques only result in inhomogeneous distributions of dopants, where the dopant is only dispersed in a very shallow subsurface region. The only way for dopants to impact the semiconductor matrix partial crystal field is by a weak, close-range contact with the matrix states. A few effective doped TiO_2 with prolonged band-to-band visible-light absorption edges, the creation of high-energy Ti, O, or dopant atoms are required. This method of doping results in homogenous doping at the atomic level because the quantity of dopant atoms included can be significantly increased and the dopant atoms are evenly distributed throughout the whole matrix. Figure 43 presents a summary of the previous analysis. For the implementation of homogenous doping, high-energy ion-implantation procedures are effective for transition-metal dopant atoms with large atomic numbers, but less effective for non-metal atoms.

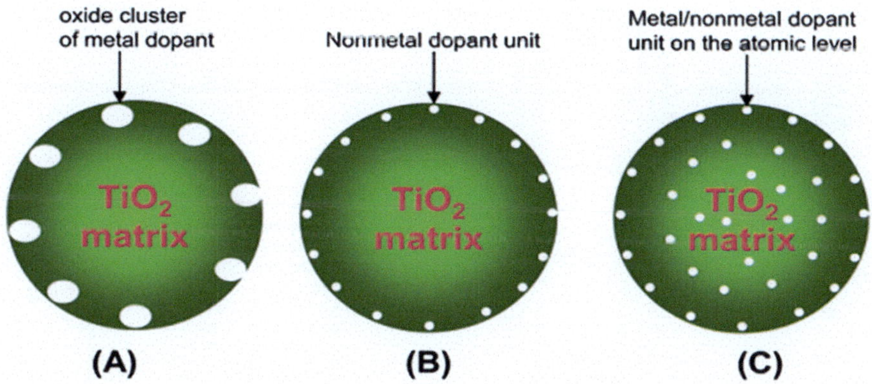

Fig. 43 Metal oxide clusters transition with Doped TiO_2 (**a**), surface non-metal dopants (**b**) and bulk metal/non-metal dopants (**c**)

While homogeneous nitrogen doping can occasionally be achieved through physical sputtering deposition, only films can be created, and the yields are still rather low. Consequently, it is scientifically and technically required to develop a straightforward and broadly applicable doping technique.

As seen in Fig. 44, two potential approaches are put up to achieve a uniform distribution of dopants throughout the whole semiconductor matrix. One is to reduce the size of the TiO_2 particles such that homogenous doping can be achieved with only a short diffusion length for dopants. The alternative is to create a gallery channel, which will enable homogenous doping throughout the entire particle bulk and facilitate dopant diffusion. In both circumstances, homogenous doping has the ability to occur before phase transition affects the subsurface layer [42].

4 Photogenerated Separation of Charge and Charge Recombination

One major issue limiting the overall effectiveness of the photocatalyst process is the rapid recombination of the produced electrons and holes. When electron—hole pairs recombine, photogenerated electrons return to the semiconductor's valance band with degeneracy in energy. While semiconductor photocatalysts with broad bandgaps are normally UV light active, those with tight bandgaps experience rapid recombination of photogenerated electron/hole (e^-/h^+) pairs. As a result, these drawbacks must be properly addressed in order to produce usable semiconductor photocatalytic systems (Fig. 45).

The morphology, dimensionality, and design of photocatalysts have a significant impact on the migration rate of these charge carriers. The photocatalytic activity is found to rise when the size of the photocatalysts is shrunk to the nanoscale due to the

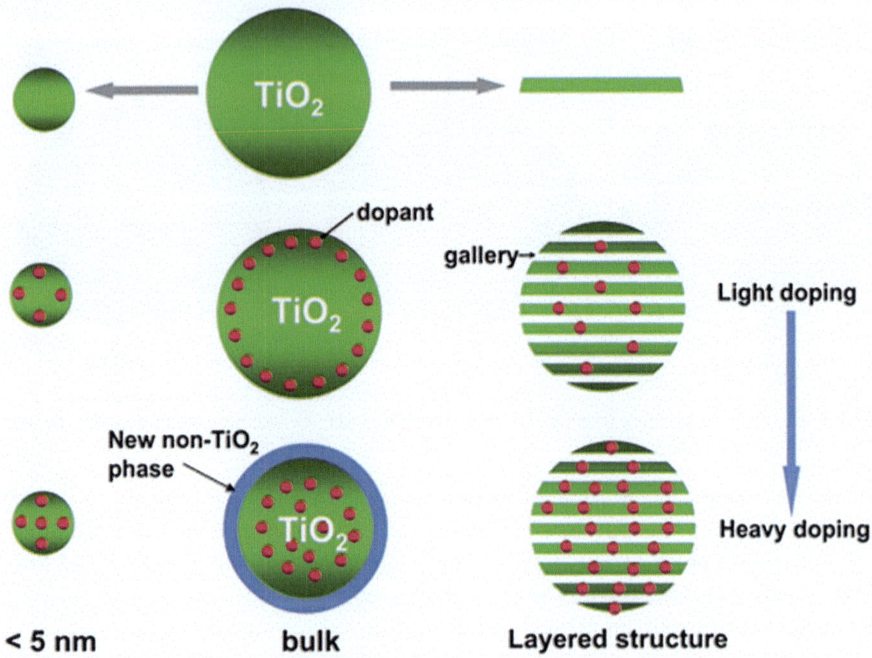

Fig. 44 Heavy doping in different TiO_2 substrates

Fig. 45 A semiconductor photocatalyst: excitation of bandgap, charge migration, charge recombination and chemical conversion

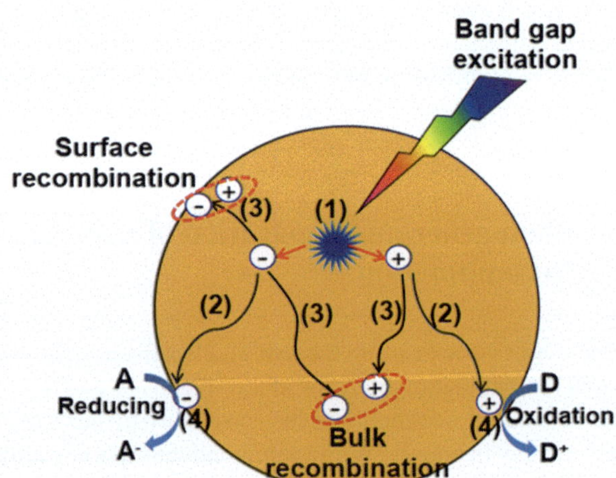

faster diffusion of the separated electrons and holes prior to recombination as well as an increase in surface area per unit mass and volume. Due to the effective and quick separation rate of charge carriers produced by the created p-n heterojunctions at the interface of hybrids, interface engineering of the nanocomposite photocatalysts can also greatly improve the photocatalytic activity. It has been demonstrated in several studies that designing a hierarchical structure properly and growing 2D nanomaterials (such as nanosheets, nanoplates, etc.) on 1D nanostructures (such as nanotubes, nanorods, nanofibers, etc.) will significantly improve the charge-separation efficiency and migration of photogenerated electrons and holes in the 2D/1D heterojunction [43].

In the schematic diagram, the energy levels of conduction band (CB) bottom and valence band (VB) top of both g-C_3N_4 and TiO_2, are labeled, which are important to understand the transfer and separation processes of photogenerated charges. The EVB levels for TiO_2 and g-C_3N_4 stand at 3.0 eV and 1.4 eV via the standard hydrogen electrode (SHE), respectively, while their ECB levels are positioned at 0.2 eV and 1.3 eV, respectively. Thus, due to suitable ECB and EVB of g-C_3N_4 and TiO_2, there is a better heterojunction expected for g-C_3N_4 and TiO_2 in nanocomposites. CB and VB of g-C_3N_4 are located well above the CB and VB of TiO_2. It means that the photogenerated electrons are easily transferred from the CB band of g-C_3N_4 to the CB of TiO_2 for the reduction process [44] (Figs. 46 and 47).

Various techniques have been employed to expand the light-harvesting zone from UV to visible and generate photocatalysts active. These ones deal with nanocatalysts, noble metal loading, ion doping, dye sensitization, fabrication of nanostructured semiconductors for photocatalysis, including nanosheet, nanotube, and nanorod, different fabrication methods for TiO2 nanotubes, including hydrothermal, anodization, and template methods, and new materials like graphene-based ones. In fact, it is crucial to develop and produce suitable heterostructures to improve the charge carrier separation in order to prevent photogenerated charge-carrier recombination on the photocatalyst. To more effectively use solar energy, several semiconductor materials with visible-light photoactivity have been developed [37].

Jia et al. demonstrated that, by control of the increase of the dipping temperature in a chemical bath deposition method, large amounts of highly crystalline 2D BiOBr nanosheets can be decorated on the outer walls of 1D TiO_{2-x} nanotubes with

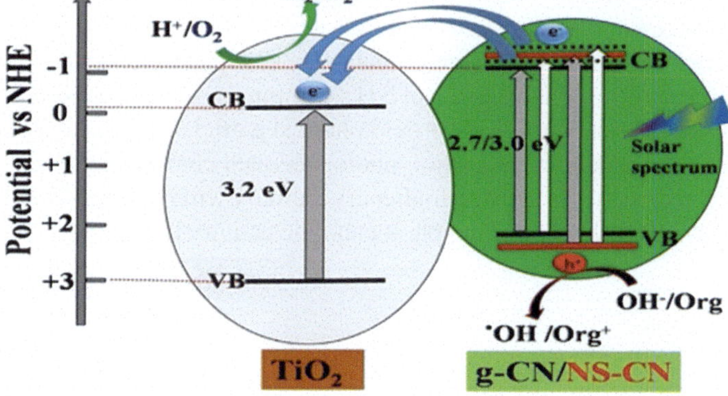

Fig. 46 Transfer and separation Charge in the TiO_2/g-C_3N_4 nanocomposite, along with the induced reactions

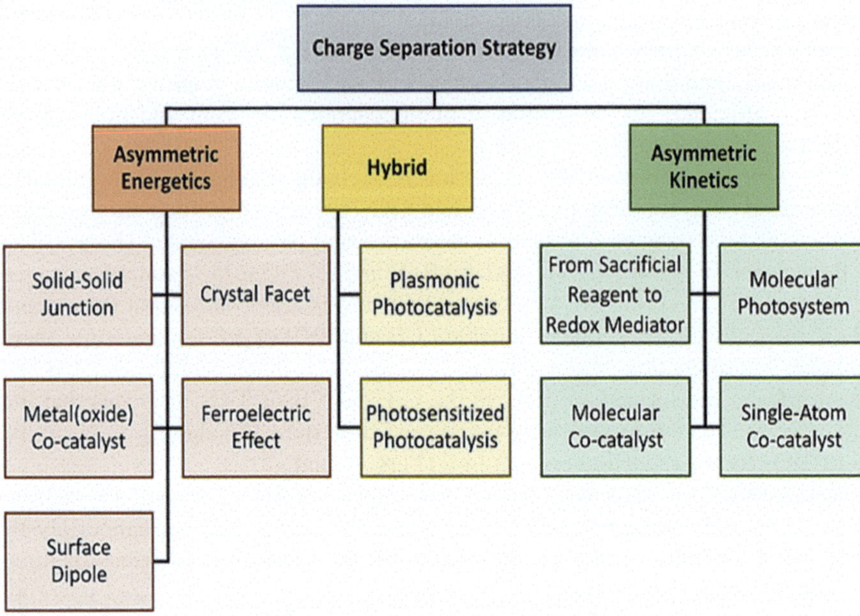

Fig. 47 Categorization of existing strategies for achieving charge separation

high photoelectrocatalytic performance in removing RhB dyes from water. The p–n junction created by the 2D/1D BiOBr/TiO_{2-x} hybrid which generates a type II band alignment, which will aid the migration of electrons from the CB of BiOBr to that of TiO_{2-x} and the movements of holes from the VB of TiO_{2-x} to that of BiOBr (Fig. 48).

Their architectural characteristics, such as morphology, crystal structure, particle size and surface area, have an impact on the separation and migration of photogenerated electrons and holes as well as the surface chemical reaction of H_2 or O_2 evolution. The enhancements to the photocatalysts' morphology can increase the incident light's reflection and scattering among the nanoparticles that make up the catalyst. For instance, by varying the concentration of NH_3 solution, Cu_2O with various morphologies, including spherical particles, porous spherical particles, cubes, and octahedral, were formed. At boundaries and flaws, photogenerated charges can be trapped and recombined very easily. As a result, higher crystallinity will result in fewer flaws and boundaries, which will enhance the photocatalytic activity [2].

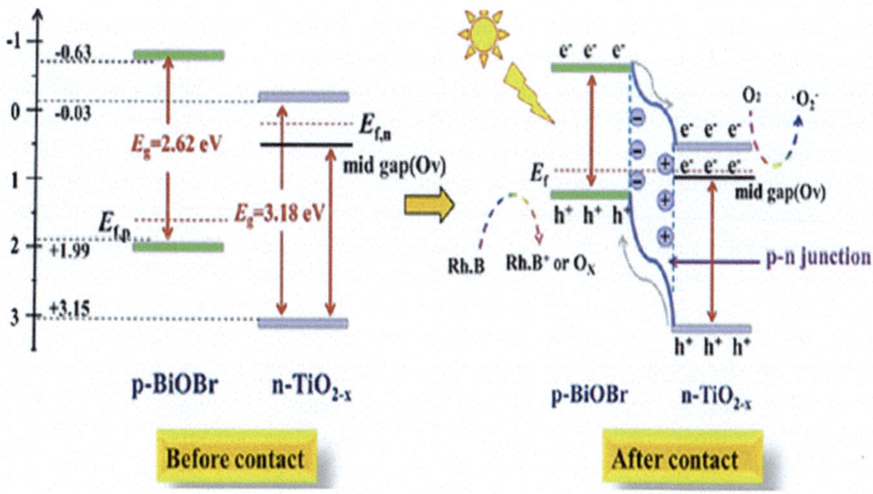

Fig. 48 a Charge transfer in the TiO_{2-x}/BiOBr. (Copyright Elsevier License Number 5313720919067)

4.1 Strategies for Enhancing the Efficiency of Charge Separation

It is vital that a successful strategy be developed to support the charge transfer and separation. According to recent study findings, separating and transporting charge carriers via semiconductor photo catalysts is a problem that can be efficiently solved by implementing a post-treatment process after sample preparation. In order to significantly improve photocatalytic water oxidation over a $BiVO_4$ catalyst, for instance, a simple and effective Cl- surface modification approach has previously been reported. As-prepared Cl-modified $BiVO_4$ produced good oxygen evolution, with an apparent quantum efficiency of 34.6 percent at 420 nm, thanks to the acceleration of photogenerated holes by Cl-. Bi_2WO_6 nanosheets that have been treated with Cl were also created to test their effectiveness as photocatalysts for oxidation.

These surface-modified catalysts with increased photoreactivity offer new perspectives and more chances to improve the efficiency of carrier separation and transport while boosting the photoreactivity of semiconductor photocatalysts. A simple approach of F-surface modification was described by Liu et al., which not only increases the effectiveness of charge separation but also quickens the charge-transfer process. The F- surface-modified TiO_2 of the TO-0.4F sample exhibits enhanced and ideal photocatalytic degradation of methyl orange (MO), photocatalytic production of the hydroxyl radical (HO.), and photocurrent response when exposed to UV light, as compared to pristine TiO_2 (denoted as TO). Here, x represents the volume of HF added in the solution. Figure 6a illustrates the photocatalytic breakdown of methyl orange (MO) under UV light. After being exposed to UV light for 40 min, the pure TO sample exhibits a high degrading efficiency of 79 percent toward MO. For TO-0.1F, however, surface modification with a modest quantity of F- results in an improved degrading efficiency of 89 percent. Furthermore, among the F-modified TiO_2 samples, TO-0.4F exhibits the maximum degrading efficiency of 95%. However, the TO-1.6F catalyst's photocatalytic degradation efficiency noticeably dropped by 51% as a result of excessive F- being present on the surface of TiO_2. In most cases, interactions between different factors lead to a catalyst's activity.

The better or lower photocatalytic performance of TO-xF in the present situation can be attributed to the surface and optoelectronic aspects since the textural and morphological characteristics of TiO_2 are almost preserved after and before surface fluorination. The reaction rate constant (k) was calculated using the fitted curves of –ln(C/C0) vs reaction time, as illustrated in Fig. 49b. The ideal k for TO-0.4F is around 1.67 times that of pristine TO, indicating that F-surface modification is advantageous for boosting TiO2's photocatalytic activity. Additionally, the colorless phenol was employed as a model pollutant to assess the photocatalytic activity of the TO and TO-xF samples in their prepared state. By producing hydroxyl radicals (.OH), the photocatalytic redox activities were further examined under UV light irradiation.

PL intensity centered at 426 nm, as shown in Fig. 49, estimates the relative concentration of .OH. As demonstrated in Fig. 6d, TO-0.4F exhibits the strongest .OH production performance among the series samples, with TO-0.1F and TO-0.4F showing stronger .OH generation than pure TO. The aforementioned findings show

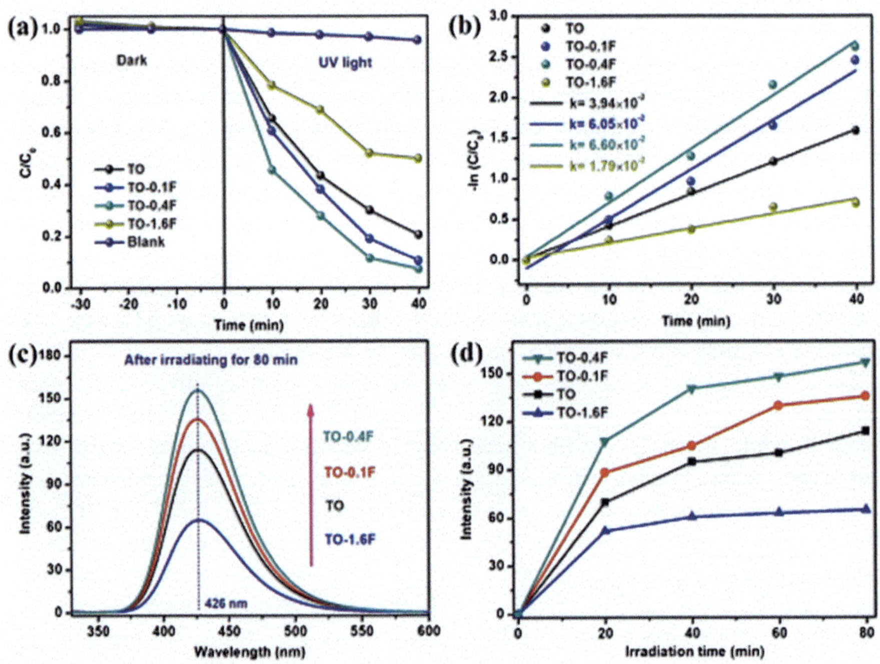

Fig. 49 **a** MO Degradation performance; **b** curves fitted of $-\ln(C/C_0)$ vs reaction time; **c** comparison of .OH radicals; and **d** profiles of .OH radicals

that when compared to pure TO, the photoinduced oxidation activity in TO-xF is increased. In most cases, the photocatalytic degradation events go forward directly through photogenerated holes or indirectly through oxidation mediated by the .OH radical. The holes in the VB are unable to oxidize the fluorine groups on the TiO_2 surface due to the F./F-couple's much greater redox potential (3.6 V). As a result, the TO-xF catalyst produces more .OH radicals to take part in the photocatalytic processes. The photogenerated holes in TO-0.4F may have a better oxidizing ability as a result of the VB downward shift, which is advantageous for the photocatalytic oxidation reactions. A potential photocatalytic mechanism using the TO-xF catalyst is suggested and is shown in Fig. 50 based on the characterization studies and band structure discussed above. As demonstrated in Fig. 50a, the photogenerated holes may be confined by coupling F species with strong electronegativity on the surface of TO, and the migration and separation of photogenerated electrons to reactive sites to take part in the redox processes are significantly enhanced. Figure 50b, c depict the F-enhanced surface trapping sites and hole transfer over the solid/liquid interface.

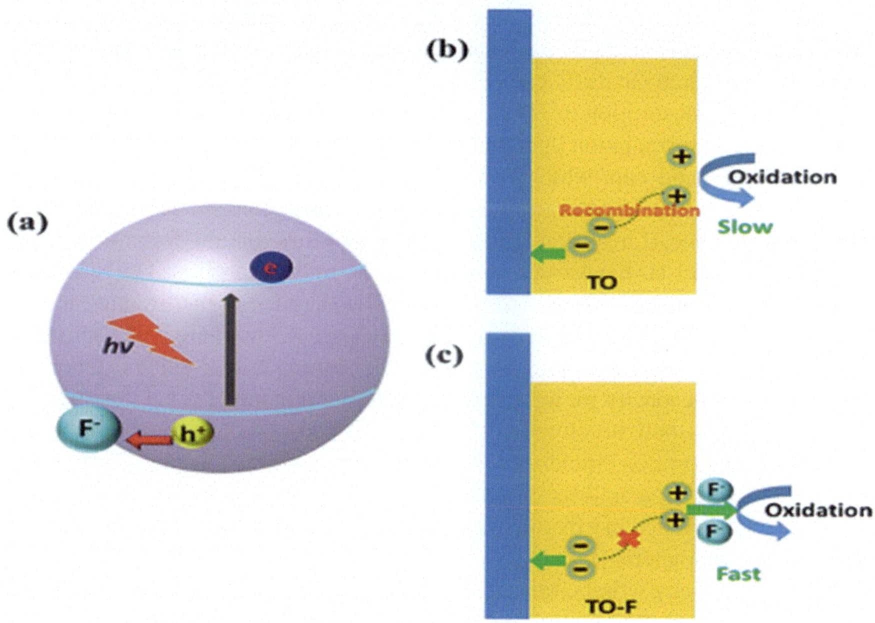

Fig. 50 **a** Mechanism of F—promoted charge separation on surface of TO-F. F—modified TO-promoted carrier charge separation and accelerated photoelectrochemical performance for **b** TO and **c** TO-F photoanodes taken from reference [45]

4.2 Improving Light Absorption

Extending the optical absorption characteristics of photocatalysts, particularly for visible-light wavelengths, is essential to achieve an improved photocatalytic performance. The photocatalytic efficiency rises as more electrons and holes are produced as a result of the photocatalysts absorbing more photon energy. Because shrinking the particles to the nanoscale increases the specific surface area, the light absorbance capacity can be greatly improved in nanostructured photocatalysts. Accordingly, the photon energy absorption by each nanostructure in 1D/2D nanocomposites with high light absorption coefficients can synergistically increase the absorption of a wide spectral range of light (UV–vis-NIR). For instance, a high specific surface area of 130.9 m2 g1 would be achieved when TiO_2 nanowires are built on MnO_2 nanosheets, which is 2.4 and 1.2 times bigger than that of TiO_2 nanowires and MnO_2 nanosheets alone, respectively. In comparison to bare TiO_2 nanowires and pure MnO_2 nanosheets, the hierarchical Z-scheme heterojunction of a TiO_2/MnO_2 nanocomposite considerably improved the full spectrum absorption of visible light from 300 to 800 nm, which improved the photocatalytic activity for the destruction of RhB dyes. Several methods to dynamically alter the energy band topologies or surface states of photocatalysts can be taken into consideration in order to further improve the photoabsorption in 1D/2D hybrid photocatalysts.

The spectrum of irradiation light that may be absorbed by photocatalyst increases with decreasing band gap, which is a crucial aspect in improving photocatalytic performance and efficiency in various solar energy conversion and harvesting applications. For instance, it has been found in various studies that the formation of 2D nanosheets on top of 1D nanostructures can increase light absorption. The externally generated 2D nanosheets can expose their complete surface to the irradiation light and absorb more photons in this type of 1D/2D design. Additionally, 2D nanostructures often have a lower band-gap energy than 1D nanomaterials, which is favorable for having a better capacity for light absorption. Regarding this, Li et al. reported the formation of core–shell-structured Bi_2O_4/BiO_{2x} hybrid photocatalysts with broad-spectrum light absorption efficiency for the photodegradation of dyes and antibiotics (Fig. 51a). These photocatalysts showed a broad absorbance range between 200 and 2500 nm, which resulted in a distinctive h + mobility, a small band gap, and superior optoelectronic characteristics.

According to many writers, raising the number of light-active sites on the photocatalyst and limiting charge-carrier recombination are the two key strategies for increasing the light absorption response. Two techniques can be used to increase the number of light-active sites: (1) designing semiconductors at the nanoscale to increase their surface area, and (2) increasing the number of highly active facets of the photocatalyst. Controlling the heterostructure's shape and construction with the right band alignment helps enhance charge-carrier separation. Additionally, some semiconductors' open pores may allow the reactants to diffuse into the photocatalyst material. Specifically, a single-step, template-free method is ued for the synthesis of meso-crystalline [37].

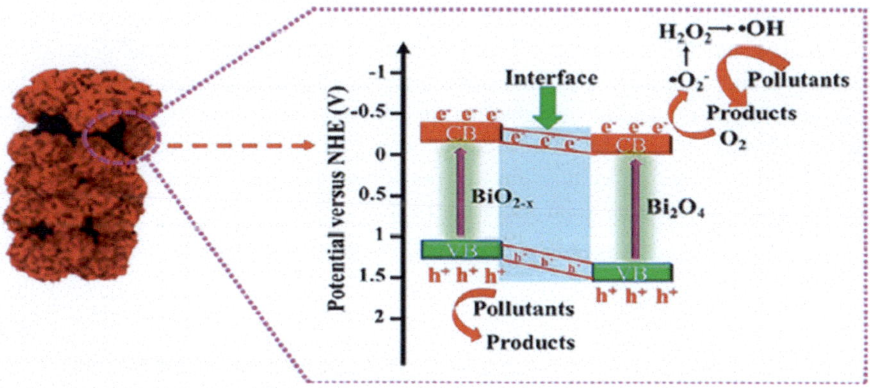

Fig. 51 Mechanism of the composite $Bi_2O_4/BiO_2{-}x$ in UV–vis–NIR irradiation

It is well known that a semiconductor's crystal structure affects how effective it is in photocatalysis. The photocatalytic effciency can be improved by optimizing the nanostructure size and morphology because photocatalytic reactions take place at the active sites on the photocatalyst surface. Increases in surface area and the percentage of crystal facets with increased photocatalytic activity both lead to an increase in the number of active sites. In general, morphological impacts on the photocatalytic activity (such as nanocrystals versus nanowires of the same material) can be understood as crystal facet-dependent reactivity or structural sensitivity.

Photocatalytic activity can be impacted by both cation and anion defects. Utilizing anion vacancies (oxygen vacancies or nitrogen deficciency) to regulate photocatalytic activity is an illustration of the method of exploiting native defects. Oxygen and nitrogen defects in g-C_3N_4 produced defects that are advantageous (O = C, NHx species, and N defects at the N-2C site), or detrimental for hydrogen evolution. Individual defects can also have positive or negative effects on the photocatalytic activity.

Cyano group (CN) is another elevant fault in g-C_3N_4 in addition to these potential flaws. Native defects, such as oxygen vacancies, can enhance exciton dissociation, which in turn contributes to enhanced charge carrier transfer and, as a result, photocatalytic activity. They can also introduce gap states to extend the photocatalytic response to longer wavelengths (depending on the semiconductor) [46].

4.3 *Ultra Violet Light and Visible Light Photocatalysts*

The majority of semiconductor photocatalysts are only activated by ultraviolet light; however, UV light makes up just about 5% of the solar spectrum, with visible light making up the other 45% and near-infrared (NIR) making up the other 50%. To make the most of solar energy, efforts should be made to widen the light response from UV to visible and NIR light. Furthermore, because high-intensity UV light places too much energy into the environment where reactions are occurring, a greater generation of reducing and/or oxidizing agents follows, frequently leading to an increase in the synthesis of byproducts. The only distinction is that the photocatalytic cycle activates with less photon energy, which most likely enhances selectivity. Oxide catalysts have a wide band gap energy because the UV portion of the solar spectrum only makes up around 4% of the entire solar spectrum [37].

4.3.1 Improving Efficiency

The photocatalytic efficiency can be expressed by the quantum yield Φ that measures the efficiency of a photocatalytic process and which is defined as

$$\Phi = \frac{\text{Amount of reactant consumed or product formed in the system}}{\text{Amount of photons absorbed by the photocatalysts at wave length}\lambda}$$

It is the absorbed photons and not the incident ones that initiate and drive a photocatalytic process. Due to light dispersion by the semiconductor surface, it is extremely challenging to quantify the real photons that have been absorbed in heterogeneous systems. The efficiency is, therefore, typically expressed as the apparent quantum yield and it is assumed that 100% of the light is absorbed. The overall quantum yield of photocatalytic processes is also constrained by the e^-/h^+ recombination and back-donation process, which occur following charge transfer from the adsorbed species to the semiconductor surface. Additionally, the effectiveness of photocatalysis in the elimination of organic pollutants depends on the semiconductor photocatalyst and the kind of irradiation, the sensitivity of the pollutant to photooxidative degradation, and the photocatalytic pathway. The targeted pollutant is one of the most crucial of the factors just mentioned. In fact, the efficiency of a photooxidative degradation can be influenced by both the material's chemical structure and its electrical properties. Another important factor in a pollutant's capacity to undergo photooxidation is the level of substitution. [24].

5 Conclusions, Challenges, and Perspectives

Oxygen evolution is the key phase in photocatalytic water splitting, which has received a lot of interest in recent years due to the intricate four-electron reaction mechanism. Despite the fact that numerous investigations have been carried out, a systematic review is necessary to comprehensively synthesize and introduce the research on photocatalytic O_2 evolution. The development of technology for solar-driven water splitting on a multi-GW scale is and will become a key. It is necessary to produce new chemically stable semiconductors with band gaps between 1.5 and 2.0 eV and adequate carrier lifetimes. High-throughput experimentation and successful expert cooperation are needed for this. It also necessitates a readiness to "fail quickly" and learn from prior errors, such as the 40-year attempts to create a successful solar absorber from a wide-bandgap material like TiO_2. Finding and analyzing novel absorber materials might be a better use of these efforts.

Any future renewable energy system that uses chemical fuels as energy carriers faces its biggest difficulty in the sustainable production of hydrogen. Due to its advantages of abundant resources, environmental friendliness, etc., photocatalytic water splitting has steadily come into focus in the societal context of improving the sustainable development of energy resources. Water oxidation, which determines the velocity of the water splitting reaction, is the main bottleneck that limits the efficiency of this process. Accordingly, effective solutions, such as cocatalyst loading, heterojunction construction, doping and vacancy formation, as well as other tactics, are based on the fundamental problems that impede the photocatalytic performance of photocatalysts,

Although a number of oxygen evolution photocatalysts have been created during the past ten years, there is still a long way to go before there are any real-world industrial uses. The majority of photocatalysts for oxygen evolution still require the presence of cocatalysts and sacrificial agents, which unquestionably raises the cost to industrial applications. They also generally have low efficiencies or poor physicochemical stability. More work will be required in the future:

Currently, techniques for improving the performance of O_2 evolution are mostly focused on increasing the absorption capacity and charge separation effectiveness, with little attention paid to studies of the surface catalytic reaction. In reality, the adsorption of reactants and the reaction activation energy are intimately related to the reactive sites of photocatalysts. For instance, creating surface flaws can unmistakably improve the photocatalysts' active sites. Therefore, it is anticipated that one of the research goals in the future will be the investigation of the catalytic active sites for O_2 evolution. Such fuels are most conveniently transported, stored, and used when they are in a liquid or gaseous state. While some of these chemical fuels do not contain carbon or nitrogen, all of them do. Water splitting with sunlight is the main difficulty since it is the only globally accessible, abundant, and sustainable source of hydrogen.

It has long been researched as a viable method of clean, large-scale fuel production to split water using solar energy and semiconductor photocatalysts to produce hydrogen and oxygen. In general, when a photocatalyst is altered with an appropriate cocatalyst, total water splitting can be accomplished. Therefore, it is crucial to create both cocatalysts and photocatalysts. Significant advancements have been made in water splitting photocatalysis during the past five years, particularly in the creation of cocatalysts and the associated physical and chemical changes. Due to its advantages of abundant resources, environmental friendliness, etc., photocatalytic water splitting has steadily come into focus in the societal context of improving the sustainable development of energy resources. Water oxidation, which determines the velocity of the water splitting reaction, is the main bottleneck that limits the efficiency of this process.

The majority of photocatalysts for oxygen evolution still require the presence of cocatalysts and sacrificial agents, which unquestionably raises the cost to industrial applications. They also generally have low efficiencies or poor physicochemical stability. More work will be required in the future. The current research focus is still on the creation of effective photocatalysts for water oxidation. Therefore, high-performance perovskites for O_2 evolution are anticipated to be produced in the future via rational design strategies based on the crystal structure and band structure. Currently, techniques for improving the performance of O_2 evolution are mostly focused on increasing light absorption capacity and charge separation effectiveness, with little attention paid to studies of the surface catalytic reaction.

In reality, the adsorption of reactants and the reaction activation energy are intimately related to the reactive sites of photocatalysts. For instance, creating surface flaws can unmistakably improve the photocatalysts' active sites. Therefore, it is anticipated that one of the research goals in the future will be the investigation of the catalytic active sites for O_2 evolution. Substantial work has been put into developing active sites on photocatalysts and understanding reaction mechanisms in the search for visible-light responsive photocatalysts, which has resulted in significant advancements in the field of heterogeneous photocatalysis for water splitting.

References

1. Olabi, Abdelkareem AG, Ali M (2022) Renewable and sustainable energy reviews, vo 1158
2. Kong D, Zheng Y, Kobielusz M, Wang Y, Bai Z, Macyk W, Wang X, Tang J (2018) Recent advances in visible light-driven water oxidation and reduction in suspension systems. Mater Today 21(8):897–924. https://doi.org/10.1016/j.mattod.2018.04.009
3. Afroz K, Moniruddin M, Bakranov N, Kudaibergenov S, Nuraje N (2018) A heterojunction strategy to improve the visible light sensitive water splitting performance of photocatalytic materials. J Mater Chem A 6(44):21696–21718. https://doi.org/10.1039/c8ta04165b
4. Lin S, Huang H, Ma T, Zhang Y (2021) Photocatalytic oxygen evolution from water splitting. Adv Sci 8(1):23–25. https://doi.org/10.1002/advs.202002458
5. Rajput S, Averbukh M, Rodriguez N (2022) Energy harvesting and energy storage systems. Electronics (Switzerland) 11(7):2–5. https://doi.org/10.3390/electronics11070984
6. Wang Z, Li C, Domen K (2019) Recent developments in heterogeneous photocatalysts for solar-driven overall water splitting. Chem Soc Rev 48(7):2109–2125. https://doi.org/10.1039/c8cs00542g
7. Xu Y, Li A, Yao T, Ma C, Zhang X, Shah JH, Han H (2017) Strategies for efficient charge separation and transfer in artificial photosynthesis of solar fuels. Chemsuschem 10(22):4277–4305. https://doi.org/10.1002/cssc.201701598
8. Nawaz A, Goudarzi S, Asghari MA, Pichiah S, Selopal GS, Rosei F, Wang ZM, Zarrin H (2021) Review of Hybrid 1D/2D Photocatalysts for light-harvesting applications. ACS Appl Nano Mater 4(11):11323–11352. https://doi.org/10.1021/acsanm.1c01014
9. Kawawaki T, Kawachi M, Yazaki D, Akinaga Y, Hirayama D, Negishi Y (2022) Development and functionalization of visible-light-driven water-splitting Photocatalysts. Nanomaterials 12(3):344. https://doi.org/10.3390/nano12030344
10. Yang X, Singh D, Ahuja R (2020) Recent advancements and future prospects in ultrathin 2d semiconductor-based photocatalysts for water splitting. Catalysts 10(10):50. https://doi.org/10.3390/catal10101111

11. Nadeem MA, Khan MA, Ziani AA, Idriss H (2021) An overview of the photocatalytic water splitting over suspended particles. Catalysts 11(1):1–25. https://doi.org/10.3390/catal11010060
12. Martin DJ (2015) Investigation into high efficiency visible light Photocatalysts for water reduction and oxidation, 149
13. Murillo-Sierra JC, Hernández-Ramírez A, Hinojosa-Reyes L, Guzmán-Mar JL (2021) A review on the development of visible light-responsive WO3-based photocatalysts for environmental applications. Chem Eng J Adv 5. https://doi.org/10.1016/j.ceja.2020.100070
14. Dutta V, Sharma S, Raizada P, Thakur VK, Khan AAP, Saini V, Asiri AM, Singh P (2021) An overview on WO3based photocatalyst for environmental remediation. J Environ Chem Eng 9(1):105018. https://doi.org/10.1016/j.jece.2020.105018
15. Xu Y, Zangari G (2021). Tio2 nanotubes architectures for solar energy conversion. Coatings 11(8). https://doi.org/10.3390/coatings11080931
16. Zhou Y, Zhang S, Ding Y, Zhang L, Zhang C, Zhang X, Zhao Y, Yu G (2018) Efficient solar energy harvesting and storage through a robust Photocatalyst driving reversible redox reactions. Adv Mater 30(31). https://doi.org/10.1002/adma.201802294
17. Jeong SY, Song J, Lee S (2018) Photoelectrochemical device designs toward practical solar water splitting: a review on the recent progress of BiVO4 and BiFeO3 photoanodes. Appl Sci (Switzerland) 8(8). https://doi.org/10.3390/app8081388
18. Lau D, Song N, Hall C, Jiang Y, Lim S, Perez-Wurfl I, Ouyang Z, Lennon A (2019) Hybrid solar energy harvesting and storage devices: the promises and challenges. Mater Today Energy 13:22–44. https://doi.org/10.1016/j.mtener.2019.04.003
19. Chinh T, Nguyen C (2018) Novel strategies to develop efficient titanium dioxide and graphitic carbon nitride-based photocatalysts
20. Yi Q, Cong S, Wang H, Zhou X, Chen J, Li K, Liu Y, Lee JM (2021) Heterostructure-induced light absorption and charge-transfer optimization of a TiO2Photoanode for Photoelectrochemical water splitting. ACS Appl Energy Mater 4(12):14440–14446. https://doi.org/10.1021/acsaem.1c03112
21. Dou Y, Zhang S, Pan T, Xu S, Zhou A, Pu M, Yan H, Han J, Wei M, Evans DG, Duan X (2015) TiO2@layered double hydroxide core-shell nanospheres with largely enhanced photocatalytic activity toward O2 generation. Adv Func Mater 25(15):2243–2249. https://doi.org/10.1002/adfm.201404496
22. Kochuveedu ST (2016) Photocatalytic and Photoelectrochemical water splitting on TiO_2 via photosensitization. J Nanomaterials. https://doi.org/10.1155/2016/407314
23. Sun B, Qian Y, Liang Z, Guo Y, Xue Y, Tian J, Cui H (2019) Oxygen vacancy-rich BiO 2–x ultra-thin nanosheet for efficient full-spectrum responsive photocatalytic oxygen evolution from water splitting. Sol Energy Mater Sol Cells 195(March):309–317. https://doi.org/10.1016/j.solmat.2019.03.030
24. Monfort O, Plesch G (2018) Bismuth vanadate-based semiconductor photocatalysts: a short critical review on the efficiency and the mechanism of photodegradation of organic pollutants. Environ Sci Pollut Res 25(20). https://doi.org/10.1007/s11356-018-2437-9
25. Shang Y, Niu F, Shen S (2018) Photocatalytic water oxidation over BiVO4 with interface energetics engineered by Co and Ni-metallated dicyanamides. Cuihua Xuebao/Chinese J Catalysis 39(3):502–509. https://doi.org/10.1016/S1872-2067(17)62943-7
26. Wei T, Zhu YN, Gu Z, An X, Liu LM, Wu Y, Liu H, Tang J, Qu J (2018) Multi-electric field modulation for photocatalytic oxygen evolution: enhanced charge separation by coupling oxygen vacancies with faceted heterostructures. Nano Energy 51:764–773.https://doi.org/10.1016/j.nanoen.2018.07.018
27. Shandilya P, Sambyal S, Sharma R, Mandyal P, Fang B (2022) Properties, optimized morphologies, and advanced strategies for photocatalytic applications of WO3 based photocatalysts. J Hazardous Mater 428:128218. https://doi.org/10.1016/j.jhazmat.2022.128218
28. Li D, Liu Y, Yang Y, Tang G, Tang H (2022) Rational construction of Ag3PO4/WO3 step-scheme heterojunction for enhanced solar-driven photocatalytic performance of O_2 evolution and pollutant degradation. J Colloid Interface Sci 608:2549–2559. https://doi.org/10.1016/j.jcis.2021.10.178

29. Shafiee HH, Sangpour P, Tabrizi NS (2019) Advanced ceramics progress identification of intermediate compound in Photodegradation of 1-Naphthol by WO 3-rGO Nanocomposites. Acerp 5(2):1–5
30. Jin GH, Liu SQ (2016) Preparation and photocatalytic activity of fluorine doped wo3 under uv and visible light. Dig J Nanomater Biostruct 11(4):1179–1188
31. Wei Z, Wang W, Li W, Bai X, Zhao J, Tse ECM, Phillips DL, Zhu Y (2021) Steering electron-hole migration pathways using oxygen vacancies in tungsten oxides to enhance their photocatalytic oxygen evolution performance. Angewandte Chemie Int Edition 60(15):8236–8242. https://doi.org/10.1002/anie.202016170
32. Kment S, Riboni F, Pausova S, Wang L, Wang L, Han H, Hubicka Z, Krysa J, Schmuki P, Zboril R (2017) Photoanodes based on TiO_2 and α-Fe2O3 for solar water splitting-superior role of 1D nanoarchitectures and of combined heterostructures. Chem Soc Rev 46(12):3716–3769. https://doi.org/10.1039/c6cs00015k
33. She X, Wu J, Xu H, Zhong J, Wang Y, Song Y, Nie K, Liu Y, Yang Y, Rodrigues MTF, Vajtai R, Lou J, Du D, Li H, Ajayan PM (2017) High efficiency photocatalytic water splitting using 2D A-Fe_2O_3/g-C3N4 Z-scheme catalysts. Advanced. https://doi.org/10.1002/aenm.201700025
34. Khurram R, Wang Z, Ehsan MF (2021) Photocatalytic response towards wastewater treatment 3:17697–17711
35. Li L, She X, Yi J, Pan L, Xia K, Wei W, Zhu X, Chen Z, Xu H, Li H (2019) Integrating CoO x cocatalyst on hexagonal α-Fe 2 O 3 for effective photocatalytic oxygen evolution. Appl Surface Sci 469:933–940. https://doi.org/10.1016/j.apsusc.2018.11.014
36. Kong L, Yan J, Li P, Liu SF (2018) Fe2O3/C-C3N4-based tight heterojunction for boosting visible-light-driven photocatalytic water oxidation. ACS Sustain Chem Eng 6(8):10436–10444. https://doi.org/10.1021/acssuschemeng.8b01799
37. Molinari R, Lavorato C, Argurio P (2020) Visible-light Photocatalysts and their perspectives various liquid phase chemical conversions. Catalysts i:1–38
38. Yang J, Wang D, Han H, Li CAN (2013) Ar300227E.Pdf. Accounts Chem Res 46(8):1900–1909. https://doi.org/10.1021/ar300227e
39. Moniz SJA, Shevlin SA, Martin DJ, Guo ZX, Tang J (2015) Visible-light driven heterojunction photocatalysts for water splitting-a critical review. Energy Environ Sci 8(3):731–759. https://doi.org/10.1039/c4ee03271c
40. Wang S, Liu B, Wang X, Zhang Y, Huang W (2022) Nanoporous MoO3−x/BiVO4 photoanodes promoting charge separation for efficient photoelectrochemical water splitting. Nano Res.https://doi.org/10.1007/s12274-022-4344-0
41. Li Z, Wang J, Chen J, Liu L, Yang X, Chen T, Chen Z, Yang M, Yan W, Fu Z, Liu M, Lu Y (2019) Realizing nitrogen doping in Bi4Ti3O12 via low temperature
42. Liu G, Wang L, Yang HG, Cheng HM, Lu GQ (2010) Titania-based photocatalysts—Crystal growth, doping and heterostructuring. J Mater Chem 20(5):831–843. https://doi.org/10.1039/b909930a
43. Parameters P (2022) Charge separation in photocatalysts
44. Raziq F, Li C, Humayun M, Qu Y, Zada A, Yu H, Jing L (2015) Synthesis of TiO2/g-C3N4 nanocomposites as efficient photocatalysts dependent on the enhanced photogenerated charge separation. Mater Res Bull 70:494–499
45. Liu X, Chen W, Wang W (2021) F-Serve as surface trapping sites to promote the charge separation and transfer of TiO_2. ACS Omega 6(51):35799–35809. https://doi.org/10.1021/acsomega.1c05891
46. Zeradjanin AR, Masa J, Spanos I, Schlögl R (2021) Activity and stability of oxides during oxygen evolution reaction—from mechanistic controversies toward relevant electrocatalytic descriptors. Front Energy Res 8:1–17. https://doi.org/10.3389/fenrg.2020.613092

The manufacturer's authorised representative in the EU is Springer Nature Customer Service Centre GmbH, Europaplatz 3, 69115 Heidelberg, Germany. If you have any concerns regarding our products, please contact ProductSafety@springernature.com

Printed and bound by CPI Group (UK) Ltd, Croydon, CR0 4YY
30/09/2026
02225317-0010